Smart Grid Test Bed Using OPNET and Power Line Communication

Jun-Ho Huh
Catholic University of Pusan, South Korea

A volume in the Advances in Computer and Electrical Engineering (ACEE) Book Series

IGI Global
DISSEMINATOR OF KNOWLEDGE

Published in the United States of America by
IGI Global
Engineering Science Reference (an imprint of IGI Global)
701 E. Chocolate Avenue
Hershey PA, USA 17033
Tel: 717-533-8845
Fax: 717-533-8661
E-mail: cust@igi-global.com
Web site: http://www.igi-global.com

Library of Congress Cataloging-in-Publication Data

Names: Huh, Jun-Ho, 1984- author.
Title: Smart Grid Test Bed using OPNET and Power Line Communication / by Jun-Ho Huh.
Description: Hershey, PA : Engineering Science Reference, [2018] | Includes bibliographical references and index.
Identifiers: LCCN 2017010835| ISBN 9781522527763 (hardcover) | ISBN 9781522527770 (ebook)
Subjects: LCSH: Smart power grids--Computer simulation. | Electric lines--Carrier transmission. | Prototypes, Engineering. | OPNET.
Classification: LCC TK3105 .H85 2018 | DDC 621.31--dc23 LC record available at https://lccn.loc.gov/2017010835

This book is published in the IGI Global book series Advances in Computer and Electrical Engineering (ACEE) (ISSN: 2327-039X; eISSN: 2327-0403)

British Cataloguing in Publication Data
A Cataloguing in Publication record for this book is available from the British Library.

Advances in Computer and Electrical Engineering (ACEE) Book Series

ISSN:2327-039X
EISSN:2327-0403

Editor-in-Chief: Srikanta Patnaik, SOA University, India

MISSION

The fields of computer engineering and electrical engineering encompass a broad range of interdisciplinary topics allowing for expansive research developments across multiple fields. Research in these areas continues to develop and become increasingly important as computer and electrical systems have become an integral part of everyday life.

The **Advances in Computer and Electrical Engineering (ACEE) Book Series** aims to publish research on diverse topics pertaining to computer engineering and electrical engineering. **ACEE** encourages scholarly discourse on the latest applications, tools, and methodologies being implemented in the field for the design and development of computer and electrical systems.

COVERAGE

- Digital Electronics
- VLSI Design
- Microprocessor Design
- Sensor Technologies
- Optical Electronics
- Chip Design
- Power Electronics
- Programming
- Electrical Power Conversion
- Algorithms

IGI Global is currently accepting manuscripts for publication within this series. To submit a proposal for a volume in this series, please contact our Acquisition Editors at Acquisitions@igi-global.com or visit: http://www.igi-global.com/publish/.

The Advances in Computer and Electrical Engineering (ACEE) Book Series (ISSN 2327-039X) is published by IGI Global, 701 E. Chocolate Avenue, Hershey, PA 17033-1240, USA, www.igi-global.com. This series is composed of titles available for purchase individually; each title is edited to be contextually exclusive from any other title within the series. For pricing and ordering information please visit http://www.igi-global.com/book-series/advances-computer-electrical-engineering/73675. Postmaster: Send all address changes to above address.

Titles in this Series

For a list of additional titles in this series, please visit:
https://www.igi-global.com/book-series/advances-computer-electrical-engineering/73675

Transport of Information-Carriers in Semiconductors and Nnodevices
Muhammad El-Saba (Ain-Shams University, Egypt)
Engineering Science Reference • ©2017 • 677pp • H/C (ISBN: 9781522523123) • US $225.00

Accelerating the Discovery of New Dielectric Properties in Polymer nsulation
Boxue Du (Tianjin University, China)
Engineering Science Reference • © 017 • 388pp • H/C (ISBN: 9781522523093) • US $210.00

Handbook of Research on Nanoelectronic Sensor Modeling and Applications
Mohammad Taghi Ahmadi (Urmia University, Iran) Razali Ismail (Universiti Teknologi Malaysia, Malaysia) and Sohail Anwar (Penn State University, USA)
Engineering Science Reference • ©2017 • 579pp • H/C (ISBN: 9781522507369) • US $245.00

Field-Programmable Gate Array (FPGA) Technologies...
Julio Daniel Dondo Gazzano (University of Castilla-La Mancha, Spain) Maria Liz Crespo (International Centre for Theoretical Physics, Italy) Andres Cicuttin (International Centre for Theoretical Physics, Italy) and Fernando Rincon Calle (University of Castilla-La Mancha, Spain)
Engineering Science Reference • ©2016 • 306pp • H/C (ISBN: 9781522502999) • US $185.00

Design and Modeling of Low Power VLSI Systems
Manoj Sharma (BVC, India) Ruchi Gautam (MyResearch Labs, Gr Noida, India) and Mohammad Ayoub Khan (Sharda University, India)
Engineering Science Reference • ©2016 • 386pp • H/C (ISBN: 9781522501909) • US $205.00

Reliability in Power Electronics and Electrical Machines Industrial Applications...
Shahriyar Kaboli (Sharif University of Technology, Iran) and Hashem Oraee (Sharif University of Technology, Iran)
Engineering Science Reference • © 2016 • 481pp • H/C (ISBN: 9781466694293) • US $255.00

For an enitre list of titles in this series, please visit:
https://www.igi-global.com/book-series/advances-computer-electrical-engineering/73675

IGI Global
DISSEMINATOR OF KNOWLEDGE

701 East Chocolate Avenue, Hershey, PA 17033, USA
Tel: 717-533-8845 x100 • Fax: 717-533-8661
E-Mail: cust@igi-global.com • www.igi-global.com

Table of Contents

Preface

The core concept of Smart Grids is to intelligently manage electric power supply in accordance with the changes in both power output and demands by integrating the IoT technology with Power Grid. The elements such as the grid itself, communication technology and networked smart-devices are organically combined to save energy and enhance the quality of electricity, which in turn let the consumers to save their electricity bills. However, one of the biggest problems researchers and power companies are facing is that there are almost no expandable test beds available currently. Since so many kinds of elements are complexly involved in the smart grid systems, it is quite difficult to construct an analytical model for the complex system and conduct direct experiments on a actual system. For these reasons, experiments are usually carried out on the simulations or the scale-downed test beds. Thus, In this book, we have tested and evaluated the performance of the simulated model in the PLC-based data-link layer, comparing it to the existing Zigbee and WiMax-based models.

Also, to compensate the transmission losses caused by the characteristics of the model, another simulation model using the Reliable User Datagram Protocol (RUDP) in the transport layer was implemented and evaluated. As a result, it was possible to construct an expandable test bed which included both of above mentioned layers. And then, a scenario for the Advanced Metering Infrastructure (AMI), which is one of the smart grid's application areas, was created and the performance of the test bed was evaluated by implementing an intelligent power management agent model. Moreover, a security model that detects Distributed Denial-of-Service (DDoS) attacks was mounted on AMI. It was observed that the AMI had responded intelligently against the fluctuations in both power supply and demands. In this book, an expandable test bed has been designed and implemented by constructing and evaluating the simulation models involving respective elements. The results showed that the test bed based on the AMI worked flexibly and we have validated our research results.

Introduction

The key to a successful management of Smart grids is to intelligently supply the electric power following the changes in power output and demand by integrating the IoT technology with the power grid. This section explains the objectives and necessities of Smart grids and describes the research environment and methods involved with the Smart grid test bed.

The Smart Grid and the Internet of Things (IoT) have become major issues in the Republic of Korea (ROK) recently and the Zigbee-, Power Line Communication (PLC)-and WiFi-related technologies are being tested in the Smart Grid-based systems. Both IoT and Smart Grid are increasingly getting attentions from the IT industry and expected to be the most notable technologies in the future to come.

The wireless sensor network technology is one that establishes an autonomous network by deploying wireless sensor nodes in an application environment and collects the information from these nodes wirelessly to utilize them for monitoring and controlling purposes, aiming to realize a ubiquitous environment where communications can be established anytime anywhere conveniently.

The wireless sensor network technology is essential for the IoT systems and can be considered as a requisite in constructing a network that carries out sensing operations between things and information processing/exchanging without any or least human interventions. For the IoT systems, it is important to decide with what kind of wireless technology the network devices should be connected. The wireless networking technologies currently available are the Zigbee, NFC, ANT, Wi-Fi, Bluetooth and RFID-based ones.

Zigbee is a wireless communication technology that complements demerits (i.e., high costs and power consumption rate) of Bluetooth and being mainly used for the wireless remote controllers, keyboards, home automation, as well as in managing various facilities. For this reason, this technology was used for the Smart Grid Test Bed complex in Jeju-island, the ROK.

PLC is a common name for the communication mode which uses electric power lines. The PLC technology uses high-frequency signals to transmit information through power lines. Frequency signals can be separated from the power line with a dedicated power line modem and then transmitted to the terminal devices. PLC has two modes: High-speed Power Line Communication and Low-speed Power Line Communication. High-speed PLC uses the frequency bands from 0.5MHz to 30MHz achieving the speed of 14Mbps to 1Gbps while Low-speed PLC utilizes the frequency bands from 10KHz to 450KHz, exerting the speed less than 9.6Kbps to control electronic devices mainly (Huh & Seo, 2014).

The power lines are originally designed for power transmission rather than for data transmission. Its channel characteristics are not ideal as they entail noises and serious signal attenuations during transmission (Hrasnica, Haidine & Lehnert, 2004; Xiao-Sheng et al., 2011).

To improve reliability of PLC, it is possible to consider improving the physical layer of the system designed in relation to channel estimation and selection, filtering design, power distribution, etc. (Xiao-Sheng et al., 2011; Yao, 2013).

Currently, PLC is being used for configuration of home-networks or factory automation systems. Owing to the improvements in both stability and transmission speed resulting from recent digital PLC technology, its field of application is expanding and designs are becoming more sophisticated (Tsuzuki, Areni & Yamada, 2012).

Also, the PLC technology is being studied widely as it is considered as a sustainable low-cost communication solution that adequately utilizes existing power distribution networks and to be suitable for the Smart Metering System. The instability in a noisy channel environment should be considered while developing a PLC-based solution through simulations and modeling. Modeling of PLC's physical characteristics by using different simulation tools has been studied in the past and still remains as a major research subject (Tsuzuki, Areni & Yamada, 2012).

1. RESEARCH OBJECTIVES AND NECESSITIES

Recently, the government of the ROK is preparing a mid-and-long-term promotional plan for the negawatt market that deals with conserved energies as the crux of their new energy business. In this plan, the ESS, EMS, and Micro Grids and Smart Grids which maintain small-scale power systems of various types of distributed power supply are the main axes. The Korean government is expecting to find some new convergent business models that would create profits and jobs with this plan.

Smart Grid is a technology that allows an efficient power usage by collecting information concerning the power consumptions and the condition of power line through integration of IT with Power Grids (Huh & Seo, 2014; Hrasnica, Haidine & Lehnert, 2004; Xiao-Sheng et al., 2011).

Meanwhile, Smart Grid is divided into five major areas: the Smart Power Grid, which is an intelligent power grid established through integration of IT technology; the Smart Place, a living environment where two-way communications between power supplier and consumer are possible; the Smart Electricity Service that provides a power transaction service and maintains the Total Operation Center (TOC); the Smart Transportation which is achieved with power charging technology and its infrastructure; and the Smart Renewable that is achieved through quality improvement of new renewable energies and their stable links in the Power Grid.

One of the major problems for the network researchers and operators involved in the Smart Grid technology is that they can hardly find reliable and scalable Test Beds today.

Since it is often impossible for them to conduct experiments directly on a network or build an analytical model for the complex system, they have no choice but to rely on the simulations on a reduced test bed. However, in most cases, designing a reduced-scale experiment that can replace a large-scale experiment with a fair degree of precision is highly unpractical as the results obtain from an alternative experiment can be falsifying and it is possible that the errors would not be found until the protocols or the applications are actually used in an internet network.

Thus, in this book, an expandable platform technology has been devised for the test bed experiments by simulating the Zigbee- and WiMAX-based Smart Grid environments (the Korean Smart Grid models) and the PLC-based environment popular in some foreign countries with the same scenario on their Physical and Data Link Layers. For the Transport Layer, Reliable User Datagram Protocol (RUDP) was proposed instead of TCP and UDP.

Also, after developing two scenarios as platform technologies for a PLC system, an AMI application embedded with an intelligent agent has been proposed to complement the existing PLC technology. Intelligent agent is a program that periodically collects information or performs certain services without any interventions by the user. Normally, such an agent program uses the parameters provided by the user and collects information in the field where the user's interest lies on. Then, it supplies information on a daily basis or an appointed time(s). This agent is sometimes referred as a 'robot' or 'bot' for short. AMI is an advanced form of AMR, which is a uni-directional metering technology and refers to the infrastructure that implements various optional or additional services through bi-directional data communications between power company and consumers. It also is a mutual information provision tool between them.

2. RESEARCH ENVIRONMENT AND METHODS

The research environment and the scope of network simulation have been examined for a Smart Grid test bed and it was determined that the OPNET Modeler was appropriate for the simulations.

This section is organized as following: the research environment in Section 2.1; the guidelines to select a network testing platform in Section 2.2; and finally, OPNET Modeler in Section 2.3.

2.1 Research Environment

The simulation environment for a performance evaluation including hardwares and softwares are shown in Table 1.

The Smart Grid test bed has been constructed with OPNET Modeler in Physical, Data Link, and Transmission Layers, and also, Microsoft Visual

Table 1. Simulation environment

Working Environment Computer Specifications	
CPU	3.0+ GHz Using x86, EM64T, x86 AMD, or AMD64 (Dual-Core)
RAM	8GB
System File Space	3GB
Working File Space	100MB
Display Resolution	1024x768
Development Environment	
OS	Windows 7 Professional (64 bit)
Compiler	Microsoft Visual Studio 2008
M&S Tool	OPNET Modeler 14.5 PL8

Studio 2008 was used as a compiler. Interworking with Microsoft Visual Studio 2008 and OPNET Modeler 14.5 PL8, all the functions necessary for the node models for OPNET have been coded and debugged with C++ language.

2.2 Guidelines for Selecting Network Testing Platform

Cost and the fidelity of experiment are the key factors to be considered when selecting a testing platform for the experiment. The cost associated with a specific platform not only includes the operational fee but also includes costs of setting up experiment environment and gathering results from it. The fidelity denotes the level of similarity between results obtained from experiments conducted on a test platform and from the target network. Generally speaking, emulation Test Beds and internet test beds have higher fidelity than most simulators. It is not always easy to0 compare the fidelity without the ground truth.

Fortunately, most platforms can offer a reasonable reliability when the experimental scenario lies within the operational range of the platform.

To determine with which testing platform the experiment should be supported, the scale of the experiment and required functionality should be considered first. The scale of an experiment is simply a scale or size of the network topology (i.e., number of nodes). The required functionality is defined as the required level of details that is sufficient to validate the network system. The required functionality has been divided into following levels:

Application-level where the experiment mainly focuses on the behavior of network applications and not highly sensitive to the underlying network connectivity.

- Routing-level where the experiment requires a specific network or topology or control on routing between nodes.
- System-level where the experiment needs a root access to the operating system.
- Hardware-level where the experiment needs to access the firmware on devices or interact with the testing machines physically.

The scale here refers to the number of nodes in the original experiment. The operational range of each platform discussed here is being illustrated in Figure 1. In addition to the scale and the required functionality, the experimenter should also consider particular characteristics of the testing platform.

Figure 1. The operational range of network testing platforms
Yao, 2013.

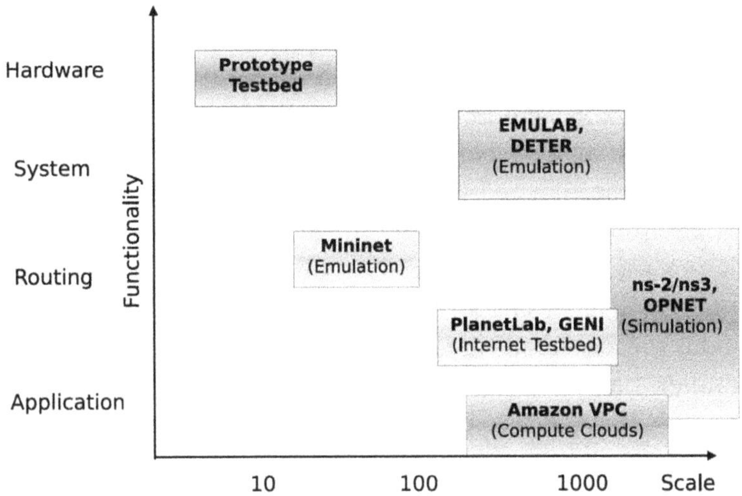

2.3 OPNET Modeler

The origin of the OPNET Modeler and IT Guru products, which are popular as the best communications and networks simulation solutions, can be traced back to a network simulator used for the project assigned to MIT by the Department of Defense in 1986. Their present forms are the result of continuous efforts and studies by the participating developers. The OPNET solution have been used and verified over 20+ years, and after its confidence being built, they are expanding their scope of application to Network Capacity Planning, Network Engineering, Network Management, etc.

OPNET Modeler is a discrete event simulator (packet-level) to design a large-scale network and a complex communication protocol. It allows an intuitive modeling of protocol in each network level as well as various operations. Because of its reliability as a network simulator, it is being used for the designing of large-scale networks and their evaluations in many countries. This modeler is also found to be an effective modeling and simulation tool for the military tactical communication field (Hrasnica, Haidine & Lehnert, 2004).

Smart Grids allows consumers to use/produce electric power efficiently and lets them save their electric bills by improving the quality of electricity and efficiently managing the power flow by organically integrating the elements such as communication technology, network smart equipment, etc.

However, one of the biggest problems researchers and power companies are facing at the moment that it is quite difficult to find a scalable test bed with which they can perform experiments flexibly.

As too many kinds of elements are complexly involved with each other in a Smartgrid system, it has been difficult to construct an analytical model for such a complex system and perform a direct experiment on an actual system so that most of the experiments have been performed in the simulated environments or the reduced-scale test beds. Thus, this section discusses a research environment along with and the test method involved. Other methodologies and their contents can be found from Reference (PlanetLab, n.d.; GENI, n.d.; Sivaneasan, So & Gunawan, 2011, 2009; Bauer et al., 2009; Galli, Scaglione & Wang, 2011, 2010; Gungor et al., 2011; Biagi & Lampe, 2010; Bumiller, Lampe & Hrasnica, 2010).

REFERENCES

Bauer, M., Plappert, W., Wang, C., & Dostert, K. (2009). Packet-oriented communication protocols for smart grid services over low-Speed PLC. *IEEE International Symposium on Power Line Communications and Its Applications*, 89-94. doi:10.1109/ISPLC.2009.4913410

Biagi, M., & Lampe, L. (2010). Location assisted routing techniques for power line communication in smart grids. *2010 First IEEE International Conference on Smart Grid Communications (SmartGridComm)*. IEEE. doi:10.1109/SMARTGRID.2010.5622056

Bumiller, G., Lampe, L. & Hrasnica, H. (2010). Power line communication networks for large-scale control and automation systems. *IEEE Communications Magazine, 48*(4).

Galli, S., Scaglione, A., & Wang, Z. (2010). Power line communications and the smart grid. *2010 First IEEE International Conference on Smart Grid Communications (SmartGridComm)*. IEEE. doi:10.1109/SMARTGRID.2010.5622060

Galli, S., Scaglione, A., & Wang, Z. (2011). For the grid and through the grid: The role of power line communications in the smart grid. *Proceedings of the IEEE, 99*(6), 998–1027. doi:10.1109/JPROC.2011.2109670

GENI. (n.d.). *Global Environment for Network Innovations (GENI)*. Retrieved from: http://www.geni.net/

Gungor, V. (2011). Smart grid technologies: Communication technologies and standards. *IEEE Transactions on Industrial Informatics, 7*(4), 529-539.

Hrasnica, H., Haidene, A., & Lehnert, R. (2004). *Broadband power line communications networks: Network design*. New York: Wiley. doi:10.1002/047002092X

Huh, J.H., & Seo, K. (2014). *PLC-based smart grid home network system design and implementation using OPNET simulation*. Academic Press.

PlanetLab. (n.d.). Retrieved from: http://www.planet-lab.org/

Sivaneasan, B., So, P. L., & Gunawan, E. (2009). A simple routing protocol for PLC-based AMR systems. IEEE Region 10 Conference, 1-5.

Sivaneasan, B., So, P. L., & Gunawan, E. (2011). A simple routing protocol for PLC-based AMR systems. *IEEE Transactions on Power Delivery*, *26*(4), 2613–2620. doi:10.1109/TPWRD.2011.2161000

Tsuzuki, S., Areni, S. A., & Yamada, Y. (2012). A feasibility study of 1gbps plc system assuming a high-balanced DC power-line channel. *2012 IEEE International Symposium on Power Line Communications and Its Applications*, 386-391. doi:10.1109/ISPLC.2012.6201334

Xiao-Sheng, L., Liang, Z., Yan, Z., & Dianguo, X. (2011). Performance analysis of power line communication network model based on spider web. *2011 IEEE 8th International Conference on Power Electronics ECCE Asia*, 953-959.

Yao, W. M. (2013). *Increasing scalability in network simulation and testbed experiments* (Ph.D thesis). Purdue University, West Lafayette, IN.

Chapter 1
Related Research

ABSTRACT

This chapter presents related and current research on Smart Grids, Test Beds, AMI (Advanced Metering Infrastructure), PLC (Power Line Communication), and other emerging fields within OPNET and Power Line Communication. The author emerges in new research trends to provide adequate information on Smart House, Smart Grid communication, Smart Grid design challenges, and PLC and the Smart Grid. This chapter will pave the way for the information in upcoming chapters.

1. TEST BED

Simulator and emulation Test Bed users often experience difficulty in generating realistic network topologies and background traffic in their experiments. These issues can be avoided by using test beds that are available on the actual Internet, such as PlanetLab and Federations of Test Beds via the Global Environment for Network Innovations (GENI).

With real Internet links included in the Test Bed, experimental networks will have real Internet background traffic and are global in scale. However, these internet test beds are not isolated from operational networks and security experiments on these test beds are difficult. Such test beds also lack repeatability (i.e., reproducibility) as the background internet traffic is not controlled by the user.

As for the Test Bed for a Smart Grid, networking was performed on the Physical and the Data Link Layer of the OSI 7 Layer, as indicated in the

DOI: 10.4018/978-1-5225-2776-3.ch001

section in Figure 1. The Test Bed experiment was conducted with Zigbee-based network which is popular in the ROK while both the WiMAX and the PLC-based networks have been implemented with OPNET for the experiments.

In the Transmission Layer, RUDP was tested and presented as the most suitable protocol that has major benefits of both TCP and UDP. Meanwhile, for the Application Layer, an AMI application has been proposed for the PLC by implementing two scenarios. At the same time, an intelligent agent was used for the AMI application to have it tested in the PLC network.

2. SMART GRID

Figure 2 shows the recent trend of Smart Grid system in the ROK. The Korean government has announced the 'Smart Grid Road Map' and plans to invest 2.75 billion won by 2030 hoping to vitalize relevant industries, starting with construction of a demonstrational complex in Jeju-Do. Their initial Smart Grid model had aimed to advance the element technologies but recently, they are seeking to invigorate the Smart Grid system business internationally by developing several promising business models and more sophisticated convergent systems. The purpose of such a Smart Grid system is to intellectualize the Power Grid first to let the supplier of electric power to adjust the level of power supply and its rates and, on the part of consumers, to control their usage by understanding the price of electricity and their own

Figure 1. Smart grid test bed

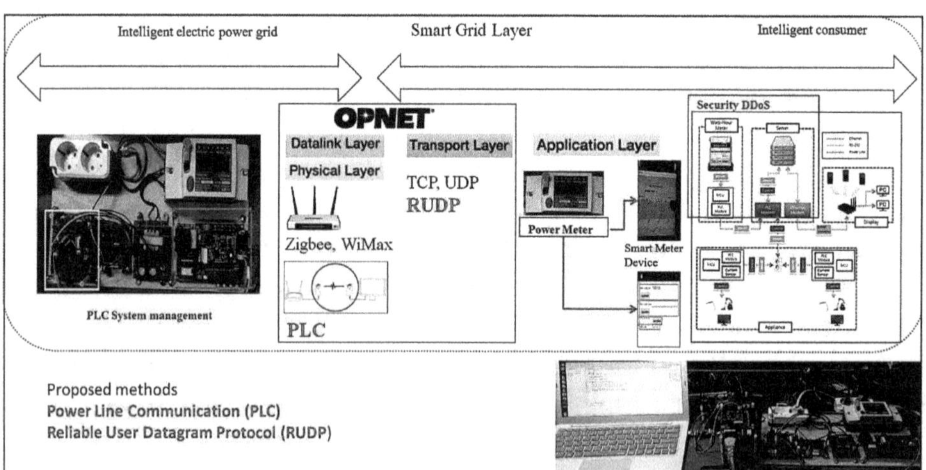

Figure 2. A trend on the smart grid in ROK

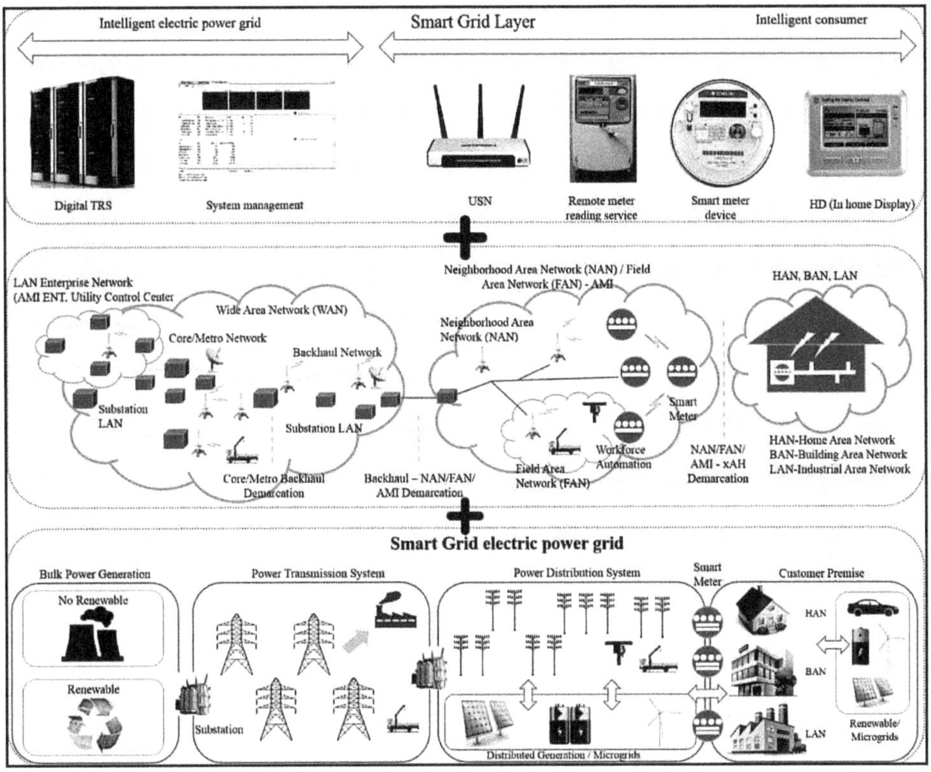

usage patterns through the smart meters connected to internet. One of the typical methods used in the Smart Metering systems is a Cable TV (CATV) network-based remote metering method.

Figure 3 shows present Smart Grid construction status overseas. Although the technologies and markets for the Smart Grid are rapidly developing, centering around the US, the policies related to them assume different forms depending on each nation's power infrastructure and industrial structure. The US, where there used be a power loss rate as high as 7% and frequent blackout events, has already announced the vision 'Smart Grid 2030' in 2003 and is actively proceeding with the development of the Smart Grid-relate technologies and supportive policies by selecting five major technical fields to be focused on and establishing the Energy Independence and Security Act (EISA) in 2007.

Meanwhile, different from the US, EU nations are focusing on increasing the consumption rate of new renewable energies following their CO_2 reduction

Figure 3. Statuses of smart grid constructions in various countries

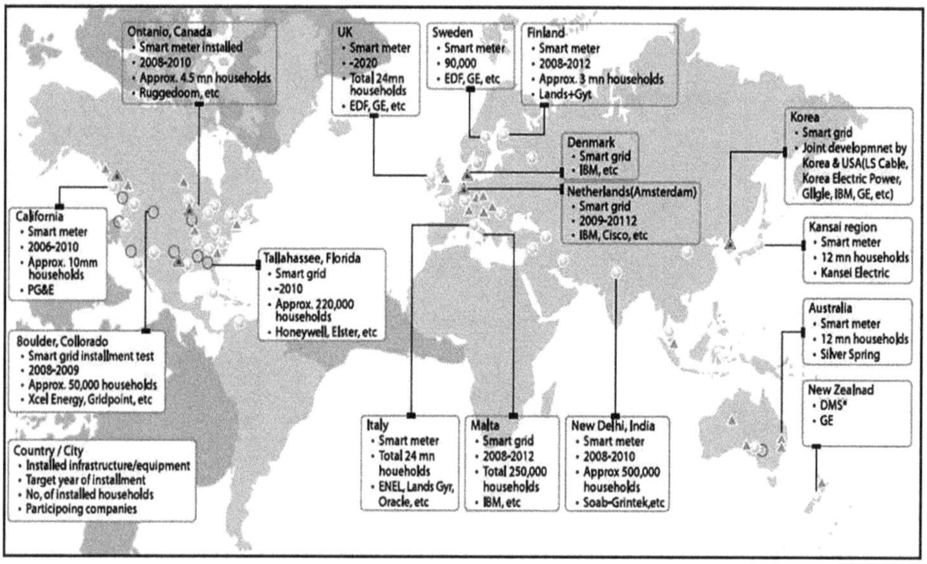

policy resulting from their concerns over rapid climate changes. They aim to increase the consumption rate from 12% (as of 2010) to 20% by 2020 and seek to build the Smart Grid systems strategically to integrate disparate renewable energies and increase grid's efficiency. In China, in order to solve the power availability imbalances between regions, they are pursuing an integrated strategy with which the extra-high voltage Power Grid will be constructed as a backbone network and four large-scale hydraulic, thermal, nuclear and renewable energy power stations will be constructed on it to improve power transmission efficiency. Even though they have initiated the Smart Grid construction plans as part of their major national electrical power business plan in rather late year of 2010, it is expected that demands for the products related to the high-Voltage Grids, Smart Meters, digital substations and distribution networks will be significantly increased considering their market size.

2.1 AMI

AMI (Advanced Metering Infrastructure) is a more sophisticated infrastructure of AMR, a unilateral Automatic Meter Reading system. It offers a bi-directional communication function between power company and consumers. It also is a means of information exchange for mutual awareness between them

(PlanetLab, n.d.; GENI, n.d.; Sivaneasan, So & Gunawan, 2011). Various types of decentralized power supply systems and intelligent power distribution systems are included and it supports highly developed time-based tariff systems such as Time-of-Usage (ToU), Critical Peak Pricing (CPP) and Real-Tme Pricing (RTP). With AMI, it is possible lead users to participate in an active energy conservation program based on Demand Response (DR). Also, since it generates data related to power use patterns of consumers through bi-directional communications between meters and power company, the provision of additional services such as load anticipation, load control and power quality monitoring will be possible.

Smart Meters is a core device of AMI and assumes the functions of both electronic power meter and communication network terminal. It allows monitoring of power usage status of home appliances through a short-distance wireless communication technology such as Zigbee.

Backhaul is used for the remote communications and collects power usage information and send them to the management system through wireless communication networks such as PLC or GSM/ GPRS network.

IHD (In-Home Energy Display) provides a household power consumption monitoring function for power equipment and helps control them with an alarm function, thereby acting as a kind of hub for the smart-energy environment.

2.2 PLC

PLC is a common name for the communication mode which uses power lines. This technology utilizes high-frequency signals and sends data over power line. Frequency signals can be separated from power line through dedicated power line modem(s), and then transmitted to the terminal devices. Power Line Communication has two modes: High-speed PLC and Low-speed PLC. The former uses the frequency band from 0.5MHz to 30MHz, achieving the speed of 14Mbps to1Gbps while the latter uses the frequency band from 10KHz to 450KHz, exerting the speed less than 9.6Kbps. This is mainly used to control electronic devices.

The Power Line Communication media was originally designed for power transmission rather than data transmission. For data communication, PLC's channel characteristics are not ideal because of obvious noises and serious signal attenuations. To improve the reliability of PLC, the design of its physical layer should be considered. For example, channel estimation and selection, filtering design, power distribution, etc. (GENI, n.d.; Sivaneasan, So & Gunawan, 2011).

Nowadays, PLC is being used for the configuration of Home-Networks or factory automations. Owing to the improvements in both the stability and the transmission speed, resulting from recent digital power line communication method, it's field of application is getting wider and holding the limelight (Kabalci, Kabalci & Develi, 2012).

Meanwhile, PLC has been studied widely lately for it is considered as a low-cost communication solution using power distribution networks and for the Smart Metering system. The instability in a noisy channel environment, requires careful simulation and modeling for the Power Line Communication based solution. Modeling of Power Line Communication's physical characteristics using different simulation tools has been studied in the past and still remains as a research subject.

3. RESEARCH TREND

This section is organized as follows: The Smart House are described in section 3.1; the Smart Grid Communications in section 3.2; the Smart Grid Design Challenge in section 3.3; the Power Line Communication and the Smart Grid in section 3.4; and finally, the Smart Grid Information and Communication Infrastructure in section 3.5.

3.1 Smart House

RFID (Radio Frequency IDentification) technology used as the control signals is being applied for the environmental control systems around the world and these control systems usually have external sensors and use voice recognition technology. To control appliances within the Smart House, Carol Rus et al. (2008) proposed voice recognition. Corcoran et al. (2002) whereas Sheng Li et al. (2008) proposed the Universal Plug-n-Play (UPnP) for the wireless home networks which control the equipment with user's PDA, mobile phones or wearable devices. The system user is able to transmit requests with his/her own voice or user interface to the home server to initiate control so that it is not necessary to pre-set voice commands and operation areas. Additionally, Hwang et al. (2009) presented RFID-based multiuser access control algorithm that can be applied to the Smart Homes where UPnP is in service. Access situation can be monitored automatically with the RFID tags carried by the user(s). One of the demerits of this method is that many RFID detectors have to be installed at various places.

Helal et al. (2005) and Liau et al. (2008) developed the Wireless Smart Floor technology which detects the location of the residents by installing the pressure sensors on the floor. Recently. the Brain Computer Interface-based smart environment control system is gaining popularity

3.2 Smart Grid Communications

The concept of establishing communications over the existing Power Lines (PLs) started in early 1900s when the first several patents had been applied (Schwartz, 2009). This technology has been further developed to be used for the remote metering and the power load control, where the single carrier NarrowBand (NB) (i.e., low frequency bands ranging from a few bps to a few kilobits) was used for the audio communications (Dostert, 2001; Ferreira et al., 2010). Following rapid development of the technology, the BroadBand (BB) Power Line Communication systems operating in the high-frequency bands (2-30MHz) have emerged in the market achieving the data transfer rate up to a 200Mbps. During last few years, communication market has turned its attention to the NB- Power Line Communication which has higher data transfer rates based on the multi-carrier schemes that operates in the bandwidths of between 3-500KHz. The PLC technology can be also used for the broadband internet access for the general population, broadband LAN service to be used at home, business sites and mobile environments, as well as providing a command and control function for automation purpose and remote metering (Latchman & Yonge, 2003; Pavlidou et al., 2003; Galli, Scaglione & Dostert, 2003; Biglieri et al., 2001).

The primary benefit of using the Power Line Communication technology is that existing Power Grids used for the Power Line Communication systems are quite extensive and pervasive compared to the other methods which employ wired or wireless technology. This adds more value to the Power Line Communication systems as any equipment or devices connected to the power lines can be operated without additional installation works.

Despite of its promising capabilities in present and future applications, the Power Line Communication technology is not being considered as the best option in the market yet. Even so, various construction plans of the Smart Grid in many countries to achieve an efficient management and distribution of power are favorable to the Power Line Communication system. The Smart Grid requires a variety of advanced networking technologies so that the PLC technology may secure a firm foothold in the Smart Grid and market.

The aim of this research book is to explicitly clarify the functionality of the Smart Grid and the benefit of using the Power Line Communication technology in the Grid.

3.3 Smart Grid Design Challenge

It has become clear that traditional means of energy generations will not meet the rapid growth of energy demands around the world. The other major concern is that many governments in the advanced nations strongly propose the necessity and urgency of containing CO_2 gas emissions to prevent the greenhouse effect and worsening climate changes. The investment in the Power Grid modernization in the last decade was not sufficient enough to prevent aforementioned problems and to adapt into a new energy generation system where renewable energy sources will be used. The new type of sustainable power generation systems involves Distributed Energy Resources (DERs), decentralized energy storage systems in response to the often fluctuating wind or photovoltaic power generations, Plug-in (Hybrid) Electric Vehicles (PHEV) which could cause a momentary high power consumption in the grids including Micro Grid, and Demand Side Management (DSM) and Demand Response (DR) where consumers participate actively (Heydt et al., 2001).

Matching the power generation with its demand at the sophisticated level requires integration of safe protective and control technologies that can ensure stable, continuous and efficient power transfer within the grid and none of these aspects can be disregarded in modern power distribution networks (Garrity, 2008; Ipakchi & Albuyeh, 2009; Wu, Moslehi & Bose, 2005). Usually, Power Grids are designed to be controlled through a conventional centralized cyber-infrastructure model commonly called as Supervisory Control and Data Acquisition (SCADA). To complement this model, a novel concept of Smart Grid was developed. The Smart Grid offer the cyber-physical infrastructure that includes wide-area monitoring, bi-directional communications, and improved control capability which complements existing technological limitations in the Power Grids. In the Smart Grid, establishment of stable and flawless communications is considered as a basic but an important requirement such that the designs for the physical, data and network communication layers are being consistently upgraded and pursued for better results.

However, establishment of an ideal Smart Grid is yet to be realized. Tomsovic et al. (2005) assert that although there are many fast-available communication methods, the computation needed for such real-time control is still very complex and poorly understood. They additionally remark that, for instance, while DR and load shedding would potentially yield economic benefits and energy savings, the correct implementation of DR and DSM applications to maximize system savings under stability constraints has not been determined clearly (Hopper, Goldman, Bharvirkar & Engel, 2007; Faruqui et al., 2009). As an example, in Zeman et al. (2008), they describe the situation where power demand soars and the power supply is reduced in response to such phenomenon to protect power distribution network. This solution may protect the network but will cause instability. Small-scale Test Bed experiments and simulations are often conducted with cautious limited environment to seek out expected or unexpected network of Grid failures but these alone cannot cover the fundamental problems emerging from the global-scale networks. Most of technical discussions regarding the Smart Grid's ideal communication method lack the specific parameters for monitoring and control functions.

Additionally, an optimal condition in subsystems does not ensure overall network or Grid optimality. Controlling and sensing in a large-scale cyber-physical system, discussed in detail in the communication optimization section, can be a quite complicated and tricky issue. For example, the results obtained from observing the stability of networked control systems can only be recognized when the observation has been performed under the restrictive and constrained terms such as a perfectly known system, a single link with zero latency and so on (Ishii & Francis, 2002; Baillieu, 2002; Fagnani & Zampieri, 2003; Tatikonda & Miller, 2004). To design a communication model and test its efficiency in terms of scalability and control distribution, distinguishing the Smart Grid's data or information source will be essential, preceding other factors. The Power Grid itself can be regarded as the source in some cases. When managing, aggregating or prioritizing the network traffics, it is important to consider the fact that the dynamics in any interconnected systems are quite coupled and dependent. Nevertheless, this is often disregarded. For the Power Line Communication system, one can comprehend better about the grid as an information delivery system by characterizing the grid as an information source because the Grid itself will perform as a physical network medium in the system.

3.4 PLC and the Smart Grid

The question of Power Line Communication's actual role in a Smart Grid has not been definitively answered yet. That is, some argues that Power Line Communication can be an excellent measure for many applications whereas others are somewhat critical and support wireless technology as a better and proven choice. Although majority of critics agree that the Smart Grid can use various communication methods (e.g., fiber optics, wireless or wireline), they still express their concern over Power Line Communication's undefined standardization protocols and its low data rates. Another demerit of Power Line Communication system presented were the expensive Power Line Communication modems and Electro Magnetic Compatibility (EMC) problems, which have been dealt with recent years.

Meanwhile, the cost Power Line Communication system establishment is comparable to setting up a wireless system as the Power Lines are already available for use. Another current advantage are that the Power Line Communication technology has gone through many Test Bed experiments and gained credibility. It's been judged that the technology has matured enough for extensive and wider applications, especially for the Smart Grids for transmission and distribution functions. By supporting AMR and AMI applications, Power Line Communication started to penetrate market successfully.

On the other hand, there are two major aspects can restrict the growth of Power Line Communication in the market. First one is use of inadequate and incomplete Power Line Communication system just to take initiative in the market for commercial reason. Some examples can be found in the Power Line Communication vendors who promote Power Line Communication-based Smart Grid applications with BB- Power Line Communication modems that are not technically appropriate for the applications. These modems have been developed for the Home and Building Area Networks (HAN/BAN) or Internet access purpose to begin with so that their operation ranges are too limited for Smart Grids.

The other obstacle for the Smart Grids is 'standardization'. The Power Line Communication industry shifted its position from having none of BB- Power Line Communication standards at all to adopting multiple non-interoperable technologies certified by some of Standards Developing Organizations (SDOs), which included TIA-1113, ITU-T G.hn, IEEE1901 FFT-OFDM, and IEEE 1901 Wavelet-OFDM (Galli et al, 2010).

For this reason, using the BB- Power Line Communication modems exclusively for the Smart Grid applications raises the question of which standard to use and of incompatibility. These problems also affect the use of multiple non-interoperable technologies that adopt multi carrier HDR NB- Power Line Communication standardized in IEEE P1901.2 and ITU-T G.hnem. It is evident that such situation brings about much confusion for both the market and users, as well as delaying of deployment. These might have been the unforeseeable obstacles but it is fact that the industry has overlooked the problems and was optimistic that they can fully manage such situations.

One of hopeful situation can be found in the PHY/MAC-agnostic coexistence scheme defined in CENELEC EN 50065 (1992), which certified some of NB- Power Line Communication standards after their publication and allowed NB- Power Line Communication solutions to be used in the market for the last two decades.

However, at the same time, there are other skeptics who point out that such a coexistence would lead to a delay in establishing a single global Power Line Communication standard as non-standardized solutions could become pervasive in the market. Such pro and con debates further delayed the global Power Line Communication deployment but the confusion was larger in US, such that US National Institute of Standards and Technology (NIST) came forward to deal with the situation addressing it the problem in the Priority Action Plan (PAP), removing the obstacles and guiding the Power Line Communication-based Smart Grid applications in the right direction (PAP-15, n.d..; Su & Galli, 2010). It is hoped that their recommendation will bring more clarity in the market and industry.

4. SMART GRID INFORMATION AND COMMUNICATION INFRASTRUCTURE

To enable flawless data flow from sensors to Smart Meter, as well as from the Smart Meter to the Data Center, a proper Smart Grid information and communication infrastructure is essential (Güngör et al., 2011). Further, to meet communication requirements of components comprising a Smart Grid, an integrated, flexible, inter-operable, reliable, scalable and secure two-way communication back bone needs optimized latency, frequency range, date rate and through put specifications (Güngör et al, 2013; Ancillotti, Bruno & Conti, 2013). Especially, data transfer and storage functions must be guaranteed in a secure way to avoid malicious external cyber-attacks (Yang, Barria & Green, 2011). Various types of cryptographic protocols and algorithms, privacy-preserving billing protocols, encryption, decryption, authentication and key management schemes have been proposed and put into use by both industry and academia to reinforce Smart Grid networks and relevant equipment or devices (Fan & Gong, 2013; Li & Zhang, 2014). Smart Grid-based communications can be categorized as home area network, neighborhood area network and wide area network in terms of the covered geographic region, as described in Figure 4 (Saputro, Akkaya & Uludag, 2012).

Currently, the Power Line Communication, wireless communication, cellular communication and internet based-virtual private networks are being used for the Smart Grid infrastructures, each having its own challenges (Güngör & Lambert, 2006).

Power Line Communication transmits data over existing Power Line cables by sending the modulated carrier signals (Wang, Xu & Khanna, 2011). Its operating frequencies for the ultra- narrow band are between 0.3-3KHz, 3-500KHz for narrow band, and 1.8-250MHz for the broadband (Galli, Scaglione & Wang, 2011). Under the time-varying channel environment, the communication lost in open circuit ends of the power system, the reflections caused by impedance mismatches at discontinuities, strong low-pass behavior of the communication channel, signal attenuation and distortion, channel congestion, interference and noise issues are the major issues in the Power Line Communication system (Anatory, Theethayi & Thottappillil, 2009).

Additionally, supporting real time communication, enabling high data rates, achieving an efficient coexistence mechanism and durable security mechanism, and IP support are necessary to construct a better sophisticated Smart Grid (Yigit et al., 2014).

Figure 4. The architecture of communication layer in smart grids
Ho, Gao & Goc, 2013.

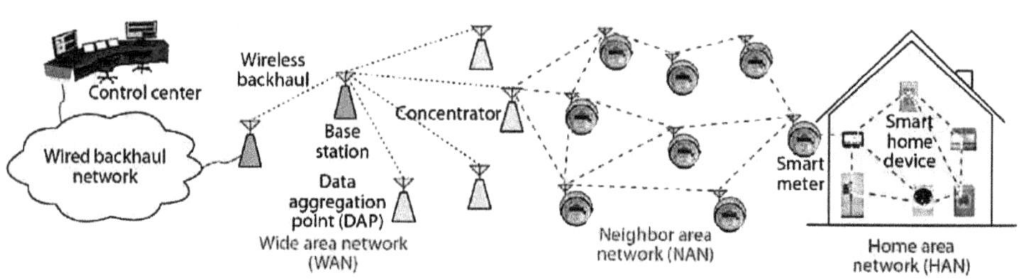

Zigbee (IEEE 802.15.4), Wi-Fi (IEEE 802.11) and WiMAX (IEEE 802.16) are the widely used wireless communication technologies in Smart Grids and they each cover the distance up to 50m, 100m and 100km with the respective data transfer rate of 250kb/s, 150Mb/sand 288.8Mb/s (Su, Qui & Wang, 2012; Ho, Gao & Goc, 2013). Most of time, the wireless communication systems or networks are affected by the interferences caused by the industrial, scientific and medical frequency bands. Wireless communication systems usually has narrow bandwidths that wireline communication systems and their operation range will be limited without additional high power supply (Mouftah & Kantarci, 2013; Gungor, Lu & Hancke, 2010). Negative impacts can originate from down-link communications, Quality of Service (QoS) differentiation and provisioning, self-healing capability, multi-casting, cluster-based routing and network design problems (Ho & Ngoc, 2013; Fan et al., 2013). Also, it has been found that the high voltage electrical equipment can slow down wireless communications or block the signals entirely. Finding the appropriate wireless equipment in the market or industry is not an easy task (Galli, Scaglione & Dostert, 2003).

Current major cellular communication technologies include GPRS, EDGE, HSPA, EVDO and LTE (US Department of Energy, n.d.). Should the cell's base station get damaged, partial or full-service interruption can be expected due to the tree-like topology of the cellular networks (Hui & Yeung, 2003). In a Smart Grid, the cellular service shared by customer markets decrease the efficiency of communication performance in an emergency situation (Gongor et al., 20011). Higher data transfer rate will be achieved by just increasing the signal power only as there is a limitation usually caused by the inter-cell interferences in the network (Akyildiz, Gutierrez-Esteves & Reyes, 2010). Also, the data and commands considered to be important for the communication will not be delivered until the prioritization or back-up process is completed within the cellular Network (Mouftah & Kantarci, 2013). Most of time, the Smart Grid utilities do not have direct control capability in the communication networks since the cellular network providers are responsible for the service they provide and therefore keep control rights (Usman & Shami, 2013). Moreover, smart antenna configuration, cross-layer optimization, signaling protocol and overhead, multi-cell environment and co-cooperative communication still remain as the subjects of debate for the device-to-device cellular communications (Mumtaz et al., 2014).

The internet-based Virtual Private Network (VPN) converts existing public network into a high-speed private network to handle its traffic more efficiently (Cohen, 2003). IPsec and MPLS technologies both are based on the overlay and peer-to-peer VPN architectures (Xu, 2011). The drawback is that overlay models require fully meshed circuits for optimal routing and it is difficult to determine inter-site circuit capacity. Peer-to-peer models also require complex filtering and all VPN routes are defined in the service provider IGP (Tan, 2003). Dissimilar to MPLS, IPsec struggles from the scalability problem when managing a large-scale VPN and cannot provide robust connections due to the unpredictable and unstable communication performance over the internet (Gongor & Lambert, 2006). Further, VPN gateway defines the policy enforcement points by encrypting all in- and out-going data for valid communication partners. This is the reason why the "covered channel resistance" occurs in the VPNs (Rossberg & Scharfer, 2011). Also, the Layer 1 VPN has both data plan and control plan connectivity as each service network often has an independent address space and separation policy to isolate itself from other VPNs (Liang & Yu, 2015). Added to this problems in the communication technologies, Smart Grids also experience deliberate cyber-attacks by industrial espionage groups and terrorists, or inadvertent compromises caused by the innocent users or equipment failures (NIST, n.d.).

The information security in a Smart Grid is a critical problem while communications are performed between generators, substations, customers and power equipment (Metke & Ekl, 2010).

The "Profile-then-detect" method which is used for the detection of DoS attacks increases the detection time necessary.

The error-prone results during the physical-layer authentication process occur because of time-varying characteristic of communication channels (Wang & Lu, 2013; Li et al., 2012). The DNP specification can easily be acquired with a nominal user fee so that the man-in-the-middle attacks by those who have the specification could result in a serious financial loss and physical danger (Liu et al., 2012).

The other problem is that the delay performance during data transfer gets worse depending on the longer key sizes in public key encryption process. The key disclosure risk caused by the short-life-span keys can be identified during the key distribution process for the symmetric key encryption (Wang & Lu, 2013; Li et al., 2012; Liu et al., 2012; Cyber Security Working Group, n.d.). The device manufacturers usually hold the initiative in establishing security measures for the host so that it is often difficult for end users to install their own security software (Knapp & Samani, 2013). The exposed application programming interface used in a big data analytic engine may cause many security breaches such as wrapping attacks, phishing schemes, meta-data spoofing, and injection attacks (Schuelke-Leech et al., 2015).

REFERENCES

Akyilidiz, I. F., Gutierrez-Esteves, D., & Reyes, E. C. (2010). The evolution to 4G cellular systems: LTE-advanced. *Physical Communication*, *3*(4), 217–244. doi:10.1016/j.phycom.2010.08.001

Anatory, J., Theethayi, N., & Thottappillil, R. (2009). Channel characterization for indoor power-line networks. *IEEE Transactions on Power Delivery*, *24*(4), 1883–1888. doi:10.1109/TPWRD.2009.2021044

Ancillotti, E., Bruno, R., & Conti, M. (2013). The role of communication systems in smart grids: Architectures, technical solutions and research challenges. *Computer Communications*, *36*(17-18), 1665–1697. doi:10.1016/j.comcom.2013.09.004

Baillieul, J. (2002). Feedback designs in information based control, stochastic theory and control. *Lecture Notes in Control and Information Sciences*, *280*, 35–37. doi:10.1007/3-540-48022-6_3

Biglieri, E., Galli, S., Lee, Y. W., Poor, H., & Vinck, A. H. (2001). BPower line communications (Guest Editorial). *IEEE Journal on Selected Areas in Communications*, *24*(7), 1261–1266.

Cohen, R. (2003). On the establishment of an access VPN in broadband access networks. *IEEE Communications Magazine*, *42*(2), 156–163. doi:10.1109/MCOM.2003.1179565

Corcoran, P. M., Desbonnet, J., Bigioi, O., & Lupu, I. (2002). Home network infrastructure for handheld/wearable appliances. *IEEE Transactions on Consumer Electronics*, *48*(3), 490–495. doi:10.1109/TCE.2002.1037033

Cyber Security Working Group. (n.d.). *Introduction to NIST IR7628 guidelines for smart grid cyber security*. Retrieved from: http://csrc.nist.gov/publications/nistir/ir7628/introduction-to-nistir-7628.pdf

Dostert, K. (2001). *Power line communications*. Upper Saddle River, NJ: Prentice Hall.

EN50065. (1992). *Signaling on low-voltage electrical installations in the frequency range 3 kHz–148.5 kHz*. CENELEC Std.

Fagnani, F., & Zampieri, S. (2003). Stability analysis and synthesis for scalar linear systems with a quantized feedback. *IEEE Transactions on Automatic Control*, *48*(9), 1569–1584. doi:10.1109/TAC.2003.816982

Fan, X., & Gong, G. (2013). *Security challenges in smart-grid metering and control systems*. Technology Innovation Management Review.

Fan, Z., Kulkarni, P., Gormus, S., Efthymiou, C., Kalogridis, G., Sooriyabandara, M., & Chin, H. et al. (2013). Smart grid communications: Overview of research challenges, solutions, and standardization activities. *IEEE Communications Surveys and Tutorials*, *15*(1), 21–38. doi:10.1109/SURV.2011.122211.00021

Faruqui, A., Hledik, R., George, S. S., Bode, J., Mangasarian, P., Rohmund, I., & Yoshida, S. et al. (2009). *A national assessment of demand response potential*. Federal Energy Regulatory Commission STAFF REPORT, The Brattle Group Freeman, Sullivan & Co Global Energy Partners, LLC.

Ferreira, H., Lampe, L., Newbury, J., & Swart, J. (2010). *Power line communications*. New York: Wiley. doi:10.1002/9780470661291

Galli, S., Koch, M., Latchman, H., Lee, S., & Oksman, V. (2010). *Industrial and international standards on PLC base networking technologies*. New York: Wiley.

Galli, S., Scaglione, A., & Dostert, K. (2003). Broadband is power: Internet access through the power line network (Guest Editorial). *IEEE Communications Magazine*, *41*(5), 82–83. doi:10.1109/MCOM.2003.1200105

Galli, S., Scaglione, A., & Wang, Z. (2011) For the grid and through the grid: The role of power line communications in the smart grid. *Proceedings of the IEEE, 99*(6), 998-1027. doi:10.1109/JPROC.2011.2109670

Garrity, T. (2008). Getting smart. *IEEE Power Energy Magazine*, *6*(2), 38–45. doi:10.1109/MPE.2007.915181

GENI. (n.d.). *Global Environment for Network Innovations (GENI)*. Retrieved from: http://www.geni.net/

Güngör, V. C., & Lambert, F. C. (2006). A survey on communication networks for electric system automation. *Computer Networks*, *50*(7), 877–897. doi:10.1016/j.comnet.2006.01.005

Gungor, V. C., Lu, B., & Hancke, G. P. (2010). Opportunities and challenges of wireless sensor networks in smart grid. *IEEE Transactions on Industrial Electronics*, *57*(10), 3557–3564. doi:10.1109/TIE.2009.2039455

Güngör, V. C., Sahin, D., Kocak, T., Ergut, S., Buccella, C., Cecati, C., & Hancke, G. P. (2011). Smart grid technologies: Communication technologies and standards. *IEEE Transactions on Industrial Informatics*, *7*(4), 529–539. doi:10.1109/TII.2011.2166794

Güngör, V.C., Sahin, D., Kocak, T., Ergut, S., Buccella, C., Cecati, C. & Hancke, G.P. (2013). A survey on smart grid potential applications and communication requirements. *IEEE Transactions on Industrial Informatics, 9*(1), 28-42.

Heydt, G., Liu, C., Phadke, A., & Vittal, V. (2001). Solution for the crisis in electric power supply. *IEEE Computer Applications in Power*, *14*(3), 22–30. doi:10.1109/MCAP.2001.952933

Ho, G. D., & Ngoc, T. L. (2013). Smart grid communications networks: Wireless technologies. In *Handbook of Green Information and Communication Systems* (pp. 115–146). USA Academic Press. doi:10.1016/B978-0-12-415844-3.00005-X

Ho, Q. D., Gao, Y., & Goc, N. (2013). *Challenges and research opportunities in wireless communication networks for smart grid.* IEEE Wireless Communications.

Hopper, N., Goldman, C., Bharvirkar, R., & Engel, D. (2007). The summer of 2006: A milestone in the ongoing maturation of demand response. *The Electricity Journal*, *20*(5), 62–75. doi:10.1016/j.tej.2007.04.007

Hui, S. Y., & Yeung, K. H. (2003). Challenges in the migration to 4G mobile systems. *IEEE Communications Magazine*, *41*(12), 54–59. doi:10.1109/MCOM.2003.1252799

Hwang, T., Park, H., & Paik, E. (2009) Location-aware UPnP AV session manager for smart home. *IEEE First International Conference on Networked Digital Technologies*, 106-109. doi:10.1109/NDT.2009.5272089

Ipakchi, A., & Albuyeh, F. (2009). Grid of the future. *IEEE Power Energy Magazine*, *7*(2), 52–62. doi:10.1109/MPE.2008.931384

Ishii, H., & Francis, B. A. (2002). *Limited data rate in control systems with networks. In Lecture Notes in Control and Information Sciences* (pp. 1–121). Springer New York.

Kabalci, E., Kabalci, Y., & Develi, I. (2012). Modelling and analysis of a power line communication system with QPSK modem for renewable smart grids. *International Journal of Electrical Power & Energy Systems*, *34*(1), 19–28. doi:10.1016/j.ijepes.2011.08.021

Knapp, E. D., & Samani, R. (2013). *Security models for Scada, ICS, and smart grid. Applied cyber security and the smart grid.* USA Waltham Syngress. doi:10.1016/B978-1-59749-998-9.00005-0

Latchman, H., & Yonge, L. (2003). Power line local area networking (Guest Editorial). *IEEE Communications Magazine*, *41*(4), 32–33. doi:10.1109/MCOM.2003.1193971

Li, C. S., Huang, Y. M., & Chao, H. C. (2008). UPnP IPv4/IPv6 bridge for home networking environment. *IEEE Transactions on Consumer Electronics*, *54*(4), 1651–1655. doi:10.1109/TCE.2008.4711216

Li, W., & Zhang, W. (2014). Simulation of the smart grid communications: Challenges, techniques, and future trends. *Computers & Electrical Engineering*, *40*(1), 270–288. doi:10.1016/j.compeleceng.2013.11.022

Li, X., Liang, X., Lu, R., Shen, X., Lin, X., & Zhu, H. (2012). Securing smart grid: Cyber-attacks, counter measures, and challenges. *IEEE Communications Magazine*, *50*(8), 38–45. doi:10.1109/MCOM.2012.6257525

Liang, C., & Yu, F. R. (2015). Wireless network virtualization: A survey, some research issues and challenges. *IEEE Communications Surveys and Tutorials*, *17*(1), 358–380. doi:10.1109/COMST.2014.2352118

Liu, J., Xiao, Y., Li, S., Liang, W., & Chen, C. L. P. (2012). Cyber security and privacy issues in smart grids. *IEEE Communications Surveys and Tutorials*, *14*(4), 981–997. doi:10.1109/SURV.2011.122111.00145

Metke, A. R., & Ekl, R. L. (2010). Security technology for smart grid networks. *IEEE Transactions on Smart Grid*, *1*(1), 99–107. doi:10.1109/TSG.2010.2046347

Mouftah, H. T., & Kantarci, M. E. (2013). Smart grid communications: Opportunities and challenges. In *Handbook of Green Information and Communication Systems* (pp. 115–146). USA Academic Press. doi:10.1016/B978-0-12-415844-3.00025-5

Mumtaz, S., Lundqvist, H., Huq, K. M. S., Rodriguez, J., & Radwan, A. (2014). Smart Direct-LTE communication: An energy saving perspective. *Ad Hoc Networks*, *13*, 296–311. doi:10.1016/j.adhoc.2013.08.008

NIST. (n.d.). *NIST Framework and roadmap for smart grid interoperability standards.* National Institute of Standards and Technology. Retrieved from: http://www.nist.gov/public_affairs/releases/upload/smartgrid_interoperability_final.pdf

PAP-15. (n.d.). *Priority Action Plan (PAP-15): Harmonize power line carrier standards for appliance communications in the home*. Retrieved from: http://collaborate.nist.gov/twiki-sggrid/bin/view/SmartGrid/PAP15PLCForLowBitRates

Pavlidou, N., Vinck, A. H., Yazdani, J., & Honary, B. (2003). Power line communications: State of the art and future trends. *IEEE Communications Magazine*, *41*(4), 34–40. doi:10.1109/MCOM.2003.1193972

PlanetLab. (n.d.). Retrieved from: http://www.planet-lab.org/

Rossberg, M., & Schaefer, G. (2011). A survey on automatic configuration of virtual private networks. *Computer Networks*, *55*(8), 1684–1699. doi:10.1016/j.comnet.2011.01.003

Rus, C., Kontola, K., Curcio, I. D. D., & Defee, I. (2008). Mobile TV content to home WLAN. *IEEE Transactions on Consumer Electronics*, *54*(3), 1038–1041. doi:10.1109/TCE.2008.4637584

Saputro, N., Akkaya, K., & Uludag, S. (2012). A survey of routing protocols for smart grid communications. *Computer Networks*, *56*(11), 2742–2771. doi:10.1016/j.comnet.2012.03.027

Schuelke-Leech, B. A., Barry, B., Muratori, M., & Yurkovich, B. J. (2015). Big data issues and opportunities for electric utilities. *Renewable & Sustainable Energy Reviews*, *52*, 937–947. doi:10.1016/j.rser.2015.07.128

Schwartz, M. (2009). History of communications carrier-wave telephony over power lines: Early history. *IEEE Communications Magazine*, *47*(1), 14–18. doi:10.1109/MCOM.2009.4752669

Sivaneasan, B., So, P. L., & Gunawan, E. (2011). A simple routing protocol for PLC-based AMR systems. *IEEE Transactions on Power Delivery*, *26*(4), 2613–2620. doi:10.1109/TPWRD.2011.2161000

Su, D., & Galli, S. (2010). *PAP 15 recommendations to SGIP on broadband PLC coexistence, PAP 15*. Tech. Rep. Retrieved from: http://collaborate.nist.gov/twiki-sggrid/pub/SmartGrid/PAP15PLCForLowBitRa tes/PAP15_-_Recommendation_to_SGIP_BB_CX_-_Final_-_APPROVED_2010-1 2-02.pdf

Su, H., Qui, M., & Wang, H. (2012). Secure wireless communication system for smart grid with rechargeable electric vehicles. *IEEE Communications Magazine*, *50*(8), 62–68. doi:10.1109/MCOM.2012.6257528

Tan, N.-K. (2003). *Building VPNs: with IPSec and MPLS*. McGraw-Hill Networking.

Tatikonda, S., & Mitter, S. (2004). Control under communication constraints. *IEEE Transactions on Automatic Control*, *49*(7), 1056–1068. doi:10.1109/TAC.2004.831187

Tomsovic, K., Bakken, D. E., Venkatasubramanian, V., & Bose, A. (2005). Designing the next generation of real-time control, communication, and computations for large power systems. *Proceedings of the IEEE*, *93*(5), 965–977. doi:10.1109/JPROC.2005.847249

US Department of Energy. (n.d.). *A survey of wireless communications for the electric power system*. US Department of Energy. Retrieved from: http://www.pnnl.gov/nationalsecurity/technical/secure_cyber_systems/pdf/power_grid_wireless.pdf

Usman, A., & Shadi, S. H. (2013). Evolution of communication technologies for smart grid applications. *Renewable & Sustainable Energy Reviews*, *19*, 191–199. doi:10.1016/j.rser.2012.11.002

Wang, W., & Lu, Z. (2013). Cyber security in the Smart Grid: Survey and challenges. *Computer Networks*, *57*(5), 1344–1371. doi:10.1016/j.comnet.2012.12.017

Wang, W., Xu, Y., & Khanna, M. (2011). A survey on the communication architectures in smart grid. *Computer Networks*, *55*(15), 3604–3629. doi:10.1016/j.comnet.2011.07.010

Wu, F. F., Moslehi, K., & Bose, A. (2005). Power system control centers: Past, present, and future. *Proceedings of the IEEE*, *93*(11), 1890–1908. doi:10.1109/JPROC.2005.857499

Xu, H. Y. (2011). The research of building VPN based on IPsec and MPLS technology. [Springer.]. *Theoretical and Mathematical Foundations of Computer Science Communications in Computer and Information Science*, *164*, 184–189. doi:10.1007/978-3-642-24999-0_26

Yang, Q., Barria, J. A., & Green, T. C. (2011). Communication infrastructures for distributed control of power distribution networks. *IEEE Transactions on Industrial Informatics*, *7*(2), 321–326. doi:10.1109/TII.2011.2123903

Yigit, M., Gungor, V. C., Tuna, G., Tangoussi, M., & Fadel, E. (2014). Power line communication technologies for smart grid applications: Are view of advances and challenges. *Computer Networks*, *70*, 366–383. doi:10.1016/j.comnet.2014.06.005

Zeman, A., Prokopenko, M., Guo, Y., & Li, R. (2008). Adaptive control of distributed energy management: A comparative study. *IEEE International Conference on Self-Adaptive and Self-Organizing Systems (SASO)*, 84-93. doi:10.1109/SASO.2008.60

Chapter 2
Design of Smart Grid Test Bed Using OPNET and PLC

ABSTRACT

In this design unit, a design to test the performances of varying models was developed for the simulations in the PLC-base data link layer. The design includes a smart home and a Smart Grid environment where a comparison between Zigbee and WiMax-based models can be performed. The Smart Grid Test Bed has been designed using OPNET and Power Line Communication is proposed in this book. It is being designed to allow test bed experiments in four layers among OSI 7 layers. This chapter is organized as follows: The Physical Layer and Datalink Layer for Smart Grid Test Bed in Section 1; the Transport Layer for Smart Grid Test Bed in Section 2; and finally, Application Layer for Smart Grid Test Bed in Section.

1. PHYSICAL LAYER AND DATALINK LAYER FOR SMART GRID TEST BED

As for the Test Bed for a Smart Grid, networking was performed on the Physical and Data Link Layer of the OSI 7 Layer, as indicated in the section in Figure 1. The Test Bed experiment was conducted with Zigbee which is popular in the Republic of Korea while both the WiMAX and the Power Line Communication networks have been designed with OPNET for the Test Bed experiments.

DOI: 10.4018/978-1-5225-2776-3.ch002

Figure 1. Physical layer and datalink layer for smart grid test bed

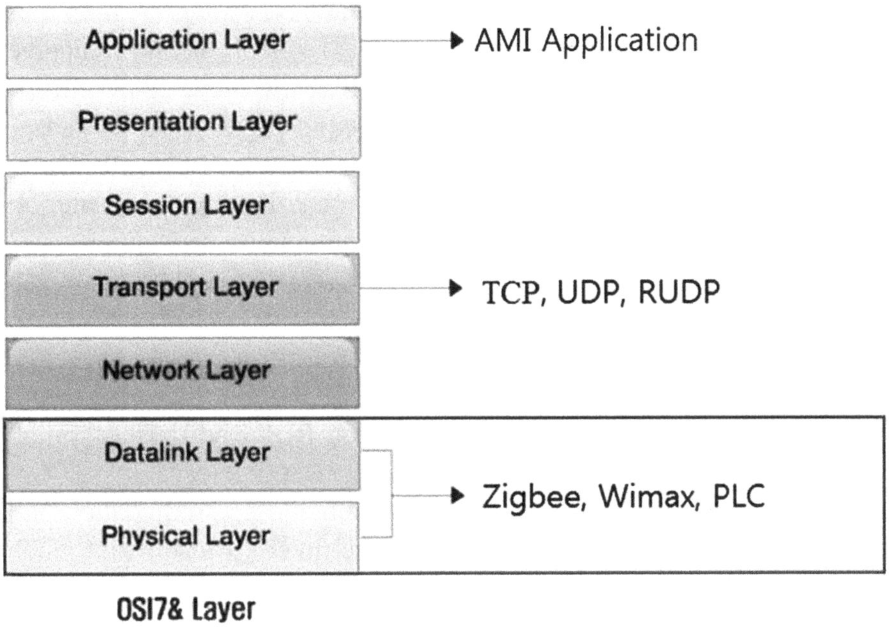

OPNET was used to implement the Smart Grid Test Bed and Power Line Communication has been proposed for the implementation where the Test Bed experiments can be conducted in four out of OSI 7 layer.

2. TRANSPORT LAYER FOR SMART GRID TEST BED

TCP demonstrates rather high packet loss rate and tends to over-respond to the ever-changing network dynamism of wireless networks which often show considerable delay variation. And also, it undergoes back off state repeatedly and enters into a slow starting mode, which in turn leading to the low through put and excessive end-to-end latency. On the other hand, UDP does not consider recovery of lost packets and the problem of out-of-order delivery, resulting in much more data losses. For these reasons, RUDP was proposed at the IETF Internet Draft. In this book, the focus was given to the minimization of data packet delay time with the aim of reducing packet losses while delivering data in due order with a proposed protocol.

Hence, by using the OPNET program under the identical environment, have conducted the comparative analysis simulations between TCP, UDP and RUDP considering 'Retransmission Function', sequence acknowledgement numbers that are used to monitor packet losses, the order of the received packets, and various timers as major testing factors. Figure 2 shows transport layer for Smart Grid Test Bed. As for the Transmission Layer, RUDP was tested and presented as the most suitable protocol that includes benefits of both TCP and UDP for the Power Line Communication networks.

2.1 Design of RUDP Architecture

Being a transport layer protocol, the purpose of RUDP is to provide a solution when UDP is not suitable because of its limitation. RUDP will be most suitable when a situation requires 'Guaranteed-order packet delivery'.

The Selective Repeat Protocol achieves reliable communications as it sends out both the acknowledgements and the timeouts to warrant reliable data transfer. RUDP was designed to manage flow control. Figure 3 describes the architecture of the RUDP.

Figure 2. Transport layer for smart grid test bed

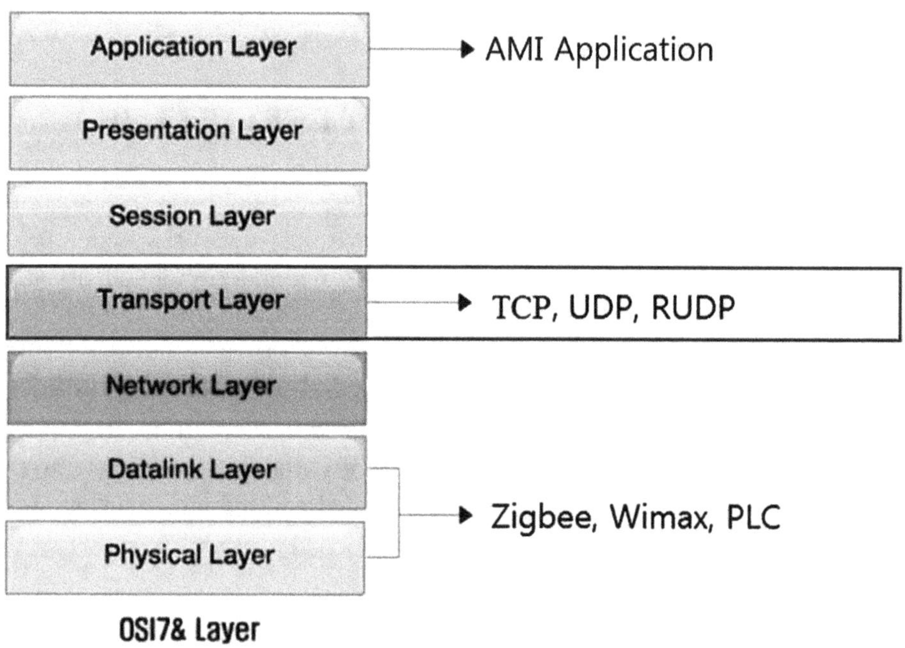

Figure 3. RUDP architecture

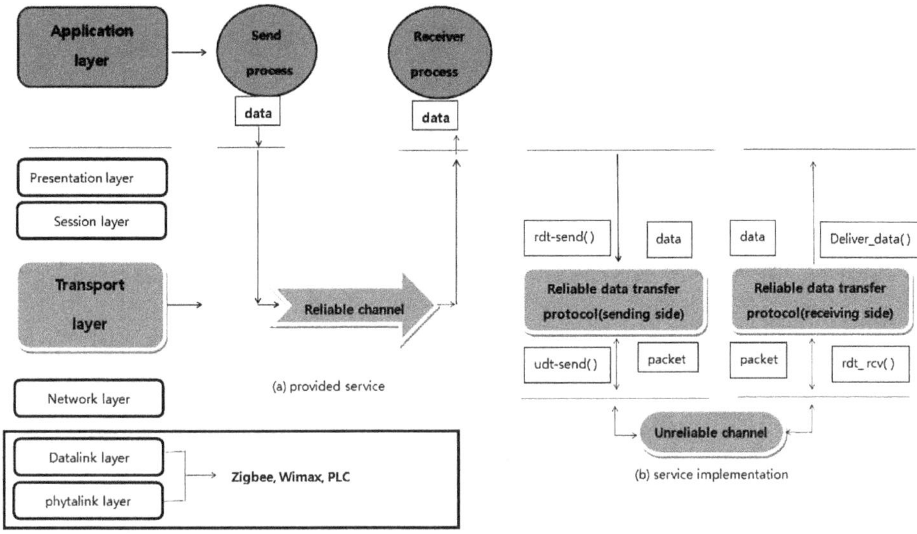

Figure 3 represents the architecture of the RUDP and Figure 3a shows the description of the service offered by the protocol. Sending/Receiving processes are performed on the application layer. By using RUDP, the data is transmitted over a reliable channel. Figure 3b is a brief description of RUDP service implementation.

The sender and receiver both maintain a certain predefined (maximum) window size to avoid communication errors where some packets could be omitted. Also, with RUDP, data can be sent through an unreliable channel but still, its reliability will be guaranteed. The architectures of the sender and the receiver sides are being discussed in other sections in this chapter later. During the implementation process, the author synchronized the shared buffers by using the counting semaphores in order to let only one thread access the buffer at a time. This was to prevent a deadlock situation. The window size can be tracked continuously with the variables ‘Base’ and ‘Next’. When a packet was sent by the sender, the variable ‘Next’ will increase by one. Then, if the Ack packet has been received by the sender, the variable ‘Base’ will also be increased by one.

By repeating these steps, one can track the number of packets in the buffer continuously. The timeouts will be handled by a timer which will be scheduled right after a packet has been sent over the channel. For the protocol, the author has simulated possible packet loss rates and the random network delays for the protocol. The packets will be lined in a queue before being carried over the socket. Each packet in the queue will be assigned with a value corresponding to the current time of the system plus the value of the network delay.

If the current time is equivalent to the value relevant to the packet, the packet will be removed from the queue and sent to the network. That is, the removed packet is to be sent over the channel after a certain time of delay. As for the sequence numbers, the author has arranged the numbers in a way that each packet will be given a specific sequence number and sent over the channel by the sender. The sender will then send the next packet which has a succeeding sequence number after receiving the acknowledgement that has the acknowledgement number equivalent to the sequence number of the packet being sent previously.

2.2 Design of Retransmission Function

Usually, the Receiver node to return an Ack packet to the Sender node once it receives a data packet from the same node but should the Ack packet has not been returned by the Receiver node, the Sender node will still retransmit the packet because of this retransmission function (1, 2). This is to warrant data reliability.

$$\text{Retransmission Timeout} = \left(\frac{n-1}{2}+1\right)RTT \tag{1}$$

$$\text{Max Retransmission Count} = Ceiling\left(\frac{Data\ Packet\ Aging}{Retransmissing\ Timeout}\right) \tag{2}$$

Here, n represents the maximum cumulative Ack counts and RTT indicates the moving average of the round-trips taken, which will be measured by using the time stamps added to a RUDP header as an extension. This performs fewer retransmissions than the ‘Go Back N Protocol’ and the ‘One-Bit Sliding Window Protocol’. The possible limitation of the Selective Repeat Protocol is that this protocol adds more complexity to both the sender and the receiver sides as each frame has to be acknowledged individually. Despite of such limitation, the author chose this protocol over two other protocols mentioned above for implementation of RUDP because of its efficiency and effectiveness. It also satisfies all the requirements for the RUDP implementation. To be more specific, the Selective Repeat Protocol allows the receiver to use only the Positive Acknowledgement (P-Ack) to acknowledge correctly received packet, including the sequence number of that packet explicitly so that RUDP will avoid the use of Negative Acknowledgement (N-Ack) to circumvent the complexity.

3. APPLICATION LAYER FOR SMART GRID TEST BED

Figure 4 shows Application Layer for Smart Grid Test Bed. Meanwhile, as for the Application Layer, an AMI application has been proposed for the Power Line Communications by implementing two scenarios for the platform

Figure 4. Application layer for smart grid test bed

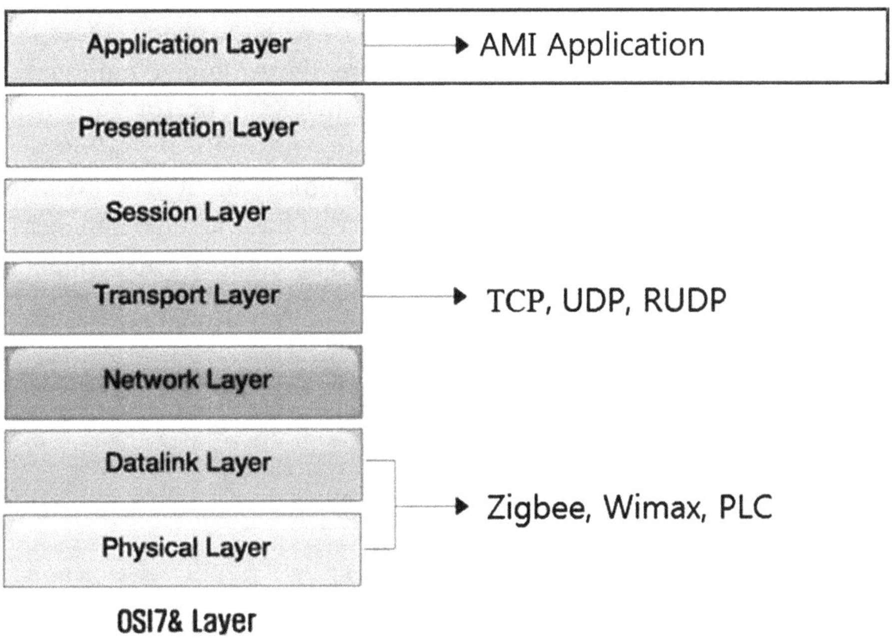

technology of Power Line Communication which uses a 220 Voltage. At the same time, the Intelligent Agent was used for the AMI application to have it tested in the Power Line Communication network.

3.1 Design of the PLC-Based Smart Grid Electric Power Control/Monitoring (Scenario 1)

Figure 5 shows a brief outline of how the mentioned system works. By setting the Arduino circuit board as a terminal device, the home server receives data and performs necessary controls. Also, the serial communication is used by the home server to communicate with the end device through Power Line. In this design, the home server aggregates and stores power use data at the Mysql database every three minutes and an Android application will be linked with this database through PHP. The same process is applied when a certain web is used instead.

The circuitry of the Smart Outlet, which can be considered as a Terminal Client, was designed as Figure 6 and largely divided into power control and measurement sections, MCU, and a communication unit.

First, MCU of Smart Outlet has been designed as Figure 7 using Arduino Uno R The Arduino board used in this design is Atmega-328-based Arduino Uno R This is a micro controller board that has a 32Kb memory unit. Terminal Client controls each module and communicates with the central sever.

through this board. As in the above picture, the Arduino circuit and the Power Line Communication modem are linked to RX/TX port of Arduino and RX/TX port of NC-E Power Line Communication respectively to allow serial communications. Since the Power Line Communication modem broadcasts to every Power Line Communication modem that has the same number, the design has been made to have their own individual number while coding the

Figure 5. The flow chart of the PLC-based Micro Grid electric power control/ monitoring system

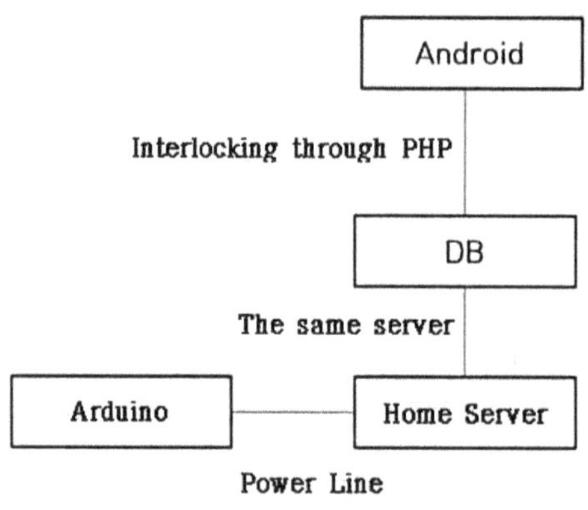

Figure 6. Smart outlet circuitry

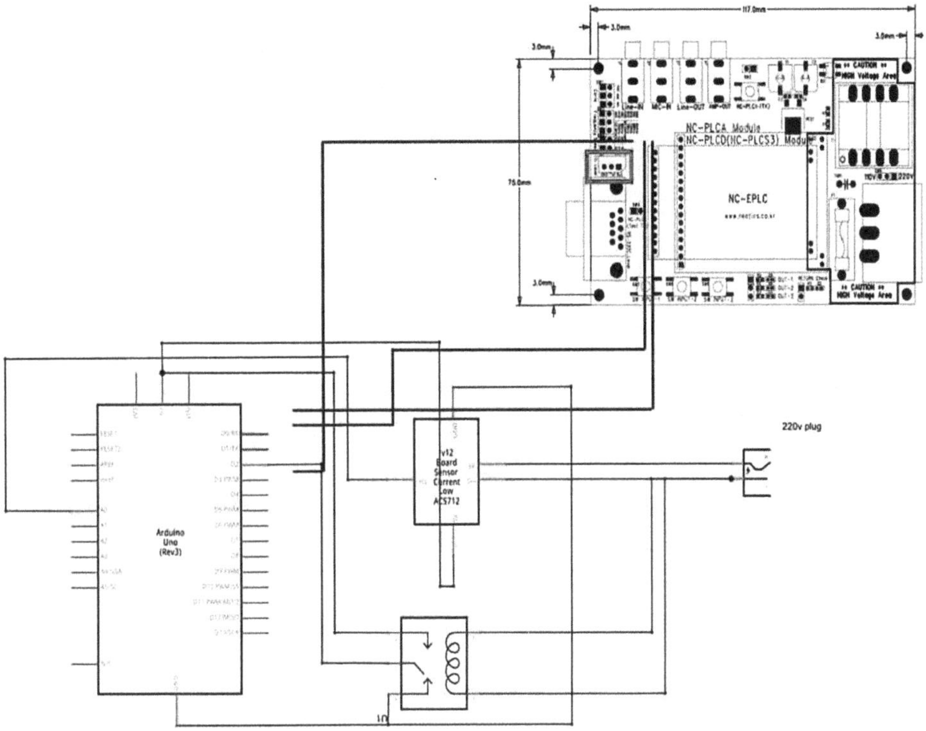

Figure 7. Arduino Uno R3

Arduino unit, so that each Smart Outlet can be distinguished by the central sever.

Figure 8 shows ACS 712. In this design, the current being used by the Smart Outlet can be measured with ACS 712 and its data is then transmitted from the home server in real-time basis. In addition, to deal with transmission errors, relevant power use amount data will be stored on MCU. RMS has been adopted to measure alternating currents and present current measurement value checked by the outlet is transmitted to the server every second to record cumulative power use amount. The source code for transmission has been designed as following.

Figure 9 shows Power control under 220V operation environment. A relay module was used for the power control purpose and the power will be switched on and off through Electric Control Service class at the home server. Actually, this module has been designed to minimize standby power, rather than to control power flow. The results of the experiment suggested that the design was suitable and the system worked properly.

3.2 Design of AMI Application (Scenario 2)

3.2.1. PLC Module

To apply Power Line Communication module, a single Data Concentration Unit (DCU) has been devised by binding it with MCU, which is to control the Power Line Communication module. This DCU plays the role of storing

Figure 8. ACS 712 (contact current measurement sensor)

Figure 9. Power control under 220V operation environment

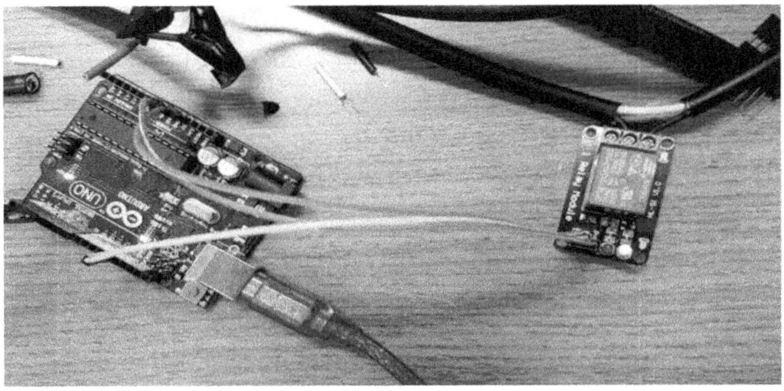

the amount of power usage measured with the electronic ammeter and of delivering it to the main server periodically. Connected to the Power Line Communication module, the usage amount is recorded by the main Server and can be sent to the Smart Phones or external displays depending upon the needs of the user. The DCU composition is shown in Figure 10.

RS-232 communications are performed between Power Line Communication module, MCU and main Server. This is a serial type interface and commonly called a serial port. For this, RS-232 communications must be established between MCU and main Server. Since both target systems (i.e., Rasberry Pi and Odriod) used in this book support RS-232 communications, not much difficulties were anticipated, and because Atmega and Arduino both have the pins available for the RS-232 communication, they were considered to be adequate for the device-to-device communications.

Figure 10. DCU composition

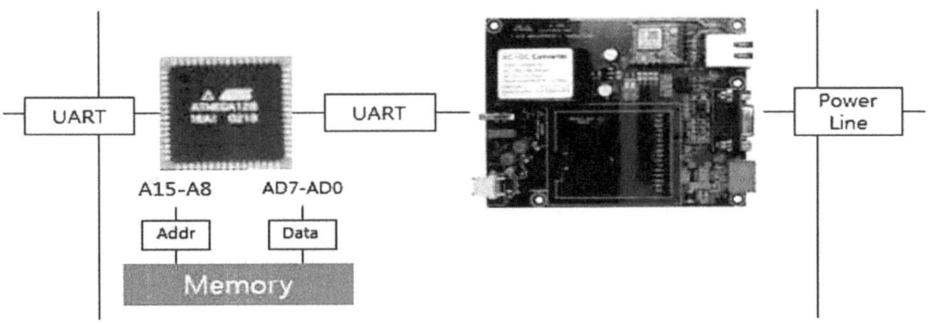

3.2.2 Digital Watt-Hour Meter

This meter has a mechanism that uses a modem communication line external the digital watt-hour meter to establish connection with the field connection device, and then it will be connected to the Power Line Communication module. The communications with the Server will be carried out through the module and Power Line.

Figure 11 shows digital watt-hour meter. The connection between the digital watt-hour meter and the field connection device can be achieved with an optical communication medium [LSIS Co., Ltd.]. This part was supplied by the manufacturer so that the original field connection device was used and it was possible to achieve a real-time monitoring function later with a Smart phone.

3.2.3 Power Line Data Communication Module

The Power Line Data Communication Module is one that is able to transmit/ receive data using the power cables (100V-230V) installed at average households. The features of this product are that first, it does not require any special cable wiring works during the installation process; second, any additional wiring and installation costs and extended construction period are unnecessary or minimal; third, interferences are infrequent and individual or simultaneous control is possible because several transmission frequencies are being used; fourth, it can be installed wherever there are power cables or wall sockets; and finally the fifth, it is a useful product which can be used for the control systems within buildings or for the product manufacturing lines. Figure 12 shows the Power Line Data Communication Module (NC-PLCD Ver 7.0).

Figure 11. Digital Watt-hour meter

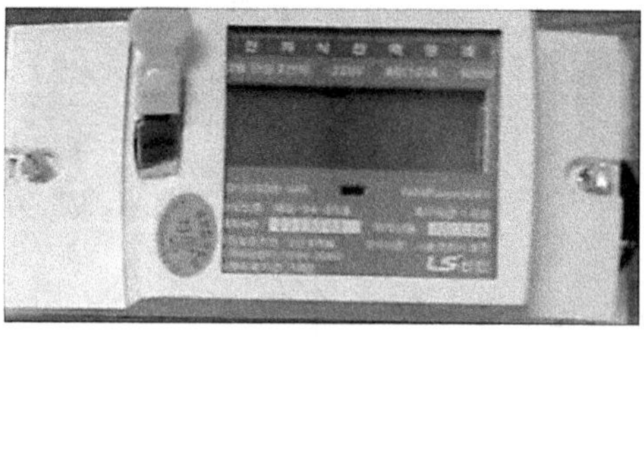

Figure 12. Power line data communication module

3.2.4 Power Line Data Communication Module Developer Unit

The PLCD developer unit is a developer's test board unit to test the basic operations and evaluate the performance of power line communication modules (NC-PLCD, NC-PLCA, and NC-PLCS3 Module). Developers can refer to this unit for the exact utilization of NC-PLCD (data communication module, NC-PLCA (audio communication module) or ND-PLCS3 (remote switch module) as well as for the connection methods for the development of circuits. This developer unit is also a circuit with which one can familiarize themselves with a basic circuit system prior to actual circuit development and learn how to use the user interface circuit which will be developed later. The features of PLCD (AC 100V- 230V) developer unit are as follows:

1. It can test NC-PLCD, NC-PLCA and NC-PLCS3
2. Since it includes the RS-232C Port and TTL Port, immediate communication tests with PC or other user interface boards are possible.
3. Connection with an AC power socket is simple as it has an exclusive connector for AC (100V- 230V) sockets.
4. Allows to set the communication band and communication speed with board's jumper setting alone, without a need of separate setting function.

5. Produced as 'one-board' form so that case-mounting and installment are convenient.
6. Its multi-voltage (100V- 230V) system makes it possible to use it for both 110V and 220V regions.

Figure 13 shows Power Line Data Communication Module developer Unit (NC-EPLC Ver 7.0). PCLD can be utilized as the Power Line Communication device for home automation (i.e., control of air conditioner, TV, refrigerator, curtains or other electric appliances) or security equipment. Also, remote control of sensors, controllers, alarms and other control units is possible and it is useful for the data communications and control operations at the places where the communication environment is not particularly suitable (e.g., factories or other noise-generating places). PLCD's additional usefulness can be observed at the places where it's been considered that the cable wiring works are difficult or impossible; work costs are high; the installment works at homes, offices or factories would incur interruption of operations; installment of expensive wireless communication devices is difficult or the surrounding environment for the installment and operation is poor; and it requires to install remote control and communication facilities with lesser costs.

Figure 13. Power line data communication module developer unit

3.2.5 Entire Functional Diagram

The entire functional diagram is shown in Figure 14. The structure is to establish communications between the watt-hour meter, main server and each electronic device by using the Power Line Communication modem.

Thus, the communication is possible without other exclusive communication lines. As an external connection, internet is used to communicate between Smart Phone and the main server. By connecting to a Smart Phone with WiFi, in-home headcount can be checked by checking the number of people

Figure 14. Entire functional diagram

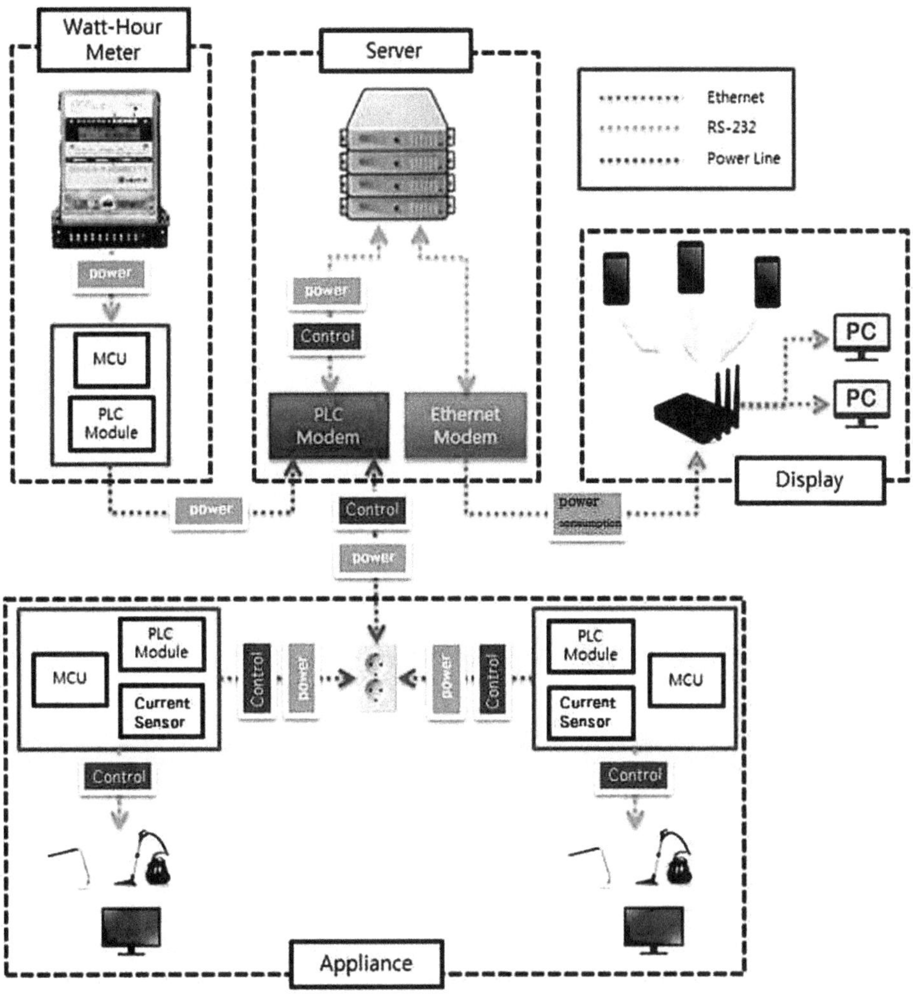

being connected to WiFi. The watt-hour meter communicates with the main server through the connection made with the Power Line Communication modem and the server stores the information received from the meter and transmits them to a Smart Phone when necessary. Each electronic device also communicates with the main server through the Power Line Communication modem and it will be controlled by the external connections or the other designated programs.

Power Line Communication routes have been assigned as following (3): For the involved cost, calculate the variable following the power flow after acquiring the variable on a certain point. Thus, where k is the number of bus i and Ni is the number of lines passing through Pi (3).

$$\text{Main PLC} = \sum_{i=1}^{k} \frac{\partial C}{\partial P_i} + \sum_{i-1}^{k} \left(\sum_{l-1}^{N_1} \mu \frac{\partial Zl}{\partial P_i} \right) + \sum_{i-1}^{k} \left(\sum_{l-1}^{N_i} \mu_l \frac{\partial W_l}{\partial P_i} \right) \tag{3}$$

As such, propose the AMI Application that deduces optimal value by calculating the total cost of entire networks, adding the costs involved in each node's network (4, 5, 6).

$$\text{PLC \#1} = \frac{\partial C}{\partial P_1} + \sum_{l=1}^{N_1} \mu_l \frac{\partial Z_l}{\partial P_1} \tag{4}$$

networks, adding the costs involved in each node's

$$\text{PLC\#2} = \frac{\partial C}{\partial P_2} + \sum_{l-1}^{N_2} \mu_l \frac{\partial Z_l}{\partial P_2} \tag{5}$$

Where is the number of lines passing through.

$$\text{PLC\#3} = \frac{\partial C}{\partial P_3} + \sum_{l-1}^{N_3} \mu_l \frac{\partial Z_l}{\partial P_3} \tag{6}$$

Where is the number of lines passing through.

In-home headcount checking algorithm is shown in Figure 15. It assigns owner's IP address to H_ip. When the user makes a connection with Wi-Fi, the IP address of the router currently being connected will be stored at N_ip

Figure 15. In-home headcount checking algorithm

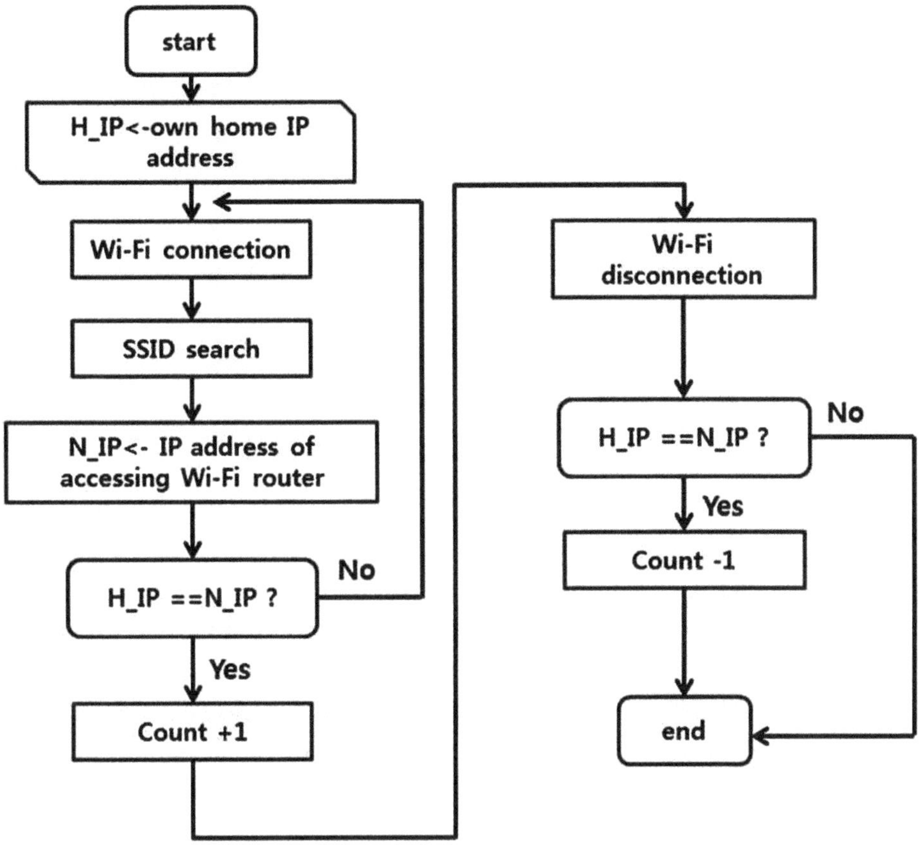

after searching SSID. Then, by comparing H_ip with N_ip, the algorithm increases the count by 1 if they are identical or go back to the initial Wi-Fi connection process if they are different. Meanwhile, an event where the Wi-Fi link was disconnected can be anticipated, and in that case, the count will be decreased by 1 if the comparison result is identical. Here, H_ip is a variable indicating stored home owner's router IP address while N_ip is a variable showing the stored IP address of the router being in access. Lastly, 'Count' indicates the number of people connected to Wi-Fi.

The In-home headcount system is shown in Figure 16. System core is composed of 4 elements:

1. Server
2. Device Manager

Figure 16. In-home headcount system core

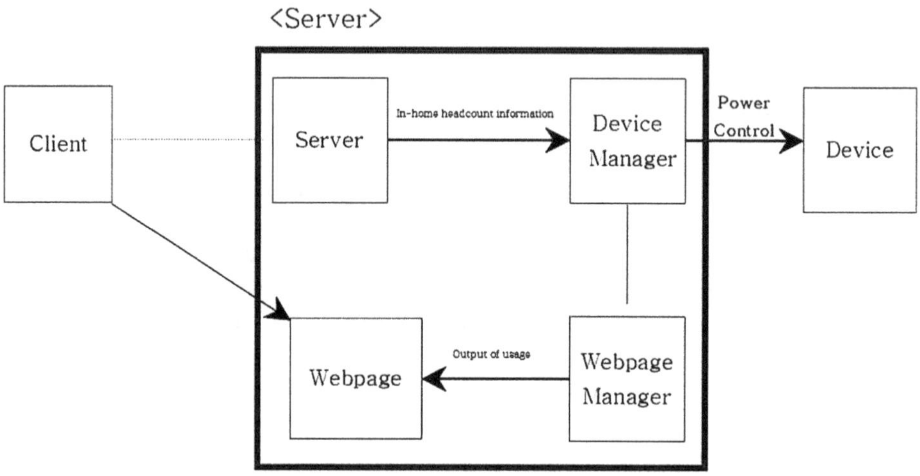

3. Log Manager and
4. Webpage Manager.

As for the direct communication with the Client, the Server first receives a signal sent by the Client after checking the current in-home headcount and control the devices through DM. If the Client sends 'In' signal, the variable value 'Home People' becomes '++', and if the signal is 'Out', it becomes '--'.

The Device Manager receives an information about the in-home headcount and issues commands to control power supply. Switch ON command will be given to the power supply if Home People ≧ 1, judging there is a person (or persons) inside. Likewise, OFF command will be sent if Home People = 0, considering that there are no people inside of home. The Log Manager has no actual influences to the system but it generates Log data after checking the recording time and successes or failures of operations every four seconds to conduct performance evaluations when needed. The Webpage Manager outputs and updates the power use amounts on a Webpage to make it easier for the client to understand the situation.

3.3 Design of Security for AMI Application

Figure 17 shows the graphic-based network intrusion detection system for the AMI that has been proposed in this book and at first, the attacker (or a zombie PC) sends a large volume of traffics towards targeted server through

Figure 17. Graphic based NIDS for AMI application

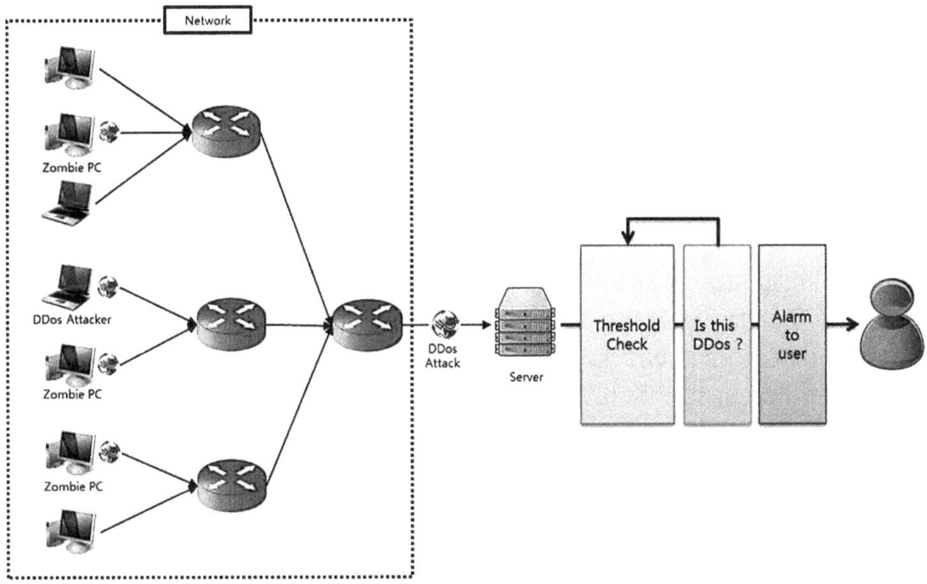

nearby router. The transmitted traffics paralyze the web server and in such a case, this solution makes a judgement about the possibility of DDoS attack by calculating traffic threshold every five seconds to warn the user.

This program's features are as follows: notify the attack events to the user when they are being initiated; real-time packet send/receive volumes are indicated with graphs; actual packet volumes are shown numerically; IPs and related information of attackers are to be displayed at the bottom of main screen; and the sources of packets are described using the Google map.

A DDoS attack event is determined by comparing the packet volume against the threshold value every five seconds and should it be determined as a DDoS attack; the user will be notified with the tray icon. The level of attack is classified as 'Suspicious', 'Warning' or 'Dangerous', depending on the difference when compared with the threshold value.

The purpose of implementing this system was to assist users to manage their servers more easily and stably by providing an intuitive interface that secures stability in server management or homepages for the entry-level operators who lack the knowledge of server management and security.

While network analysis tools are difficult to use and need professional skills, the system suggested in this book made it easier for non-professional users to utilize the tools by showing the analysis results graphically and just checking them with a few clicks. On the other hand, one of the characteristics of the program is that it has a function which warns the traffics reaching the threshold value initially set. That is, a possible DDoS attack packet(s).

This scheme has been implemented by using Jpcap library based on Java in a Window 7 development environment. Since it is possible to capture and analyze the packets that come in and out of one's own network through Jpcap library, developing the system for the network intrusion detection purpose was possible.

Analyzed pattern was used to update packet's threshold periodically. And, in order to determine if the packet is originated from the suspicious regions, message IP has been checked with the packet header information obtained. Information regarding Send/Receive packets volume and abnormal packets are handled with Java classes and can be output to user's monitor. These information (e.g. Send/Receive IP, protocol, the number of communications made and all sorts of flags) will be able to be stored as logfiles.

In adaption stage, analysis of the network installed earlier was needed. The analysis can be approached with statistical method. For instance, supposing that three kinds (A, B, and C) of incoming/outgoing sockets are to be collected for a period of 's' seconds and for 'n' times, where Cs is the number of A packets collected for the period of s seconds, then the number of packets of A can be calculate (7) as

$$T_A = \sum_{i-1}^{n} C_{s_i} \tag{7}$$

Accordingly, the mean of the number of A packets entering for s seconds (8) will be

$$M_A = \frac{T_A}{n} \tag{8}$$

Therefore, the variance for each packet type can be calculate (9, 10, 11) as

$$V_A = \frac{\sum_{i-1}^{n} \left(C_{s_i} - T_A\right)^2}{n-1} \tag{9}$$

$$V_B = \frac{\sum_{i-1}^{n}\left(C_{s_i} - T_B\right)^2}{n-1} \tag{10}$$

$$V_C = \frac{\sum_{i-1}^{n}\left(C_{s_i} - T_C\right)^2}{n-1} \tag{11}$$

The standard deviation SA can be calculated with square root of the variance VA. By using calculated variance VA and the standard deviation SA of A packets, it is possible to learn the average number of the normal packets, the range of abnormal packets and probabilities, for a period of 's' seconds. However, even if the network systems are the same, the results may vary depending on the access patterns and network conditions. Thus, the accuracy will be increased if the detection patterns (i.e., n times for s seconds) are organized irregularly across a variety of time zones. The current threshold of traffic can be found using the mean and standard deviation of already calculated for A packets.

This chapter describes a methodology with which researchers or students can design and extend a Smartgrid test bed. It is hoped that the proposed design technique will be quite useful to them. The additional test bed environments and methodologies can be found from Liu, Sigle, and Dostert (2011), Guymon, Brown, and Ramirez (n.d.), Zaballos, Vallejo and Selga (2011), Papadopoulous et al (2013), Berger, Schwager, and Escudero-Garzas (2013), Kim, Bae, and Kim (2011), Liu et al. (2010), Yan et al (2012), Aalamifar, Hassanein, and Takahara (2012), and Artale et al. (2013).

REFERENCES

Aalamifar, F., Hassanein, H. S., & Takahara, G. (2012). Viability of powerline communication for the smart grid. *Communications (QBSC), 2012 26th Biennial Symposium on*. IEEE.

Artale, G., Cataliotti, A., Cosentino, V., Cara, D. D., Fiorelli, R., Russotto, P., & Tine, G. (2013). Medium voltage smart grid: Experimental analysis of secondary substation narrow band power line communication. *IEEE Transactions on Instrumentation and Measurement*, *62*(9), 2391–2398. doi:10.1109/TIM.2013.2270924

Berger, L. T., Schwager, A., & Escudero-Garzas, J. J. (2013). Power line communications for smart grid applications. *Journal of Electrical and Computer Engineering*, *2013*, 1–16. doi:10.1155/2013/712376

Guymon, M., Brown, K. J., & Ramirez, C. (n.d.). *Smart grid over power line communication network*. U.S. Patent Application No. 13/321,842.

Kim, Y., Bae, J. N., & Kim, J. Y. (2011). Performance of power line communication systems with noise reduction scheme for smart grid applications. *IEEE Transactions on Consumer Electronics*, *57*(1), 46–52. doi:10.1109/TCE.2011.5735480

Liu, J. (2010). Application of power line communication in smart power consumption. *Power Line Communications and Its Applications (ISPLC), 2010 IEEE International Symposium on IEEE*. doi:10.1109/ISPLC.2010.5479945

Liu, W., Sigle, M., & Dostert, K. (2011). Channel characterization and system verification for narrowband power line communication in smart grid applications. *IEEE Communications Magazine*, *49*(12), 28–35. doi:10.1109/MCOM.2011.6094003

Papadopoulos, T., Kaloudas, C. G., Chrysochos, A. I., & Papagiannis, G. K. (2013). Application of narrowband power-line communication in medium-voltage smart distribution grids. *IEEE Transactions on Power Delivery*, *28*(2), 981–988. doi:10.1109/TPWRD.2012.2230344

Yan, Y., Qian, Y., Sharif, H., & Tipper, D. (2012). A survey on cyber security for smart grid communications. *IEEE Communications Surveys and Tutorials*, *14*(4), 998–1010. doi:10.1109/SURV.2012.010912.00035

Zaballos, A., Vallejo, A., & Selga, J. M. (2011). Heterogeneous communication architecture for the smart grid. *IEEE Network*, *25*(5), 30–37. doi:10.1109/MNET.2011.6033033

Chapter 3
Implementation of Smart Grid Test Bed Using OPNET and PLC

ABSTRACT

In this chapter, a design that allows testing of the performances of various models was developed with OPNET for the simulations in the PLC-base data link layer. As the model proposed earlier, the design includes a smart home and a Smart Grid environment where a comparison between Zigbee and WiMax-based models can be performed. The Smart Grid Test Bed has been implemented using OPNET and Power Line Communication is proposed in this book. It is being designed to allow Test Bed experiments in four layers among seven OSI layers. This chapter is organized as follows: the physical layer and datalink layer for Smart Grid Test Bed in Section 1; the transport layer for Smart Grid Test Bed in Section 2; and finally, application layer for Smart Grid Test Bed in Section 3.

1. PHYSICAL LAYER AND DATALINK FOR SMART GRID TEST BED

A comparative analysis of performances between Zigbee and WiMAX has been conducted by deploying twelve terminal nodes in each home network that uses either of these three communication technologies. For the Zigbee scenario, Zigbee Coordinator, six Zigbee Routers and twelve Zigbee end-devices were deployed whereas one Server, two Routers, and six base stations, and twelve terminal nodes were deployed for the WiMAX scenario to compare

DOI: 10.4018/978-1-5225-2776-3.ch003

data transfer rates between a coffee pot and the server, as an example. Figure 1, Figure 2 and Figure 3 represent Mobile WiMAX and Zigbee and Power Line Communication scenarios respectively and show complex network systems in a Smart Grid Test Bed.

Figure 4 shows scenarios of performance evaluation. The Mobile WiMAX-based node model and router model will be demonstrated in the future extended study. The four separate electric cookers are being displayed in each scenario but this is to show the readers that other appliances can be used in the system replacing the coffee pot but the same interface will be used for the system. Applicable appliances could be automated vacuum cleaners, Smart Closets, or CCTVs with mobility and intelligence, all of which are expected to be produced in the future.

1.1 Implementation of Zigbee Architecture for Smart Grid Test Bed

Zigbee refers to the standard technology for the small-scale wireless Mesh networks that feature low-cost, low-power, long battery life, higher security and low-level transmission speed. Because of its low-cost characteristic, a wide range communication is possible with low-power requirement by deploying

Figure 1. Mobile WiMAX scenario

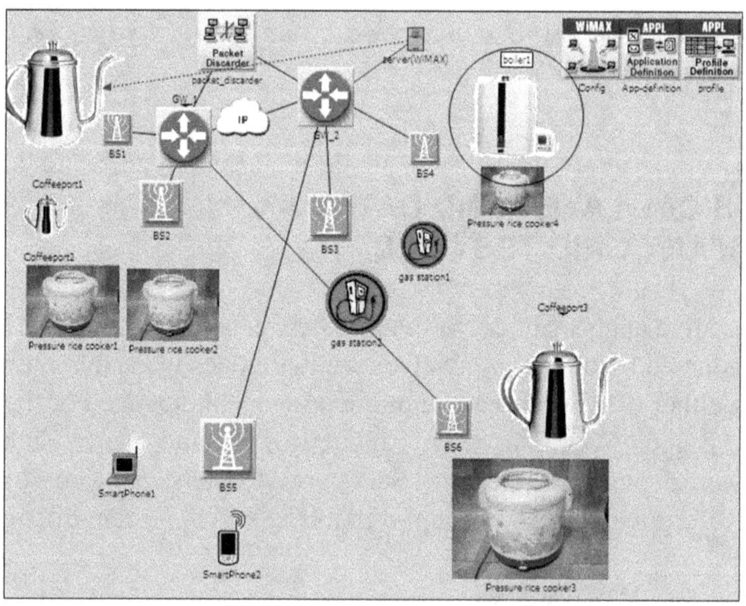

Figure 2. Zigbee scenario

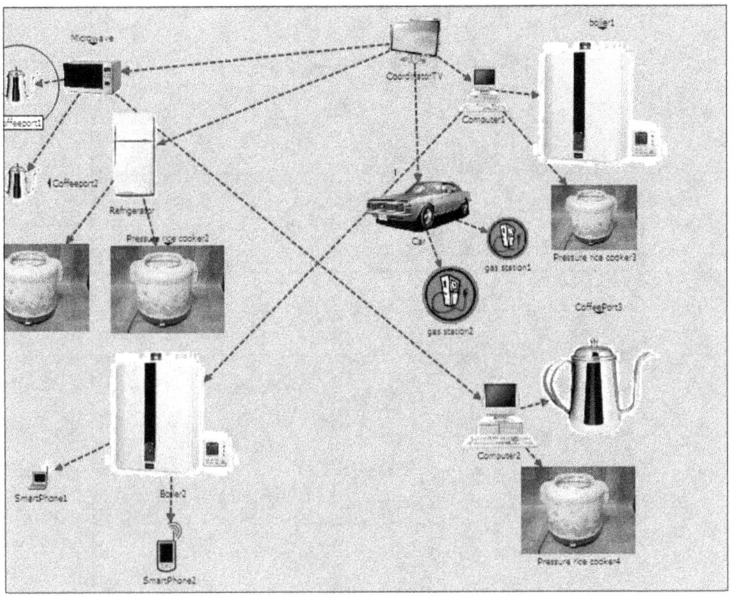

Figure 3. Power line communication scenario

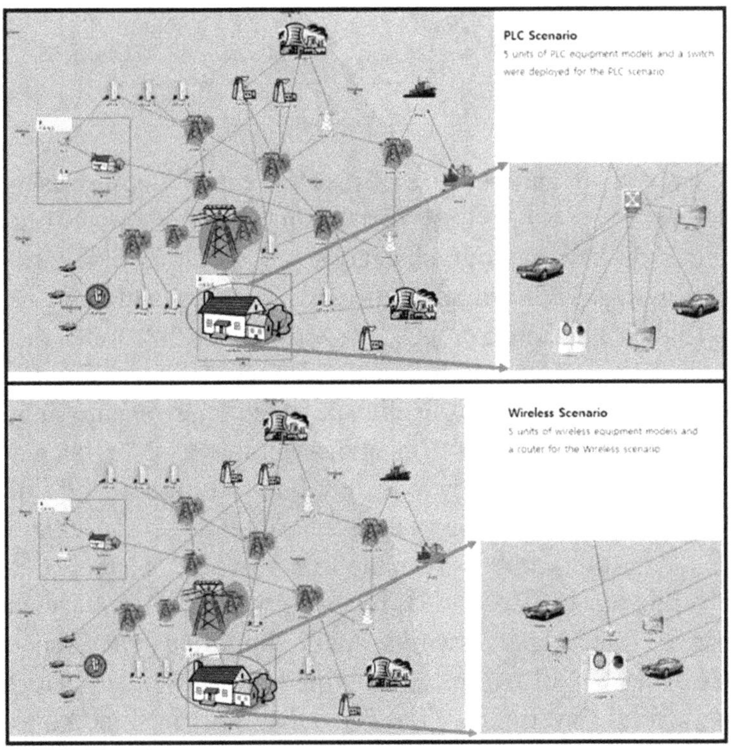

Figure 4. Scenarios of performance evaluation

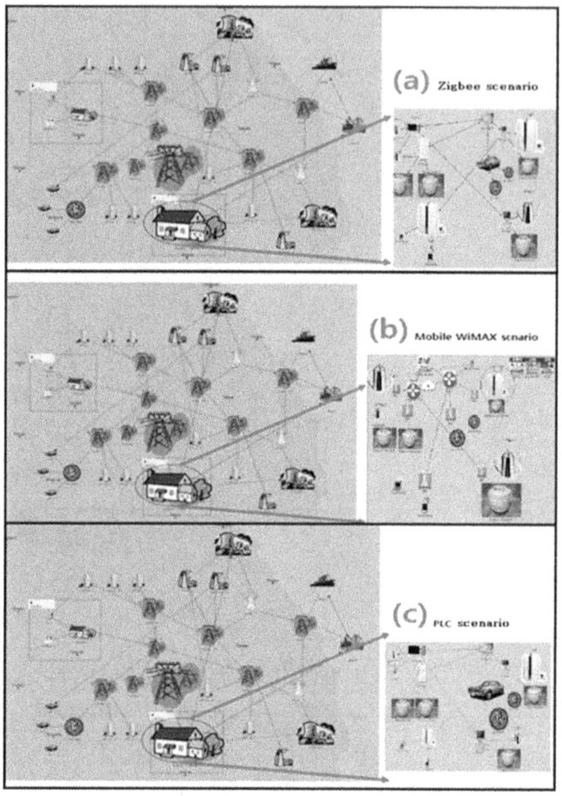

nodes on many spots and because of its efficient low-power usage, a long battery life can be expected as well. Further, since the Zigbee network provides 128bit symmetric-key cryptography, it offers higher level of security. Zigbee basically follows OSI 7 Layer and defines the physical and MAC layers with IEEE 802.15.4, in addition to proceeding with standardization of the Network layer, Application layer and ZDO for the Zigbee Alliance. The PHY Layer defined in IEEE 802.15.4 is basically subjected to a transmission speed of 250kbit/sec within the range of 10m and several layers have been defined to make the best use of its low-power feature but in this book, the experiments have been conducted according to the scenario under 2.4GHz and 250kbps conditions. Also, the access was made to a wireless channel on the MAC sub-layer defined in IEEE 802.15.4 by using the CSMA/CA mechanism.

The Star, Tree and Mesh networks will be supported in the Network Layer and the tree-type network has been configured for the simulation for this case. In the Application Layer, the FTP service that transmits one kbit-sized file have been used. As in Figure 5, Zigbee refers to a standard technology that features the low-cost and low-power functions, long battery life, security and low-level transmission speeds for the small-scale Wireless Mesh Networks. Because of its low-cost characteristic, wide-range communications can be

Figure 5. Architecture of terminal nodes in Zigbee

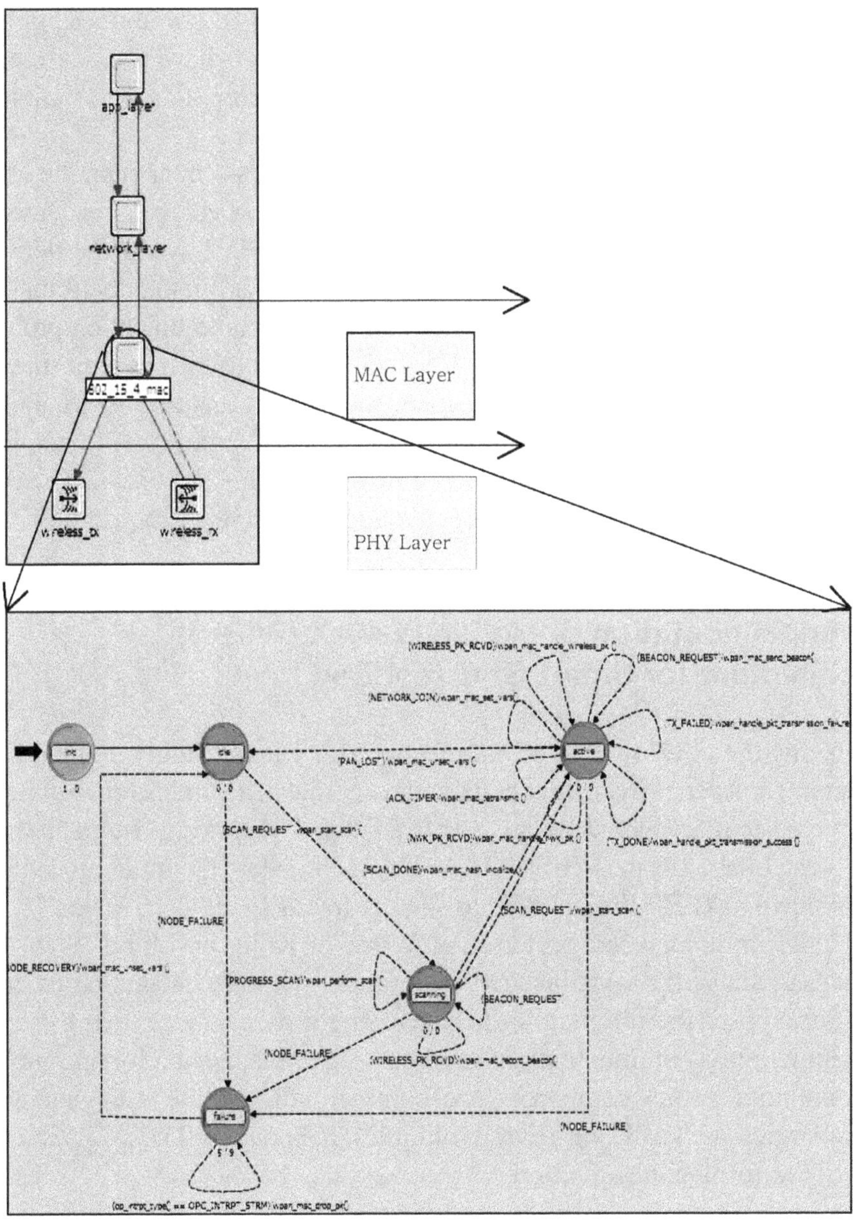

possible with a low-power by deploying a large number of nodes in many places. This low-power characteristic can result in extended battery life as the battery will be used efficiently. Additionally, a Zigbee network will have higher security as it offers the 128bit symmetric-key cipher algorithm. Zigbee basically follows OSI-7 layer and the IEEE 802.15.4 defines both the Physical and the MAC Layers. The Network and Application Layers and ZDO are defined by the Zigbee Alliance. Although the Physical Layer provided by IEEE 802.15.4 basically premises a transmission speed of about 250kbit/s within the distance of 10m and several Physical Layers have been defined to emphasize Zigbee's low-cost and low-power characteristics, simulation on the condition of 2.4GHz and 250kbit/s.

At the same time, the access was made to a wireless channel on the MAC sub-layer provided by IEEE 802.15.4 by using the CSMA/CA mechanism.

In an event that 'Idle State' receives data from the PHY Layer or the Network Layer, it will have three choices. First, 'Idle State' will make the transition to 'Failure State' if there is any problem during the communication process, and then reverse the transition to 'Idle State' again after a certain interval without trying to fix the error. Second, make the transition to 'Scanning State' which scans the surrounding nodes. The transition to either 'failure state' or 'active state' will be decided following the result of scan. The third choice is that if the transition has been made to 'Active State', 'Idle' will be used to suit the purpose.

1.2 Implementation of Mobile WiMAX (WiBro) Architecture for Smart Grid Test Bed

Following the development of internet, users are increasingly demanding a system(s) that can provide higher data transfer rate in a wireless environment. Therefore, to meet such demands, the IEEE 802.16 Working Group has been newly established by the IEEE 802 LAN/MAN Standard Committee (LMSC), also known as IEEE Project 802, to standardize IEEE 802.16 series.

This technology is being referred as 'Wibro' in Korea or 'Mobile WiMAX' overseas. Since the existing 3rd generation wireless technologies have already exposed their limitations in providing the data services to the users in current wireless communications market, both domestic and foreign mobile telecommunications companies are commercializing the 4th generation technologies such as Long-term Evolution (LTE) and WiIMAX (Wibro) in priority to resolve the situation.

1.2.1 Terminal Nodes Model in Mobile WiMAX

Figure 6 shows deployed terminal nodes in a network of Mobile WiMAX nodes. For Mobile WiMAX, a separate PHY process model will not be used on the PHY layer while using OPNET Modeler 14.5 PL8 so that most of the functions are implemented at the pipeline stage, or on the MAC layer.

On the other hand, a typical example of a function that seems to belong to PHY layer but implemented at the MAC stage instead is 'Time Division Duplex (TDD)', which is utilized in our simulation. 'Time Division Duplex refers to a function of time sharing. A single frequency is shared internally by designating it as transmission or reception frequency to enable bi-directional communications. Both the framing and transmission timing functions are needed to implement 'Time Division Duplex and being achieved at the MAC stage in OPNET model.

Figure 6. Architecture of terminal nodes in mobile WiMAX nodes

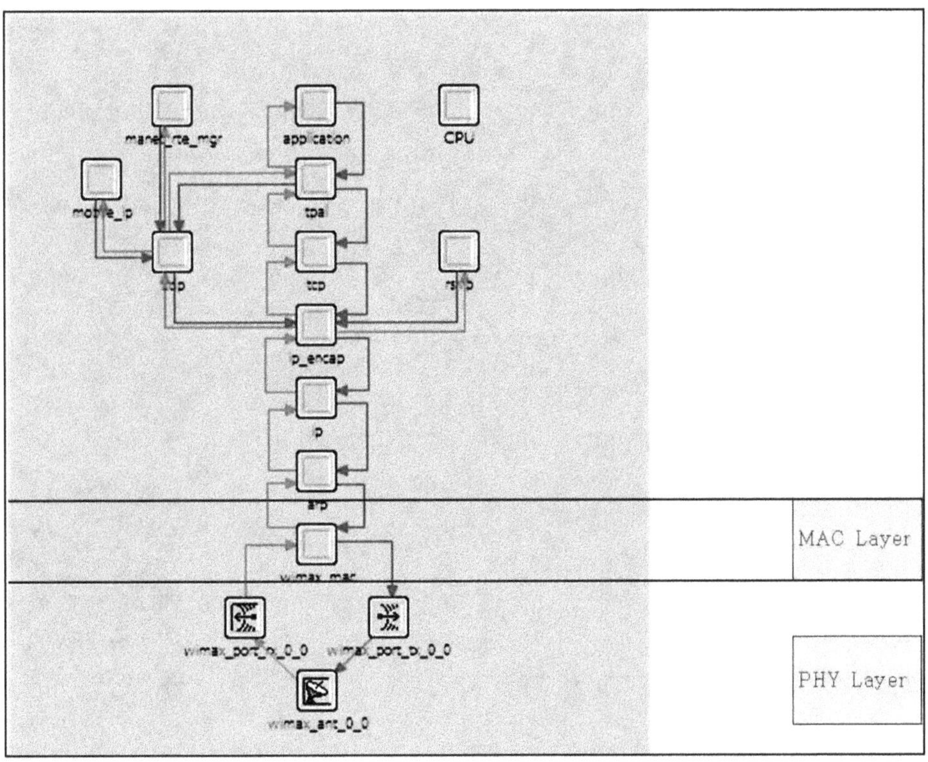

1.2.2 WiMAX_MAC Process Model

Figure 7 represents implemented WiMAX_MAC process model that shows the routes of all transmission and reception packets take. The packets received from the IP layer are of pure data packets but those from the PHY layer are mixed with the control packets. The former will be processed directly in this model but the latter will be handed over to the child-process.

The state transition, from idle state to hl_pk state, will take place once the data packet is received from the IP layer. At this state, a suitable CID will be determined based on the service class and MAC address, and the packet will be stored in this CID. CID is comprised of information concerning the bandwidth, data storage queque and QoS. The data stored in each CID will be transmitted to the PHY layer after acquiring a bandwidth and adapting to QoS while the packet received from the PHY layer will be sent directly to the IP layer after passing through the reassembly process, without acquiring any bandwidth.

Figure 7. WiMAX_MAC process mode

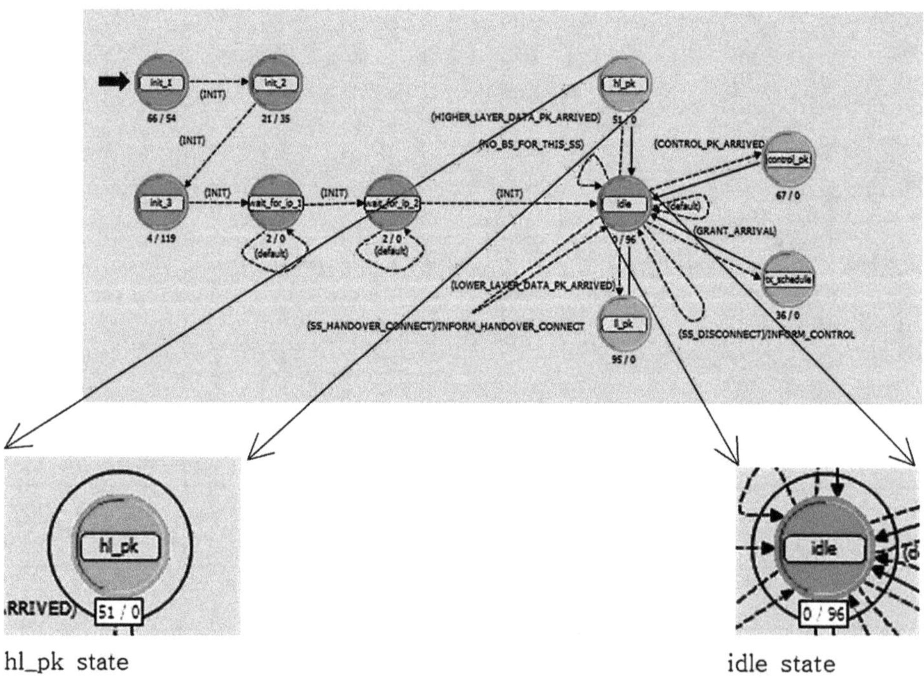

hl_pk state

idle state

1.3 Implementation of PLC Architecture for Smart Grid Test Bed

This section is organized as follows: the PLC Router node model in paragraph section 1; the PLC terminal node model in paragraph section 2; the PLC link model in paragraph section 3; and finally, PLC palette model in paragraph section 4.

1.3.1 PLC Router Node Model

Figure 8 shows a PLC Router node model which has been developed based on the route model to perform generating, discarding and relay functions, and the model supports OSPF, RIP and any other routing protocols used in

Figure 8. PLC router node model

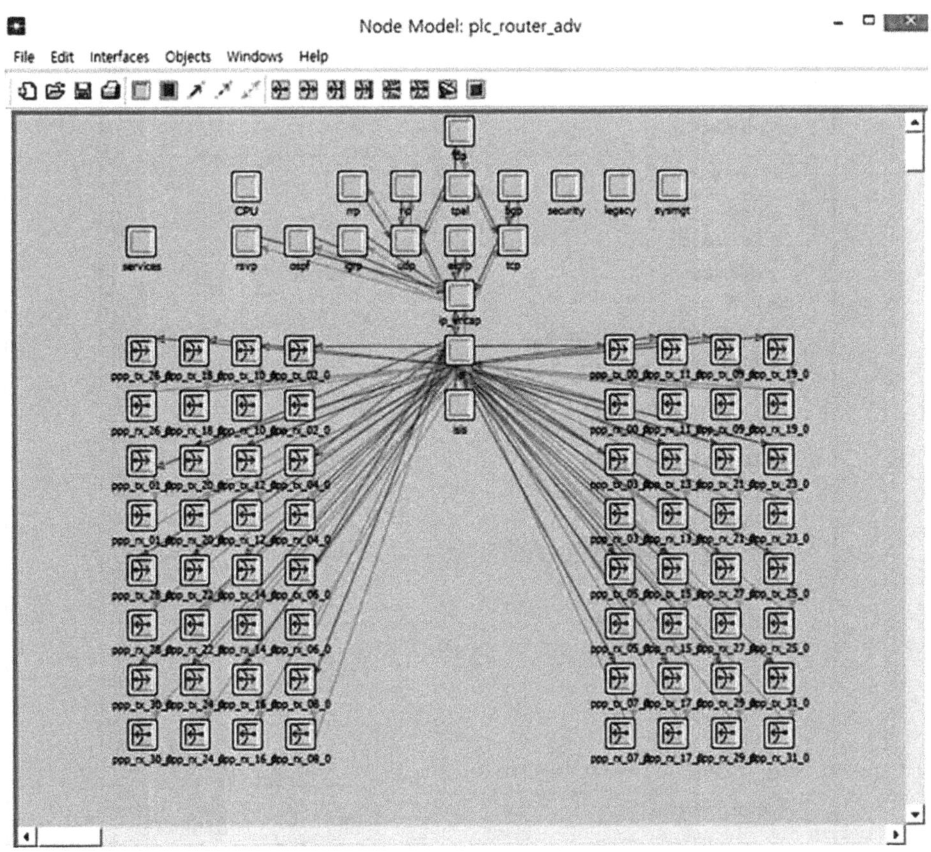

OPNET. As a linking interface with other node models, the PPP interface was used because of its extendability, which was taken into consideration while designing the process model. This design allows the linkages with 32 nodes max. and should the number of the links exceed 32, the situation can be solved by linking two or more relay nodes. Respective router node model takes a 'Start type' form and contains all the characteristics of router provided by OPNET.

Figure 9 shows attributes of Power Line Communication Router node model. The router node model includes all the attributes of routers provided by OPNET and is possible to set IP address, routing protocol, and other parameters.

As seen in Figure 10, since 32 nodes can be connected to a single router, star-form structure will be possible. In this form, however, if the router in the center fails to function, the connection between the A-network and the

Figure 9. Attributes of PLC router node model

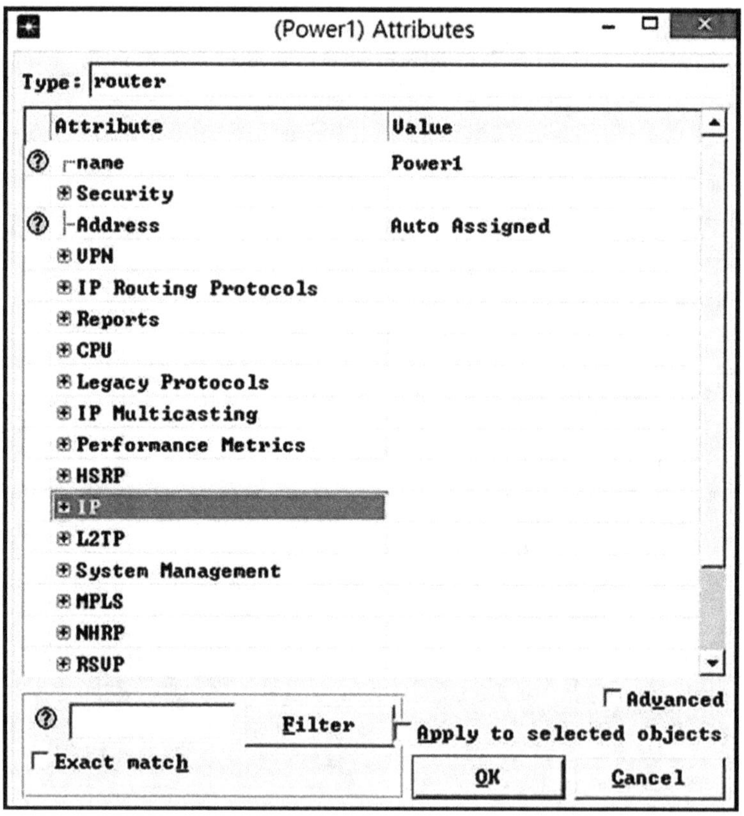

Figure 10. Power line communication router node model mechanism

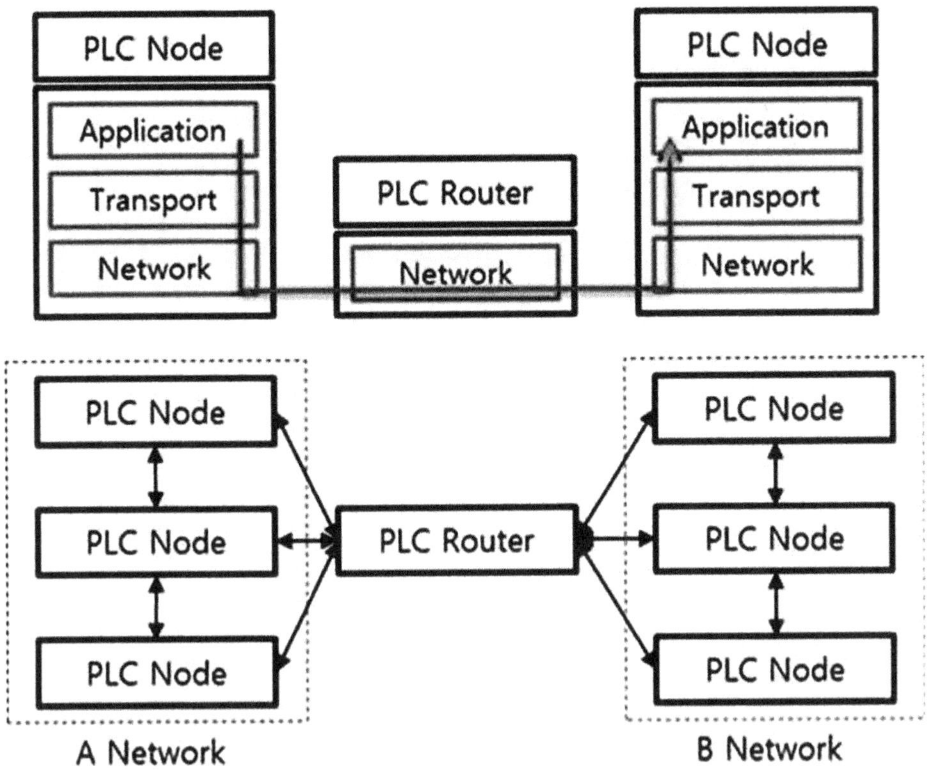

B-network becomes impossible. Even so, the connection using the router(s) is preferable because there will not be much trouble in terms of entire network level even if some of the nodes fail.

1.3.2 Power Line Communication Terminal Node Model

The Power Line Communication terminal node model can generate and discard IP packets but cannot perform the relay function. It was designed to provide a basic IP protocol and to use UDP and TCP as the transport protocols. Like the Router model, PPP interface has been used as a linking interface with other node models.

This is used as an end-to-end terminal model and can be connected to only one adjacent node. Figure 11 shows a Power Line Communication terminal node model that includes all the terminal attributes provided by OPNET, and it is possible to set IP address and transport protocol parameters, and etc.

Figure 11. Power line communication terminal node model

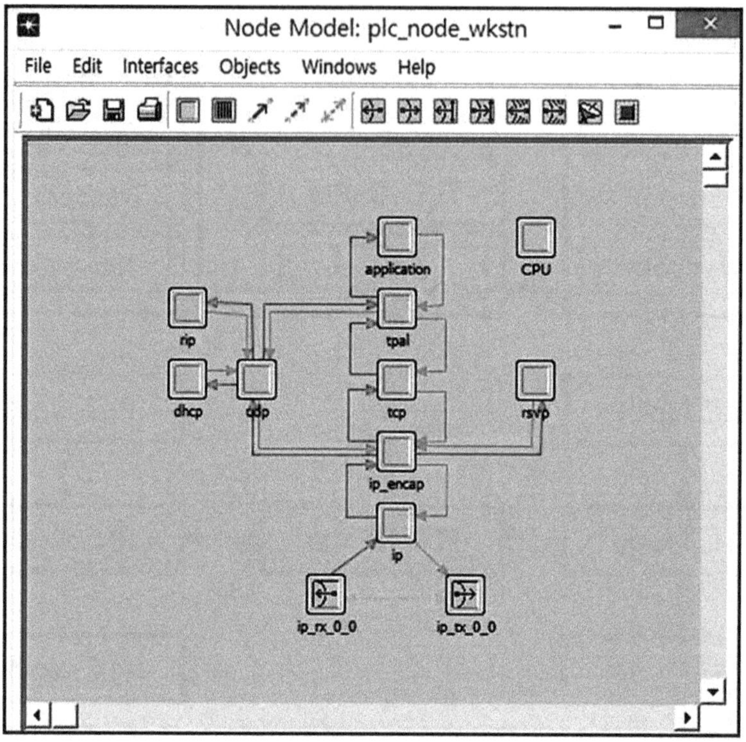

Figure 12 shows attributes of Power Line Communication terminal node model. And as shown in Figure 13, the connections between the nodes are made up with one-to-one connection so that in order to connect with remotely located node, one has to go through another node(s). Thus, ring-form connection can be established. However, in this form, failure in one node could lead to whole network failure and such phenomenon can be fixed with FDDI. Meanwhile, the terminal node model incorporates all the attributes of terminals provided by OPNET and is possible to set IP address, transmission protocol parameters.

1.3.3 PLC Link Model

The link model refers to a model that links nodes. That is, traffics flow between the nodes. This Power Line Communication link model has been implemented based on the PPP model and the default bandwidth (for data rate) was set

Figure 12. Attributes of power line communication terminal node model

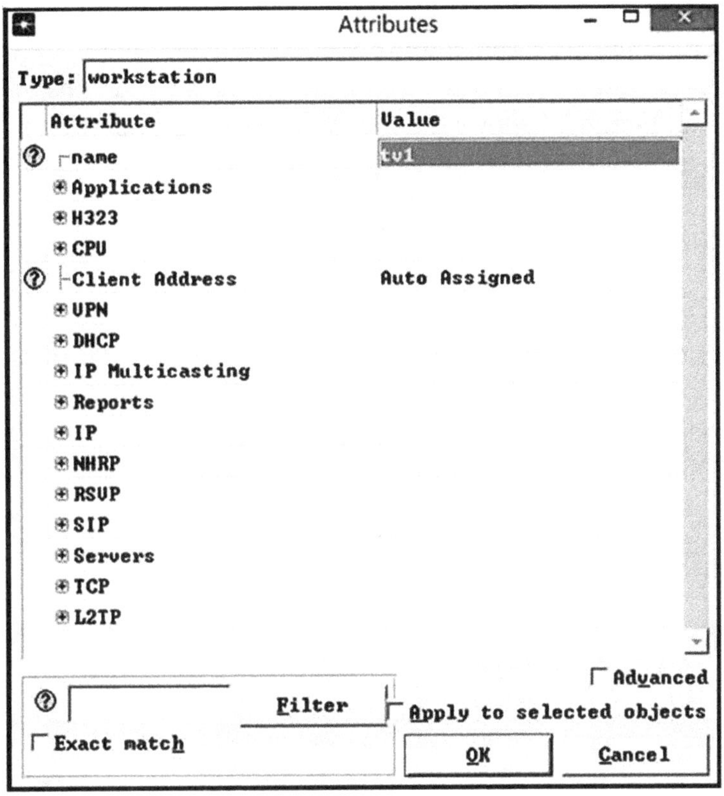

Figure 13. Power line communication terminal node model mechanism

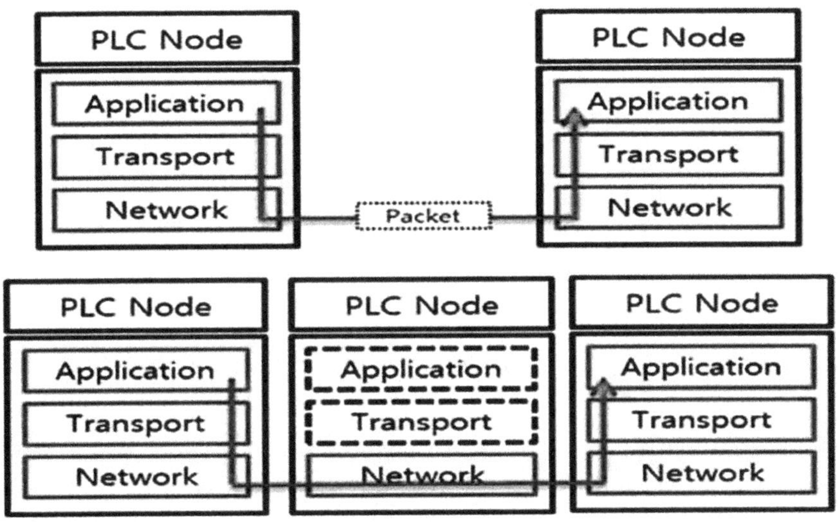

at 11Mbps. The linking speed (bandwidth) of each node may vary so that it was made the way that the user could change the linking speed if necessary.

In order to alter the bandwidth of the link model, right-click the model and then open 'Edit Attributes' to select or input desired bandwidth to set the data rate. Figure 14 shows the link model setting display and Figure 15 represents the bandwidth setting display.

1.3.4 Power Line Communication Palette Model

The purpose of the palette model is to make scenario production in OPNET easier. It is programed in a way that the user can take convenient 'Drag & Drop' steps by creating a power Line Communication palette for the nodes or link model. When arranging a node model in the scenario, one only needs to open 'Object Palette', select the node model to deploy, and then put into the scenario. Figure 16 shows implemented Power Line Communication palette model.

Figure 14. Link model setting display

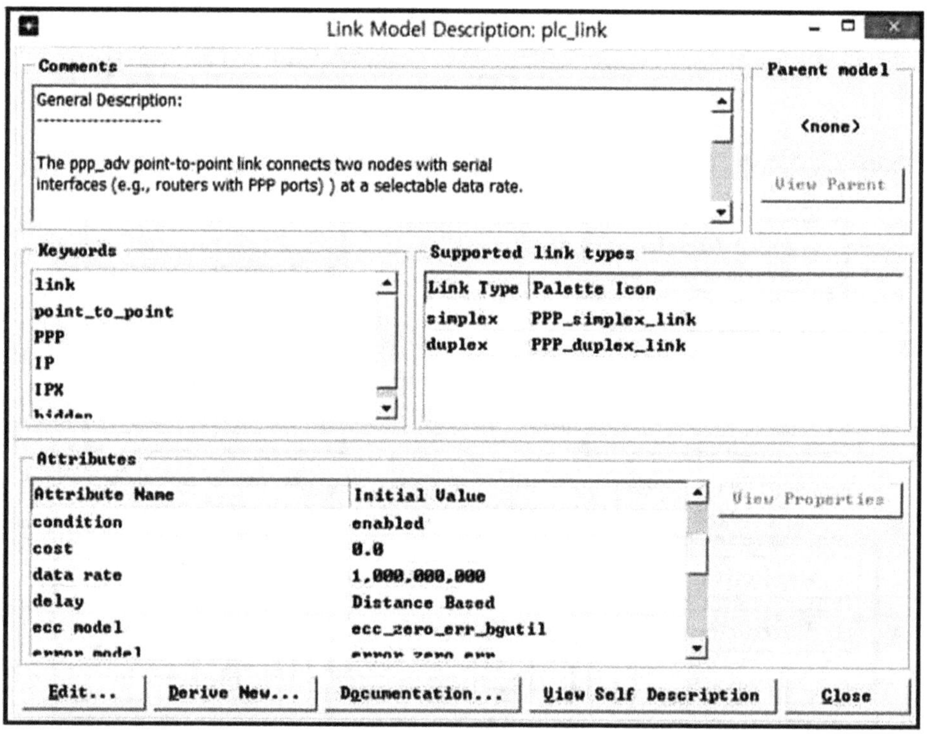

Figure 15. Bandwidth setting display

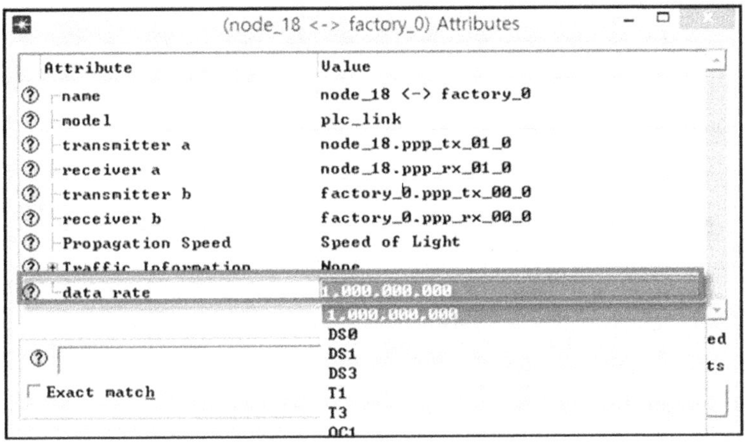

Figure 16. Power line communication palette model

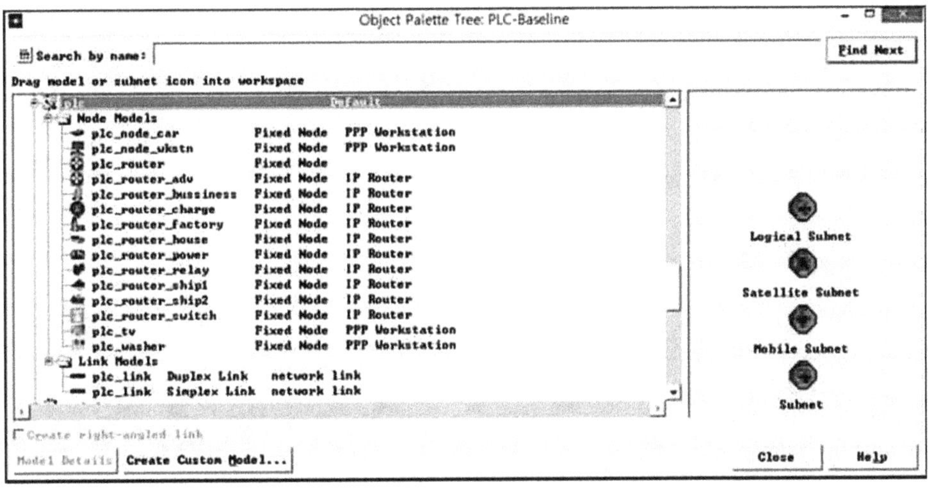

1.4 Performance Evaluation

Communications-Based technology for a Zigbee, Mobile WiMAX, Power Line Communication has been introduced in this study using OPNET.

Figure 17 shows the traffic transmission/reception rates. The average traffic generation rate of both networks was one Kpbs, equivalently.

Transmission/Reception rates may not coincide with each other in the wireless conditions but in Mobile WiMAX networks, the rates were consistent

Figure 17. Traffic transmit/receive rates

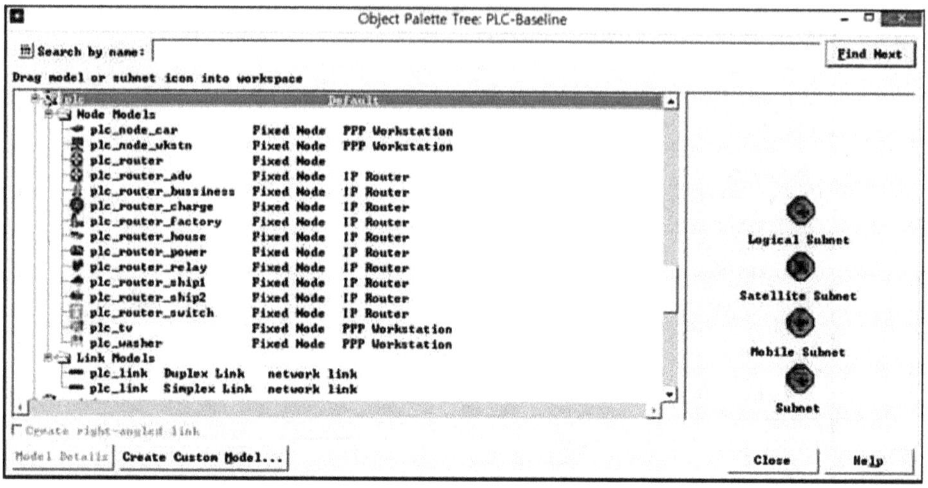

and no losses have been found due to the HARQ and ARQ functions of the Mobile WiMAX.

Figure 18 shows the results of WiMAX and Zigbee and Power Line Communication networks' End-to-End delay measurements. The delays in both Zigbee and WiMAX networks were approx.

Figure 18. WiMAX and Zigbee end-to-end delays

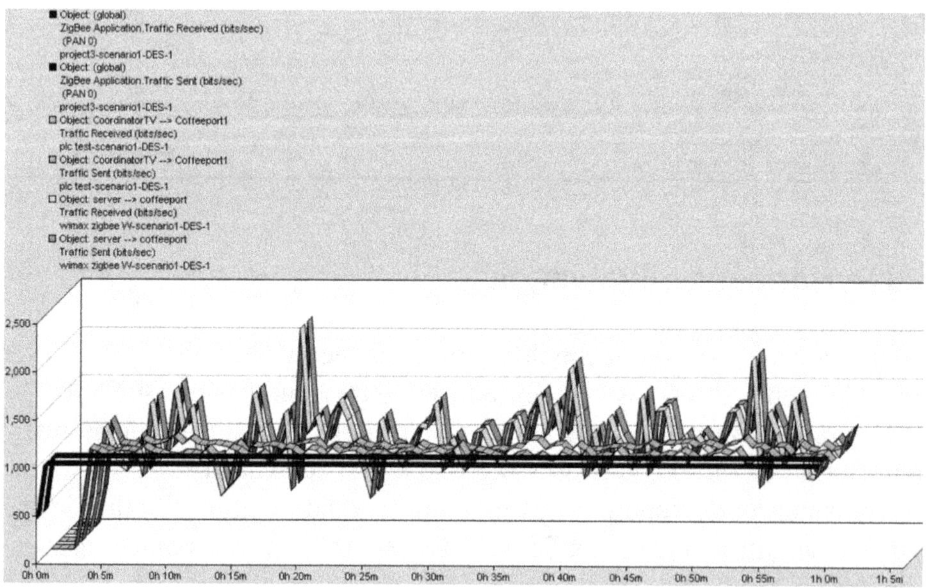

The delays were 0.00022sec, 0.006sec and 0.0035sec for PLC, Zigbe and WiMax, respectively.

Because WiMAX supports connection-oriented MAC and uses the OFDMA method, its delay is longer than that of wired communications. That is, while Zigbee is usually used for the networks where low transmission speed satisfies network's requirements and designed to be used for simple signal deliveries from the sensors, intermittent data transmissions and home networking, Mobile WiMAX was designed to perform long-distance communications similar to the mobile phone service, high-speed transmission and Quality of Service, etc.

2. TRANSPORT LAYER FOR SMART GRID TEST BED

This chapter is organized as follows: the TCP and UDP in section 1; and finally, RUDP in section 2.

2.1 TCP and UDP

2.1.1. TCP and UDP Client Node Model

The TCP and UDP Client node model has been developed based on the workstation model provided by OPNET Figure 19. Here, the Layer 2 uses MAC to perform TCP/IP communications. The statistics has been provided to measure the End-to-End delays of application traffics.

Figure 19. The structure of the TCP and UDP client node models

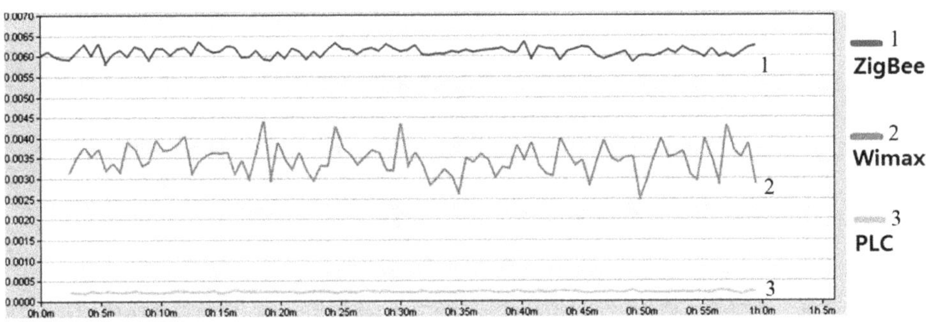

2.1.2 TCP and UDP Server Node Model

The TCP and UDP Server Node Model also has been developed based on the workstation Model provided by OPNET Figure 20 and the Layer 2 utilizes MAC to carry out TCP/IP communications. Similar statistics are given for the measurement of application traffic delays. Like the Client Model, both the IP traffics and the application traffics can be generated.

2.1.3 TCP AND UDP Router Node Model

Figure 21 shows the structure of the TCP and UDP Router Node Models. This Router Node Model has been developed based on the router model provided by OPNET and its Layer eight has four MACs to allow four Ethernet links

Figure 20. The structure of the TCP and UDP server node models

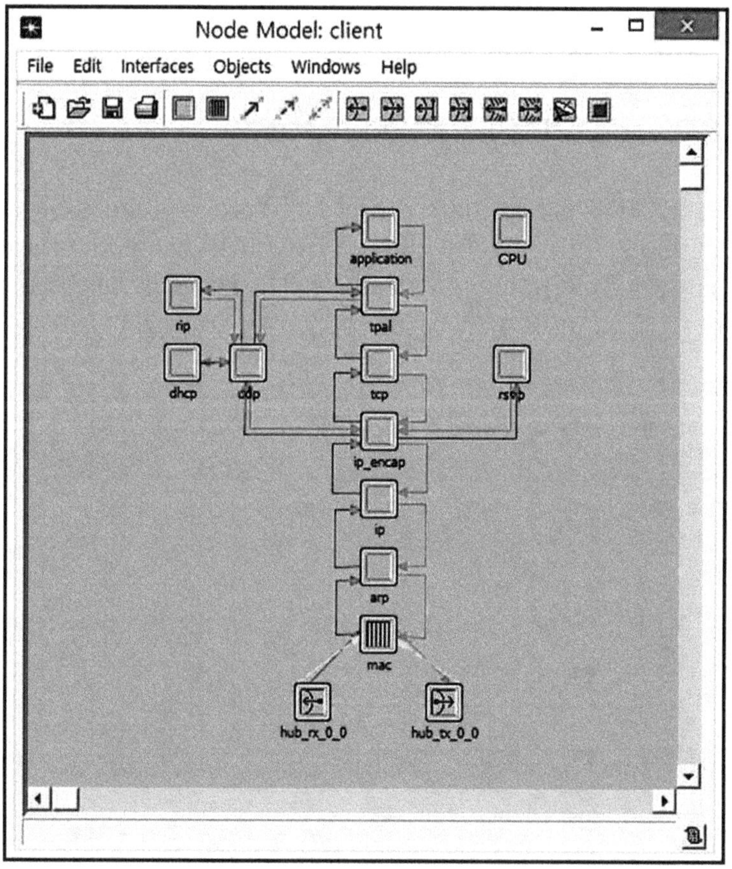

Figure 21. The structure of the TCP and UDP router node models

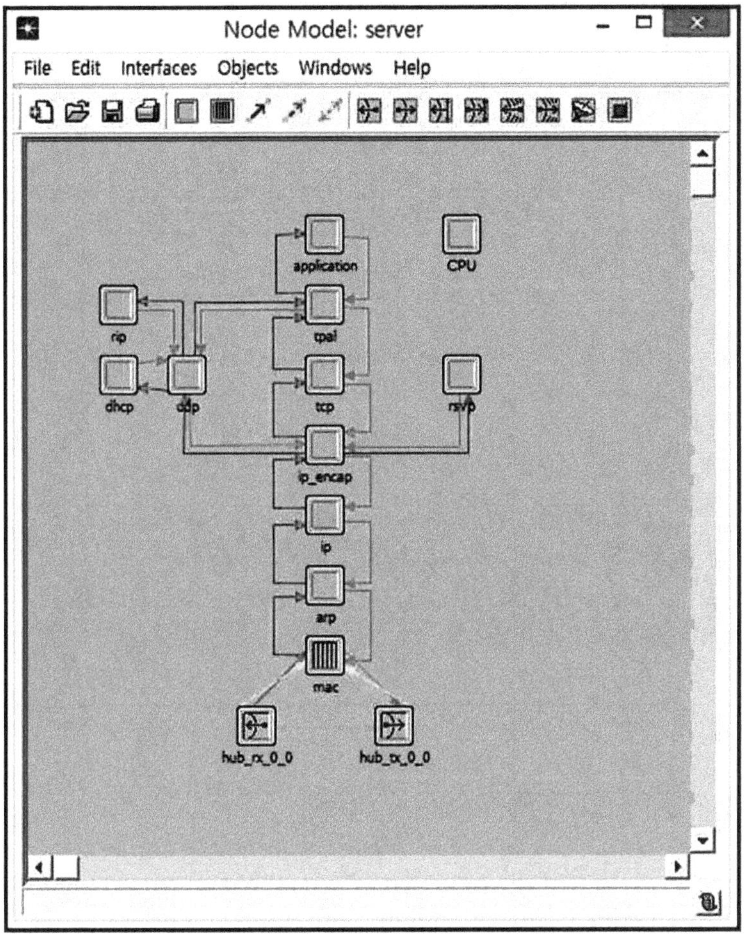

and eight PPP links to be used simultaneously. In this model, routing will be performed on the network and all the routing protocols provided by OPNET can be used.

2.1.4 TCP and UDP Switch Node Model

This TCP and UDP Switch Model has been developed based on the switch model provided by OPNET Figure 22. Its Layer 2 has eight MACs that interwork with eight links simultaneously. The traffics cannot be generated.

Figure 22. The structure of the TCP and UDP switch node models

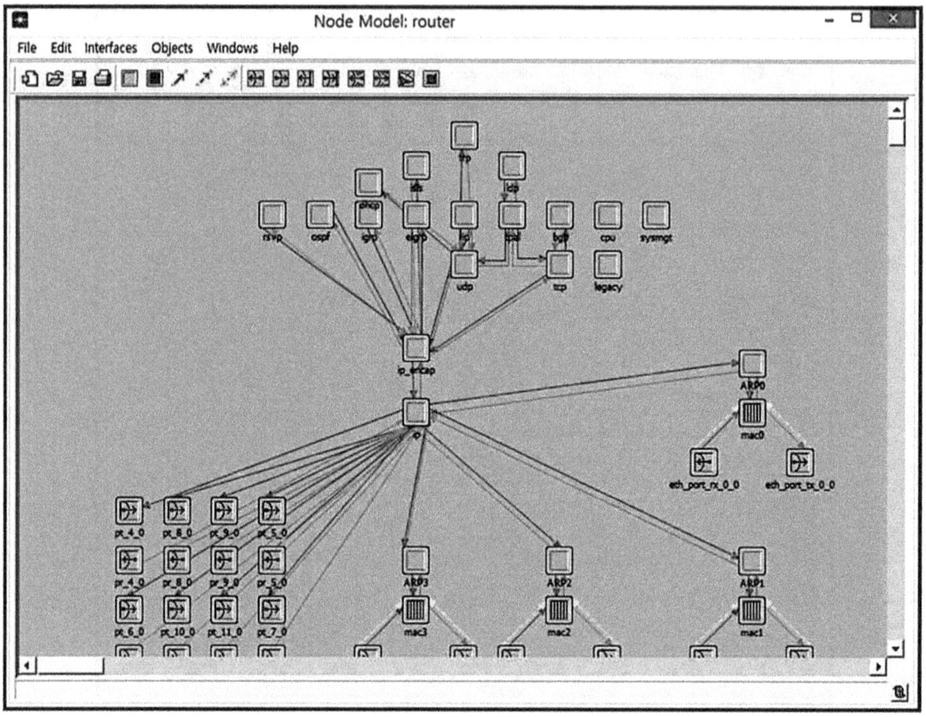

2.1.5 TCP and UDP Process Model

To measure the delays of the model, provided the statistics of the End-to-End delays for the Application Model being provided by OPNET. Figure 23 and Figure 24 are the state transition diagrams of the process models.

2.2 RUDP (Proposed Method)

2.2.1 Node and Process Model for RUDP

Figure 25 shows a RUDP node model and Figure 26 shows RUDP process model. RUDP node model has been implemented using app_UDP module on Tpal. This model, Layer 2, was developed to utilize MAC and perform TCP/IP transmission. The additional attributes in the RUDP node model are the port numbers, the number of retransmissions, time-out periods and the transmission protocol. The RUDP process model was implemented to use all of TCP, UDP, RUDP transmission protocols. For the RUDP model,

Figure 23. TCP and UDP app_demand process model

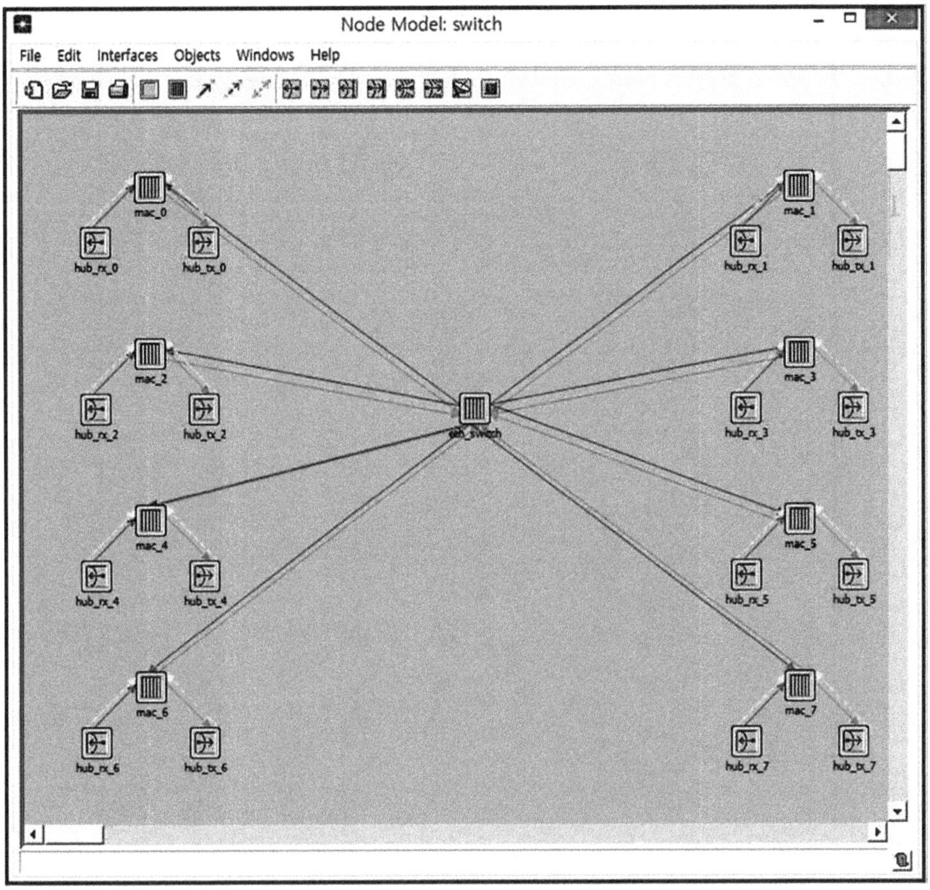

used a methodology that guarantees the packet reliability by transmitting the packets in accordance with the 'Sequence (order) Number' and by receiving the Ack packets after message deliveries have been completed. This is called the 'Selective Repeat Protocol'. If a Ack packet does not arrive within the designated time-out period, the same packet will be retransmitted until a fixed maximum number of retransmissions has been achieved. This model is expected to discard the packet if it is unable to receive relevant Ack packet at the end. The number of retransmissions and time-out periods can be defined with the 'Property Window' so that the user can run the simulations by altering them each time.

Figure 24. TCP and UDP Traffic_engine process model

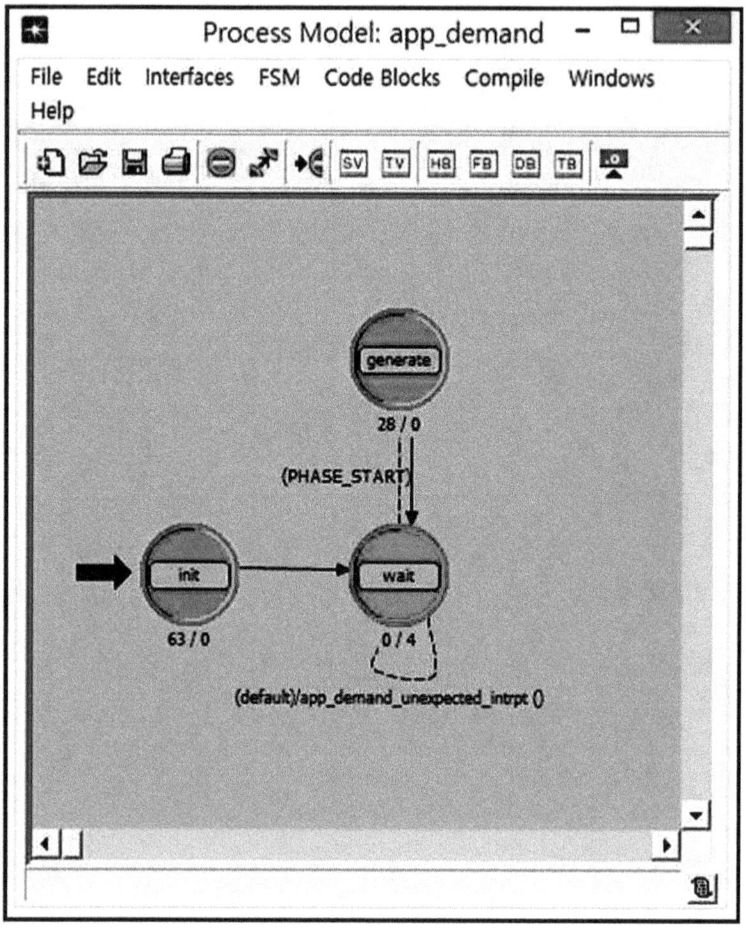

2.2.2 Demand Model for RUDP

Figure 27 shows RUDP Demand model. This model was implemented to simulate generated traffic.

The RUDP Demand model is to simulate traffic generation and, as below in Figure 28, the Start Time is the traffic occurrence time; End Time is the traffic (occurrence) ending time; Inter-arrival Time refers to the traffic occurrence cycle; and Packet Size represents the scales of traffic occurrences.

Figure 25. RUDP node model

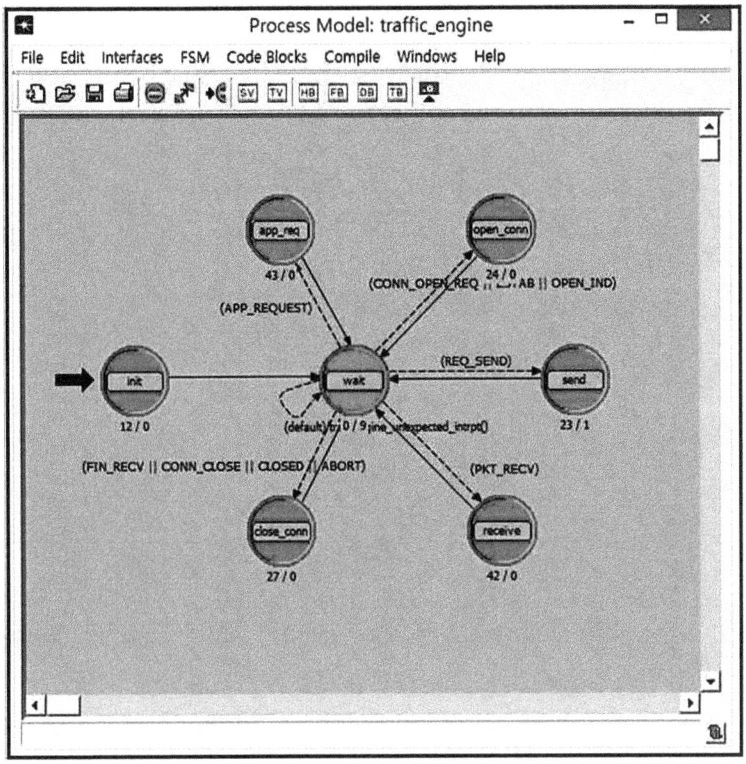

2.3 Performance Evaluation

Proposed RUDP refers to the Reliable UDP. Normally, TCP is reliable but slow in speed and UDP is faster but lacks reliability. Meanwhile, RUDP has both benefits. For example, one of the TCP's characteristics is that there is a relationship between the server and the client. That is, each needs separate connection management. On the other hand, RUDP mainly allows client-to-client communications. First of all, in a common Client/Server structure, the communications between the Clients gain their reliability by sending data through the server but the speeds are slower than UDP and the server can be overloaded. For these reasons, RUDP, which has the major benefits of TCP (Reliability) and UDP (Higher Speed), has been devised. TCP requires that one side has to be on standby assuming the role of a server but UDP can immediately initiate communications without doing so. Therefore, as the most suitable protocol for the Smart Grids, RUDP has been proposed. The

Figure 26. RUDP process model

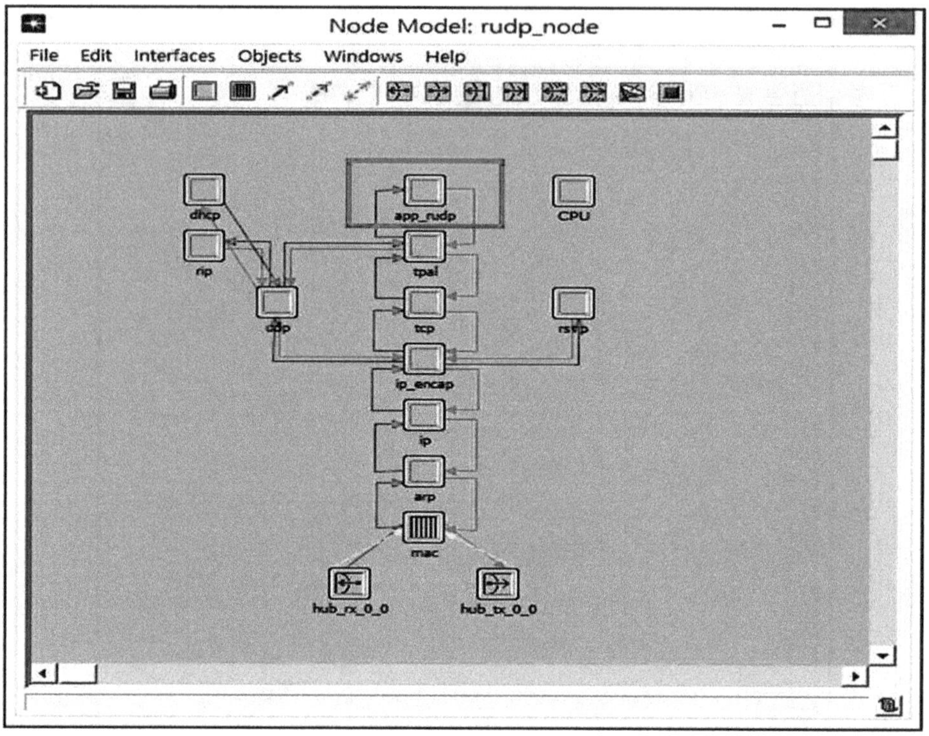

simulations demonstrated better speeds and reliability compared to UDP and TCP.

Table 1 shows the total accumulated number of received packets. 100% success rate is shown for TCP, approx. 99.86% for RUDP, and 98.57% for UDP. Comparing the success rates based on the simulation results, RUDP had a lower rate than TCP but was higher than UDP. This demonstrates that RUDP is superior than UDP in terms of reliability.

Table 2 shows the comparative results of simulations between TCP, UDP and RUDP, including their characteristics. Overall descriptions suggest that the proposed RUDP is faster than TCP and superior to UDP in terms of reliability.

Figure 27. RUDP demand model

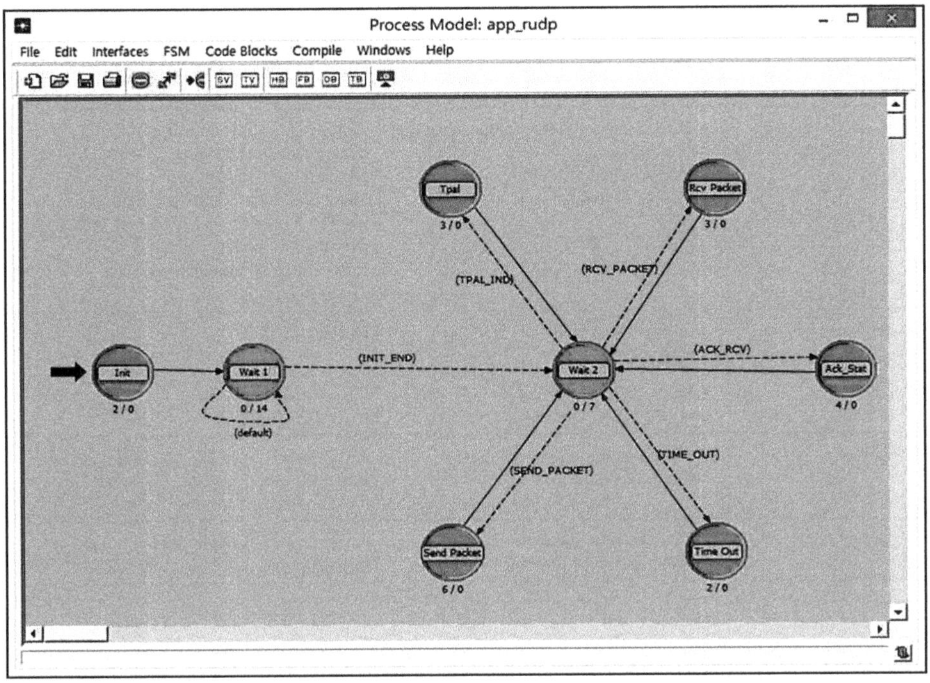

Figure 28. RUDP demand model attributes

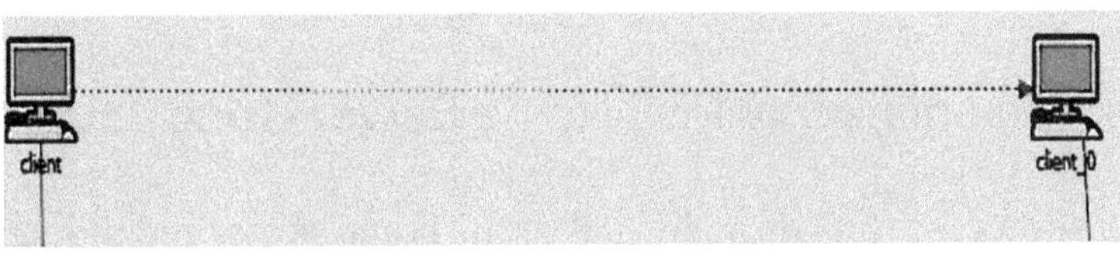

Table 1. Number of received packets

	TCP	RUDP	UDP
packets transmitted	97.22222	97.22222	97.22222
packets received	97.22222	97.08333	95.83333
Success rate	100%	99.86%	98.57%

Table 2. Comparative results

Criteria	Existing TCP and UDP	Proposed RUDP
Characteristics	• TCP- reliable but slow • UDP- low reliability	Has both advantages
Networking	• TCP- one side has to become a server and be on standby • UDP- immediate communication is possible without being on standby	Followings mainly appear in the communications between Clients: The communication reliability is usually guaranteed by sending data through server but TCP is slower than UDP and may cause overloads to the server. Instead, RUDP complements such disadvantages
Speed	• TCP - higher speed • UDP - lower speed	Faster than TCP
Reliability	• TCP - better reliability • UDP - lower reliability	Improvement has been made to have better reliability than UDP
Scalability	Both are being used for most of networking operations	Airborne networks, On-line video games. RUDP has been proposed for Smart Grids as there were no precedents.
Results	• TCP - reliable but slow • UDP - fast but less reliable	Faster than TCP and has better reliability than UDP

3. APPLICATION LAYER FOR SMART GRID TEST BED

3.1 Implementation of the PLC-Based Smart Grid Electric Power Control/Monitoring (Scenario 1)

3.1.1 Implementation of PLC Home Server Using Java

Figure 29 shows the threads and UML of the home server and this server has been implemented with Java. Home server' basic structure is described below.

The value entered every five seconds will be stored at ConcurrentHashMap as 'Key: appliance name (uniqueIndex)' and 'Value: a model that includes Time and amount (TimeWatt.class)'.The power used for a period of three minutes and the relevant String will be processed by splitting it in Carriage Return(\n) units. The results of Aggregation of the power use amounts entered from the Ardunoid circuit are displayed in the following Figure 30. A desk lamp was used for the simplicity and it was confirmed that its power consumption was measured every five seconds.

Figure 29. The threads an UML of the home server

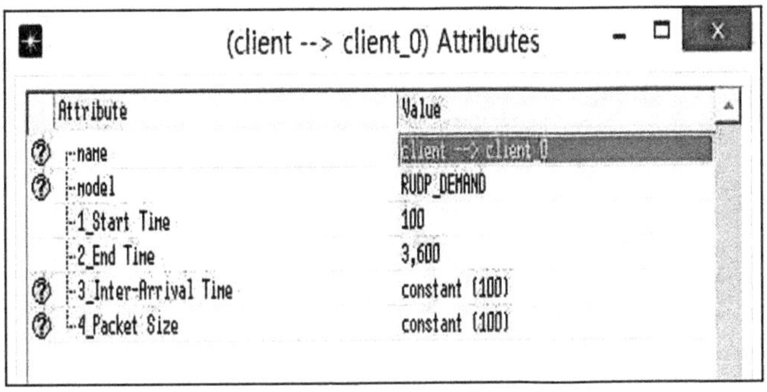

Figure 30. Power use table and GUI

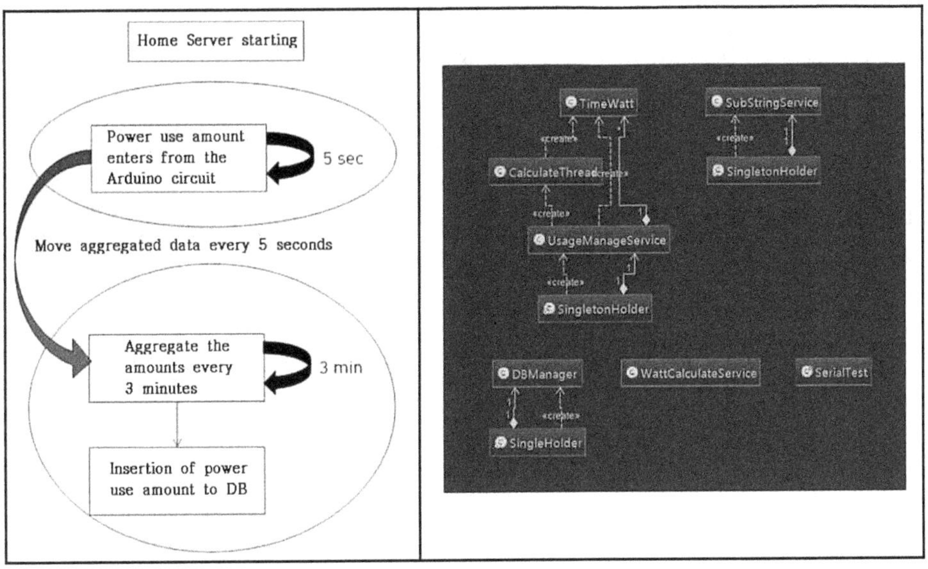

3.1.2 Implementation of Smart Outlet Circuitry Using Java

Figure 31 shows home server main class. Below is the fabrication design of Home Server which is to perform the data transmission/reception function. The main function of Home Server is to perform serial communications with the Smart Outlet for data delivery and reception.

Subsequently, delivered data will be processed and stored at the DB as in Figure 32. The power use data is transmitted every 5 seconds and stored separately following the individual index of each Smart Outlet. This power

Figure 31. Home server main class

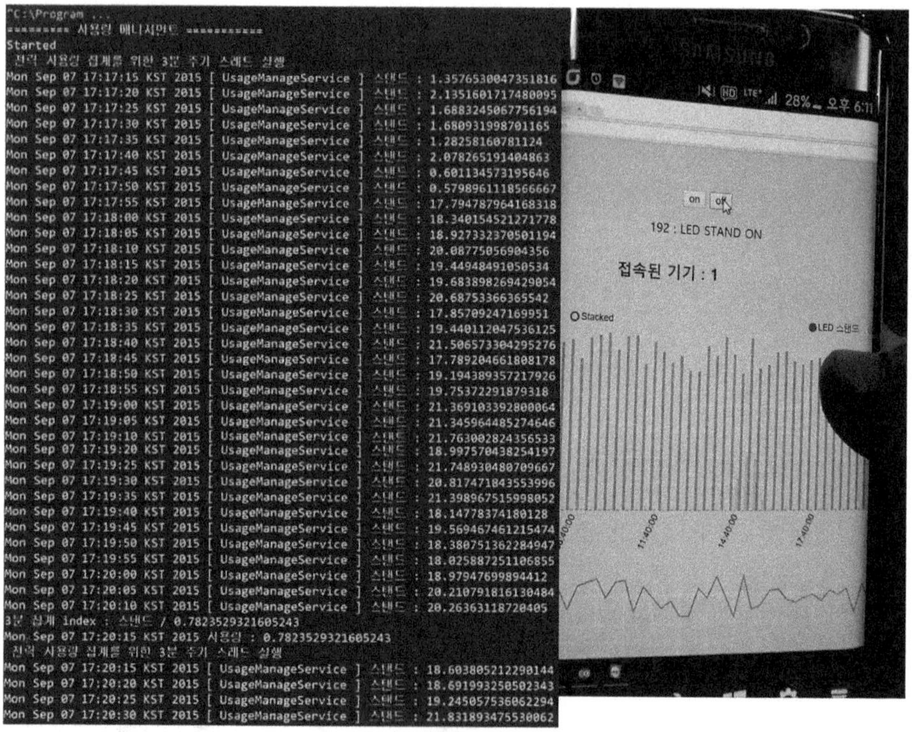

Figure 32. ManageService UML diagram for data processing

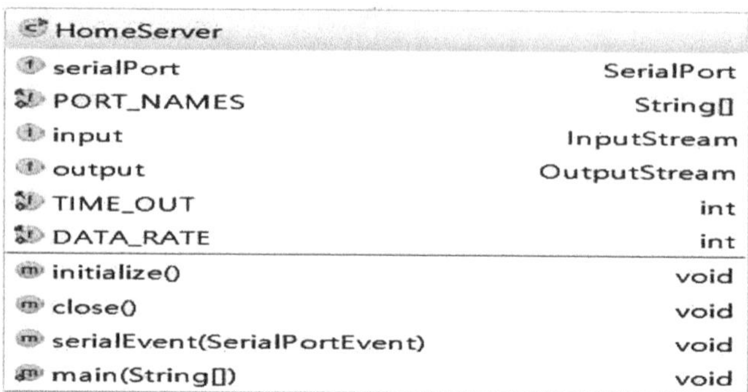

use data will be indicated with Wh unit every 3 minutes for each appliance once the thread CalculateThread has been initiated.

The database used to interlink the Smart Outlet with appliance is comprised of objectID, which is the primary key of the Smart Outlet, and an appliance name relevant to this objectID.

Figure 33 is the composition of the DB table which stores appliance names and this table will be managed with NameManageService class and NameDAO Data Acess Objects as below. These loads HashMap that stores objectID and objectName when Home Server starts its operation and lets the user to conveniently check and change the name of the appliance being connected to the Smart Outlet by interlinking with relevant HashMap when he/she checks the amount of power used.

The power use data processed with CalculateThread every three minutes in each Smart Outlet will be stored at electric_usage table through ElectricUsageDAO as shown in Figure 34. The electric_usage table includes objectID, time period (3 minute-base), appliance name and power use amount Figure 35.

Figure 33. Composition of DB table that stores appliance name

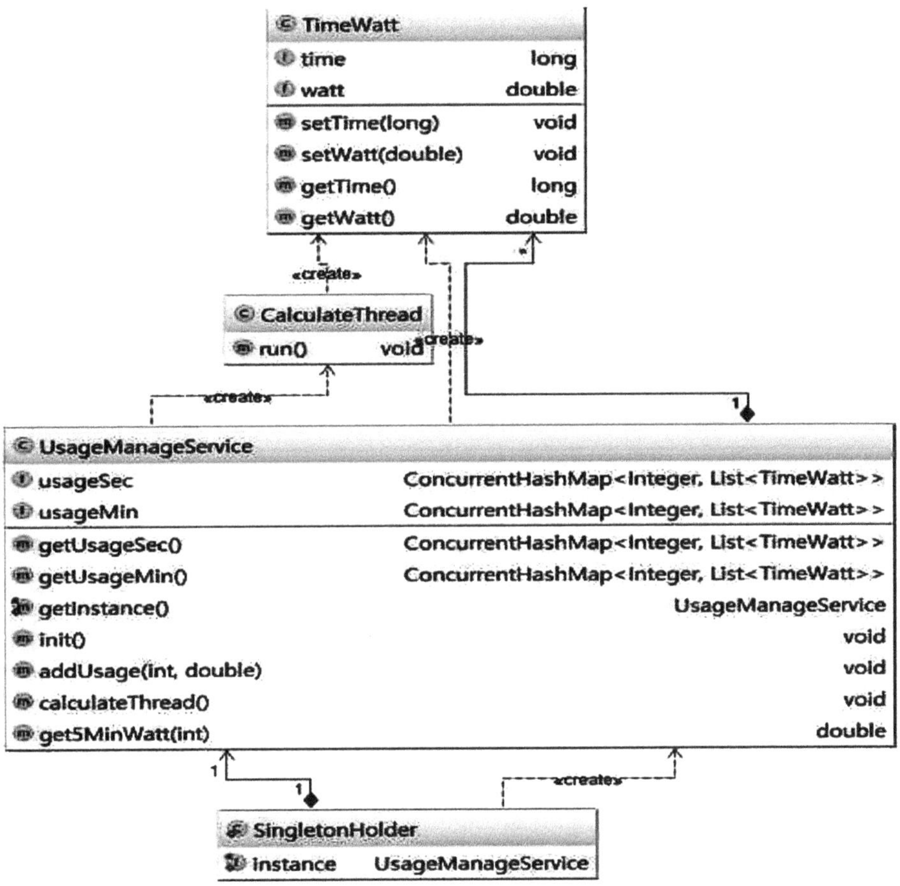

Figure 34. Composition of NameManageService class that interlinks smart outlet with relevant appliance name

Column ID	Name	Type
0	**objectId**	**INTEGER**
1	objectName	VARCHAR

Figure 35. Electrical power use DAO

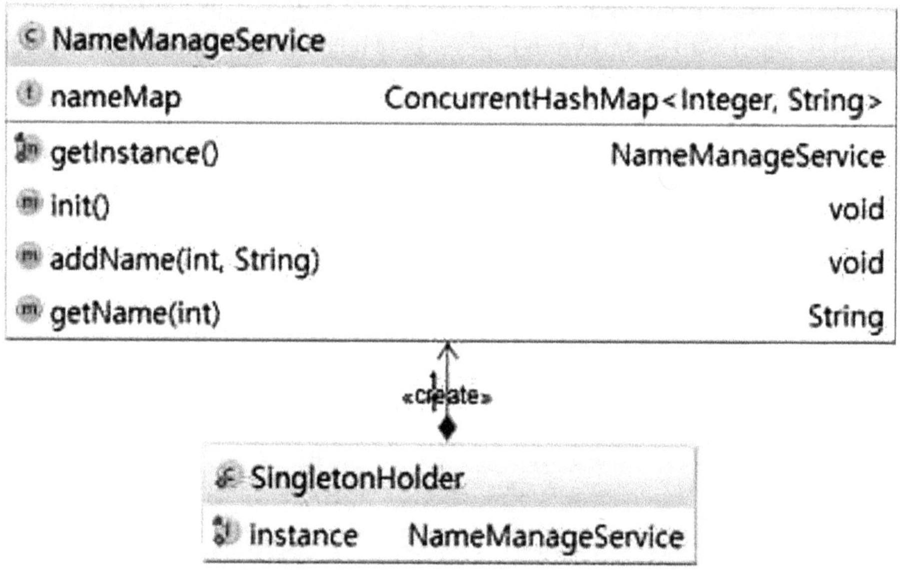

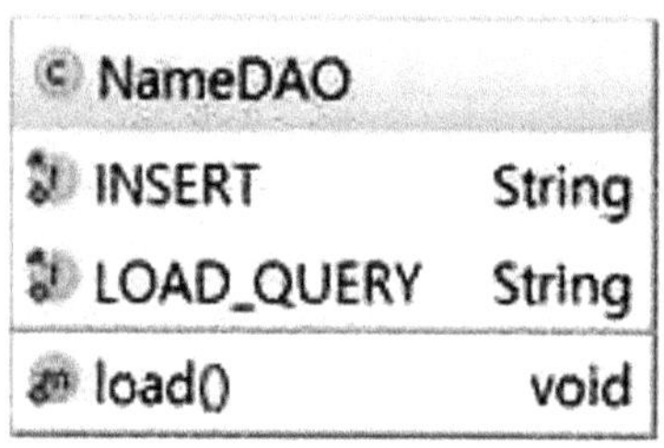

3.1.3 Performance Evaluation

As for the PLC data transmission/reception performance evaluation, communication reliability and speeds in different environments have been tested and for Home Server, the delays in the multi-thread environments were tested. For the test in the multi-thread environments of Home Server, the time required to carry out 500,000 times of IN/OUT processes for each number of thread (up to 50) has been measured. In this case, the data structure was 'Thread Safe', consisting of HashTable, synchronizedHashMap and ConcurrentHashMap. Since the data is stored at the database every 3 minutes, 288 data per appliance will be stored in a day.

Assuming that there are 100 Smart Outlets, a virtual database that includes a total of 30,000 data to measure the delay times in both PC and Mobile environments for the data loading process was created. To indicate the delay occurring at the zone between Smart Outlet and Home Server, a time difference value between StartTime of MCU and Home Sever (Java), has been calculated. The length of used power line was two meters.

The test was performed under the environment where i7-4770k cpu and 12Gb Ram have been used. The results of IN/OUT processes for 50 threads are shown in Figure 36. To induce appropriate results, 'ConcurrentHashMap', which is 'Thread Safe' and has the lowest delay time in average, was used. Since there are around 20-40 power outlets in an actual home and the r/w process will be carried out once every 5 seconds, the delay under 400 ms can be guaranteed.

Figure 37 shows the delay results when loading a database with 30,000 data. The test was carried out for both mobile and PC environments. For these environments, the time taken to load the web that has a function of checking the power consumption rate was measured. The results showed that the delays were between 0.8 to 1.2sec for PC and 0.3 to 0.4sec for mobile device.

Additionally, Figure 38 shows the delays occurring between the Smart Outlet and Home Server connected with a 2 m-length power line. The results shows that the delays are within the limits of 0.01 to 0.08 seconds, both

Figure 36. Delays occurring while performing 500,000R/W for 50threads

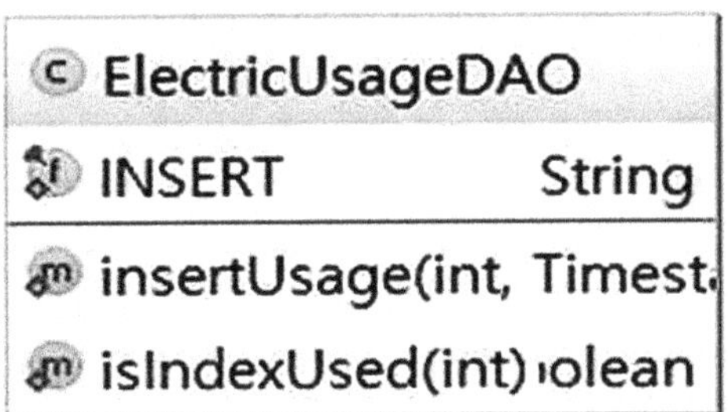

Figure 37. Loading delays in mobile and PC environment for 30,000 data

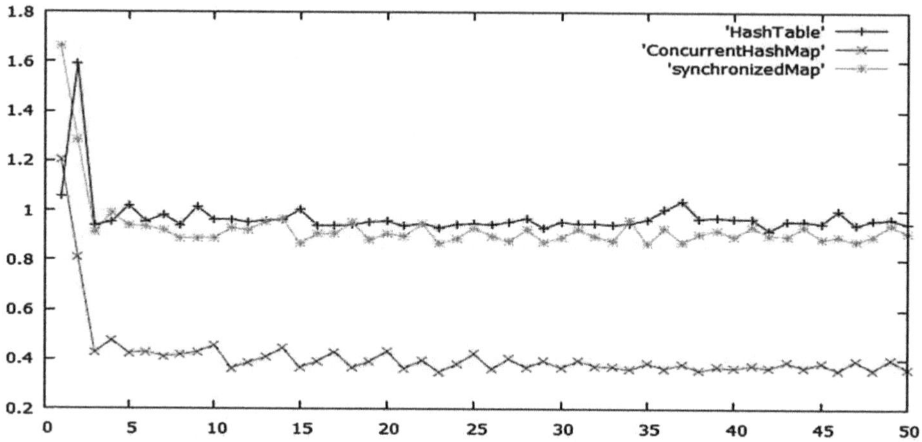

Figure 38. Delays between smart outlet and home server

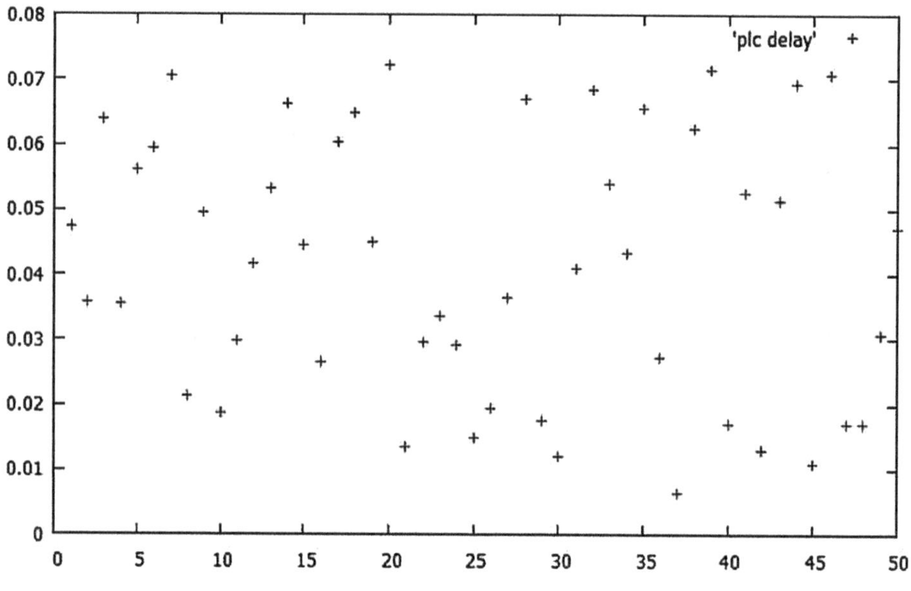

of which are negligibly small when compared with the de- lays generated between the Home Server and the web server.

With performance evaluation test results in Figure 38, it will be safe to conclude that the product, Smart Outlet, has reached the adequate level of commercialization in terms of reliability and performance under the household Power Line Communication environment.

3.2 Implementation of AMI Application (Scenario 2)

3.2.1 Developed Products

Figure 39 shows Implementation Diagram AMI Application (Scenario 2). The power usage information on the watt-hour meter is recorded with MCU (right side quadrangle). It transmits the information through the PLC module (left side quadrangle). ① and ② of Figure 39 are the sections that will be used to control communications of each device. ③ will be used for the communication support between the Server and the watt-hour meter.

In Figure 41, the information on the watt-hour meter is transmitted to the Server through the Power Line Communication module and the Server outputs this information on the webpage. The characteristic of the Client is that it is possible to monitor the power usage in realtime by accessing the webpage and the Server controls the power supply of electronic devices through Device Manager, receiving commands from the Client.

3.2.2 Implementation of AMI Application Using Intelligent Agents by Java on the Power Line Communication Network

Figures 42 and 43 are the UML of the Smart Metering Control System embedded with an intelligent agent proposed in this book. Key Classes are

Figure 39. Implementation diagram AMI application

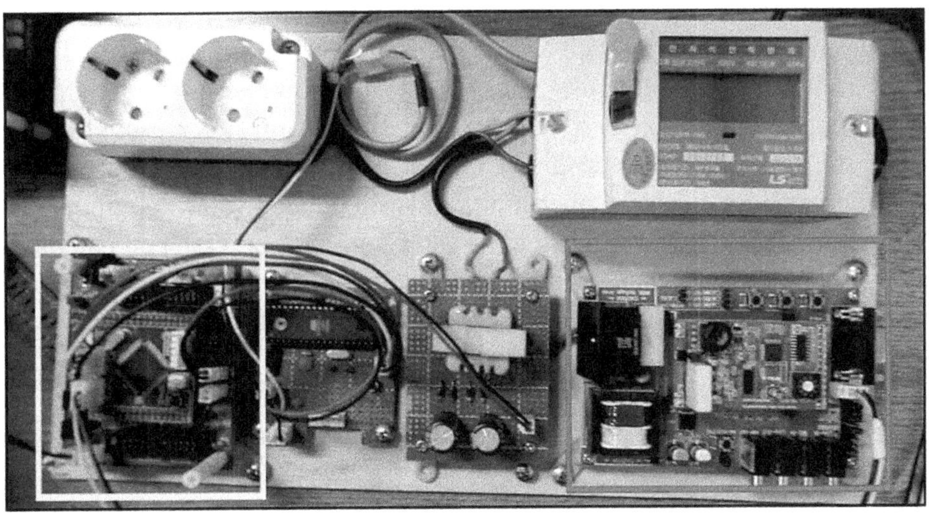

Figure 40. Server implementation diagram

Figure 41. Whole system implementation diagram

as follows: CIAgent is a base class that defines the common programming interface and behaviors of all the agents in the framework; CIAgentState object is to be created and its 1st state will be set as UNINITIATED; CIAgentTimer class implements a 'Runnable' interface, which requires that it implements run() function summoned by runnit thread; CIAgentTimer provides two basic functions to CIAgent; and CIAgentEventListener interface extends the standard Java EventListener interface used by all java.awt components. This interface registers itself using the add CIAgentEventListener() method; CIAgentEventQueue class implements a queue of CIAgentEvents. Its only data member is eventsQueue, a Vector instance; and ExampleThread class is used to update Android User Interface (UI).

BetterBuyerAgent and BetterSellerAgent use the hard-coded logic with a slightly more aggressive negotiation strategy and warrant further discussion.

Figure 42. UML of the smart metering control system using intelligent agent 1

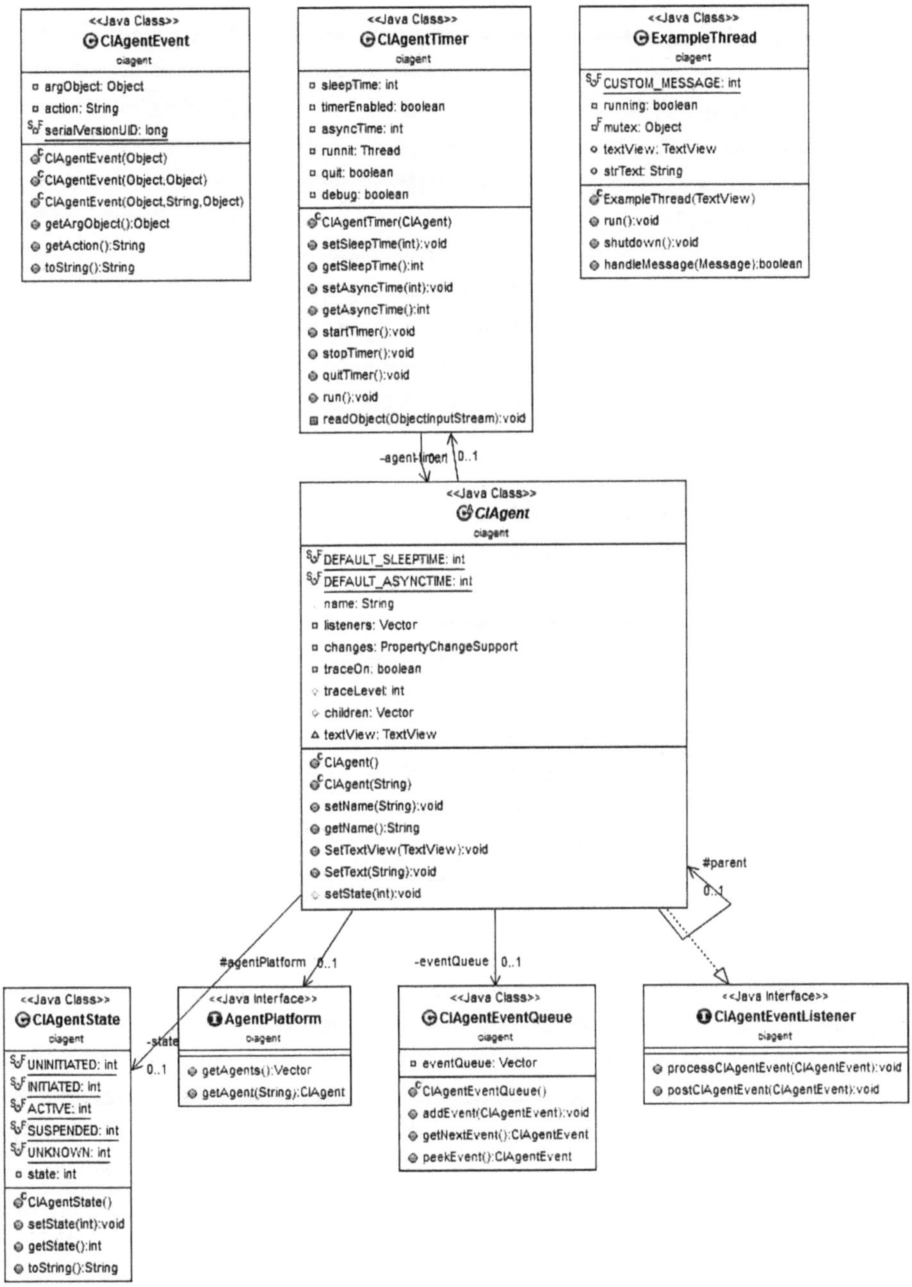

Figure 43. UML of the smart metering control system using intelligent agent 2

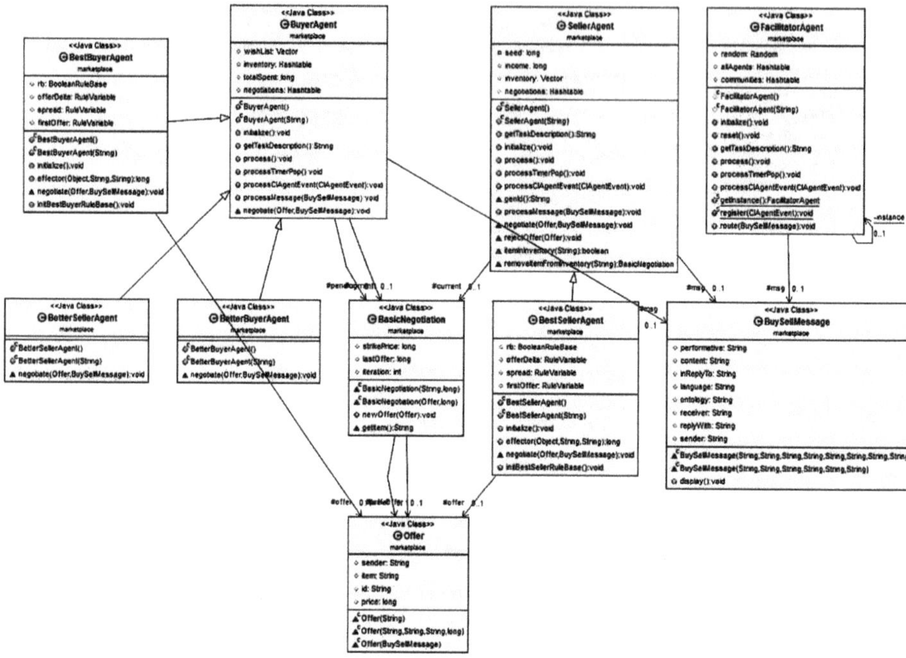

They both use BooleanRuleBase class that provides the rules for determining the prices and actions during the negotiations.

3.2.3 Power System Control Programs Using Java

In Figure 44, a functional diagram showing the functional relationships between respective Classes. Main functions are largely comprised of DeviceManager, WebpageManger, countingServer and LogManager.

The DeviceManager communicates with each device's controller through PLC and assumes the role of delivering the orders to respective devices by user's request. Device represents each device carrying out the RS232 communications through PLC physically.

Here, [(int)SenderDevice] [(int)ReceiverDevice] [(int)Unused] [(string) CommandLine] [EndOfCommand]-type protocol is used. MailBox is a central part of the DeviceManager to process commands and a sort of non-circular queue of which is to deliver the requirements from a user to each device in the form of a Mail. The DeviceManager checks the MailBox, and if there

Figure 44. Functional diagram

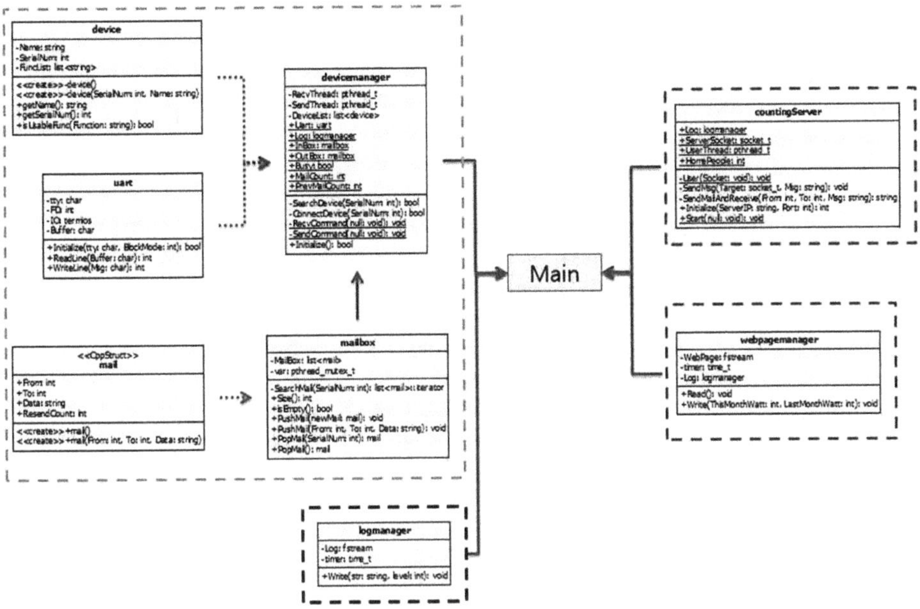

are new any requests, it delivers them to the relevant device(s), and stores the replies back to the MailBox.

Figure 45 shows DeviceManager class. In our proposed system, the DeviceManager receives the readings from the digital watthour meter and displays the power usage amount on the webpage. DeviceManager manages currently linked devices as well as continuously added devices through PLC.

Figure 46 shows Device Class assuming the role of setting the port number for each connected device. And Uart Class is to control PLC and uses 'Shall' term provided by Ubuntu 13.0.

Figure 45. DeviceManager class

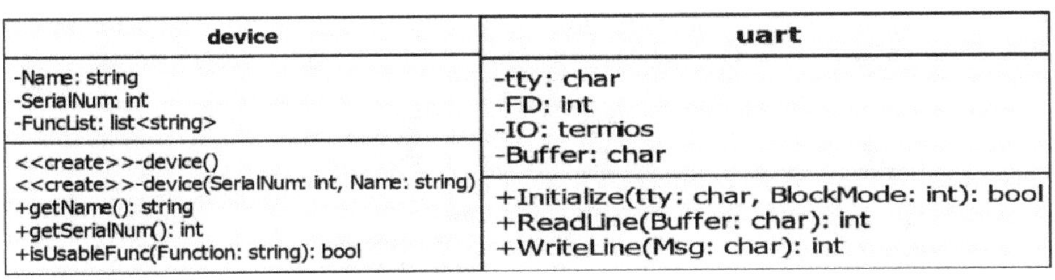

Figure 46. Device class and Uart class

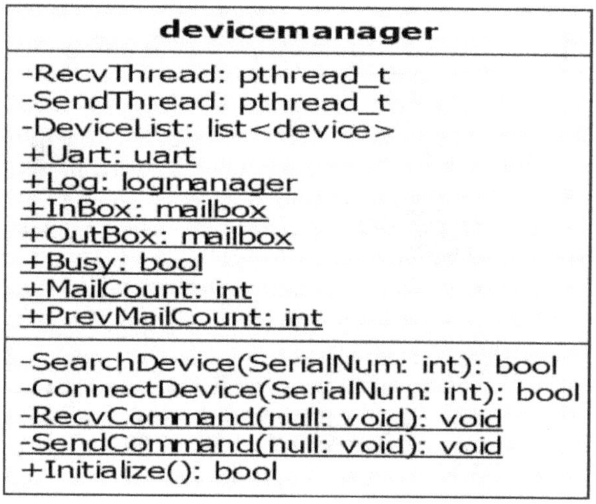

Figure 47 shows that Mail Class sets the standard for the signal received from the Client through the socket which has been left open with Mail. And MailBox is a fundamental part for the execution of commands from DeviceManager and a sort of non-circular queue for delivering the incoming requests from the user to each device in a form of Mail. This checks MailBox and if there's a new request, delivers it to the device and then stores the reply from the device in the MailBox again.

In Figure 48, this Class is to communicate with EndUser directly through Ethernet using CounterServer. The main function is to check in-home headcount. For later addition of functions, [(int)SendUser], [(int)DeviceNumber] and [(string)CommandLine]-type protocols have been developed so that individual request can be made to each device.

In Figure 49, webpage manager class assumes a function of updating current power usage on the web through Apaches. And Log manager class

Figure 47. Mail class and mailbox class

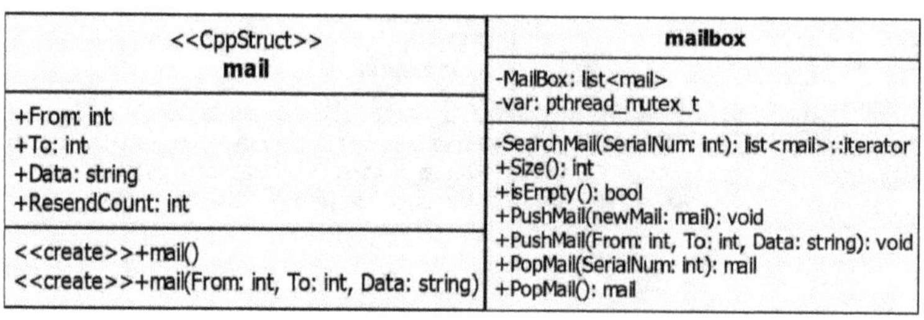

Figure 48. CountingServer

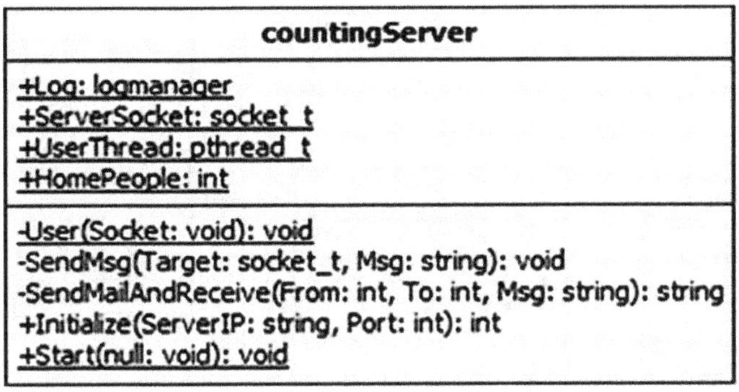

assists the administrator in checking the condition of the server by recording the information or the problems occurring on the server and generating the text-type log files.

Figure 50 is a correlation diagram of the power system control program which has been developed by using Atmega 128. SERIALNUM is a serial (unique) number of each device, and for the power controller, 2XXX series number is given and only one device can be connected. For general devices, 3XXX numbers will be assigned and number 1000(fixed) is for the server. The Broadcasting number for all of the devices is 1111.

3.2.4 User Interface of AMI Application and Test Bed Experiments Using Intelligent Agents by Java

Figure 51 shows the User interface for the Android-based applications. Four domains of the user interface have been described depending on their functions. Domain ① has the function of updating previously stored Mac address(s) by the user. The function of Domain ② is to output and update

Figure 49. Webpage manager and log manager

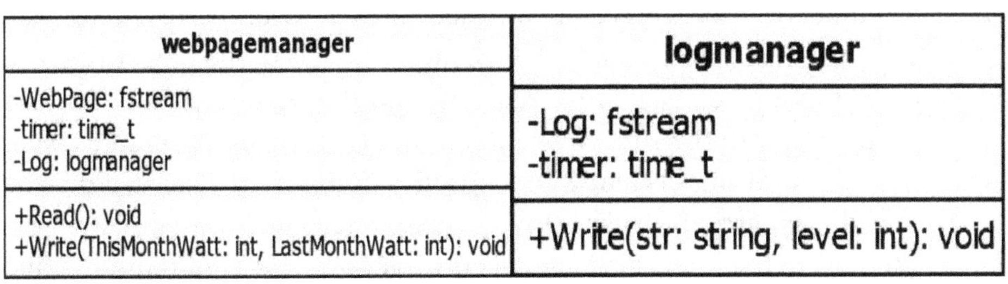

Figure 50. Relationship diagram

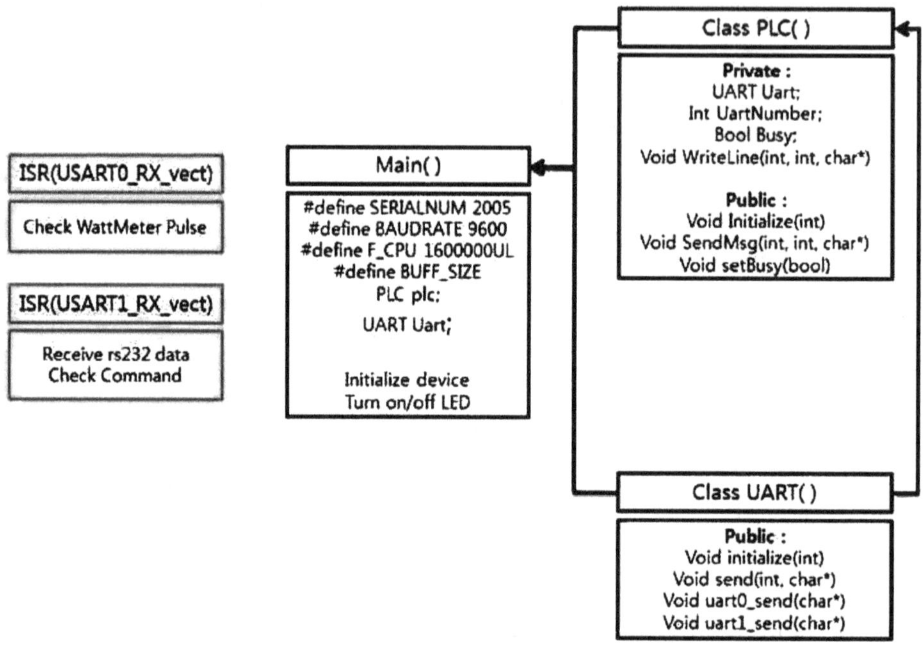

the Mac address currently accessing. In Domain ③, access to the server URL, which is to check power consumption, is made and one's own power consumption information can be checked. Finally, Domain ④'s function is to output Mac address currently being stored. Should the Mac address stored in Domain ④ matches to the Mac address being accessed in Domain ②, then it is determined that someone has entered the room. This is to eliminate the possibility of an error of determining that someone has entered the room when it is just an incident that nearby resident has accessed the same Wi-Fi network and there is actually no one at home.

3.2.5 Performance Evaluation

The communication rates have been tested in two different environments: the first environment where many devices were linked to the power line and power consumptions consistently changed; and the second, where devices were linked similar to the first environment but the consumption rates were steady. The observation was made for respective conditions. That is, when the communication is carried out using PLC, when devices are linked directly with RS232, and when the intelligent agent is involved. As seen in Figures 52 and

Figure 51. User interface of Android-based applications

Figure 52. Tests in the environments where there are no interferences

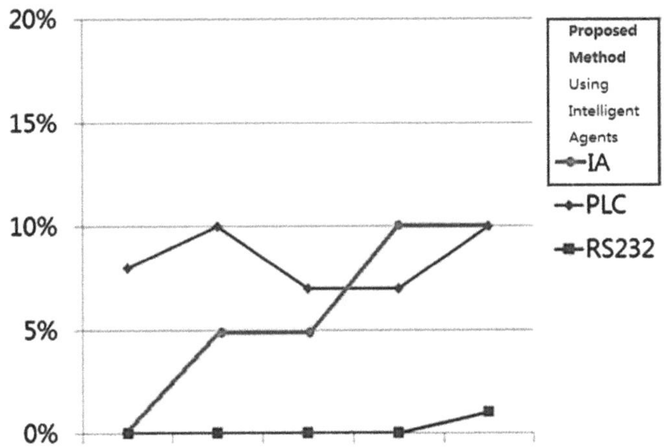

53, the PLC communication resulted in about 10% failure even though there were no interferences, and 20-35% failure when interferences were present. Meanwhile, almost 0% failure had resulted when direct linking method was used for the first two environments. Next, the round lines show the results when the Intelligent Agent has been applied and the communication failure rates in both environments are significantly lower than that of the PLC in average, showing the improvements.

The results analysis of above shows that the PLC method lacks reliability and it can be observed that the failure rate increases when there are more interferences. If the reliability level drops lower, more retransmission requests will be sent out so that the efficiency of entire system will be compromised. However, considering the nature of the PLC technology and the fact that it would be difficult to increase reception success rate unless the module is changed, the best possible simple form of communication is preferred. Additionally, a mechanism that processes re-requests and re-responds is necessary when the messages have not been delivered properly.

Therefore, for later use, the standardized protocol is essential for the effective PLC applications. An assumption was made that using TCP/IP after establishing a PLC-applied ethernet environment could be the best possible method when the generality and scalability are taken into account.

In this chapter, a model was implemented and evaluated by using Reliable User Datagram Protocol (RUDP) in the transfer layer to surmount/complement the transmission loss caused by noise, which is a typical disadvantage of

Figure 53. Tests in the environments where interferences exist

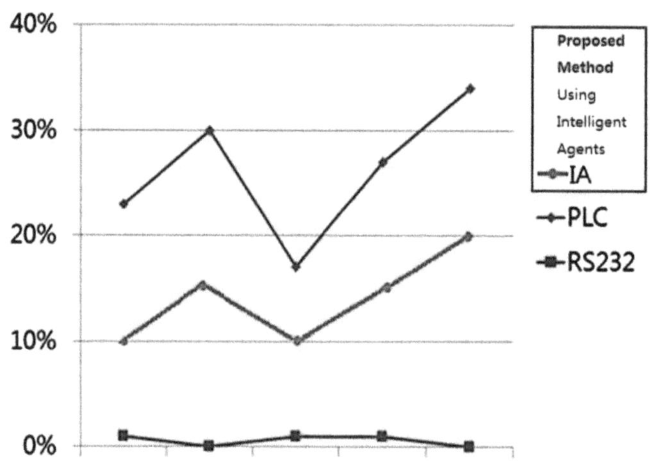

the PLC-based communications. As a result, it was possible to implement a scalable test bed in which both of above mentioned layers were included.

It is expected that this implementation technique will be helpful to the researchers and students who are interested in this field. Additional contents related to the topic can be found in Maneerung, Sittichivapak and Hongesombut (2011), Korki, Hosseinzadeh and Moazzeni (2011), Kuzlu, Pipattanasomporn and Rahman (2014), Mlynek, Misurec and Koutny (2011), Kim et al (2011), Lazaropoulos (2012), Milioudis, Andreou, and Labridis (2012), Bauer et al (2009), Della et al (2013), and Mohsenian-Radd and Leon-Garcia (2011).

REFERENCES

Bauer, M. (2009). Packet-oriented communication protocols for smart grid services over low-speed PLC. *Power Line Communications and Its Applications, 2009. ISPLC 2009. IEEE International Symposium on IEEE.*

Della Giustina, D. (2013). Automation of distribution grids with IEC 61850: A first approach using broadband power line communication. *IEEE Transactions on Instrumentation and Measurement*, *62*(9), 3272–3283.

Huh, J. H., Je, S. M., & Seo, K. (2016). Communications-based technology for smart grid test bed using OPNET simulations. Information Science and Applications (ICISA), 227-233.

Huh, J. H., Kim, N., & Seo, K. (2016). A Study of Power Line Communication-based Smart Outlet System Expandable at Home. *Journal of Korea Multimedia Society*, *19*(5), 901–909.

Huh, J. H., Otgonchimeg, S., & Seo, K. (2016). Advanced metering infrastructure design and test bed experiment using intelligent agents: Focusing on the PLC network base technology for Smart Grid system. *The Journal of Supercomputing*, *72*(5), 1862–1877.

Huh, J. H., & Seo, K. (2016, August). Smart grid framework test bed using OPNET and power line communication. In Soft Computing and Intelligent Systems (SCIS) and 17th International Symposium on Advanced Intelligent Systems, 2016 Joint 8th International Conference on (pp. 736-742). IEEE.

Kim, S., Kwon, E. Y., Kim, M., Cheon, J. H., Ju, S., Lim, Y., & Choi, M. (2011). A secure smart-metering protocol over power-line communication. *IEEE Transactions on Power Delivery*, *26*(4), 2370–2379. doi:10.1109/TPWRD.2011.2158671

Korki, M., Hosseinzadeh, N., & Moazzeni, T. (2011). Performance evaluation of a narrowband power line communication for smart grid with noise reduction technique. *IEEE Transactions on Consumer Electronics*, *57*(4).

Kuzlu, M., Pipattanasomporn, M., & Rahman, S. (2014). Communication network requirements for major smart grid applications in HAN, NAN and WAN. *Computer Networks*, *67*, 74–88. doi:10.1016/j.comnet.2014.03.029

Lazaropoulos, A. G. (2012). Towards modal integration of overhead and underground low-voltage and medium-voltage power line communication channels in the smart grid landscape: Model expansion, broadband signal transmission characteristics, and statistical performance metrics. *ISRN Signal Processing*.

Maneerung, A., Sittichivapak, S., & Hongesombut, K. (2011). Application of power line communication with OFDM to smart grid system. *Fuzzy Systems and Knowledge Discovery (FSKD), 2011 Eighth International Conference on* (Vol. 4). IEEE.

Milioudis, A. N., Andreou, G. T., & Labridis, D. P. (2012). Enhanced protection scheme for smart grids using power line communications techniques—Part II: Location of high impedance fault position. *IEEE Transactions on Smart Grid*, *3*(4), 1631–1640. doi:10.1109/TSG.2012.2208988

Mlynek, P., Misurec, J., & Koutny, M. (2011). Modeling and evaluation of power line for smart grid communication. *Przeglad elektrotechniczny*, *8*, 228-232.

Mohsenian-Rad, A., & Leon-Garcia, A. (2011). Distributed internet-based load altering attacks against smart power grids. *IEEE Transactions on Smart Grid*, *2*(4), 667–674. doi:10.1109/TSG.2011.2160297

Chapter 4
Security for AMI Application

ABSTRACT

Among many hacking attempts carried out in the past few years, the cyber-attacks that could have caused a national-level disaster were the attacks against nuclear facilities including nuclear power plants. The most typical one was the Stuxnet attack against Iranian nuclear facility and the cyber threat targeting one of the facilities operated by Korea Hydro and Nuclear Power Co., Ltd (Republic of Korea; ROK; South Korea). Although the latter was just a threat, it made many Korean people anxious while the former showed that the operation of nuclear plant can be actually stopped by direct cyber-attacks. After these incidents, the possibility of cyber-attacks against industrial control systems has become a reality and the security for these systems has been tightened based on the idea that the operations by network-isolated systems are no longer safe from the cyber terrorism. The ROK government has established a realistic control systems defense concept and in the US, the relevant authorities have set up several security frameworks to prepare for the threats. At the same time, Smart Grids are not an exception any longer. Thus, in this chapter, the security tests have been tested under the Smart Grid environment and several DDoS (Distributed Denial of Service) attack scenarios were developed for experiments.

1. INTRODUCTION

In AMI, the power usage information is transmitted to the power management system remotely through the communication infrastructure comprised of relays, concentrators centering around Smart Meters. Thus, AMI provides

DOI: 10.4018/978-1-5225-2776-3.ch004

two-way information provision, home network interworking and power control functions. It is necessary to design the system that optimizes supply and demand of electric power. And also, since AMI data is delivered through communication networks, a strong security level is required for the devicesand against Cyber-attacks to offer a variety of safe services in a Power Grid.

2. IMPLEMENTATION OF GRAPHIC BASED NETWORK INTRUSION DETECTION SYSTEM FOR AMI APPLICATION USING JAVA

Figure 1 shows the description of UML of our system and here, Tray Class will be executed first in Background once the program has started. At the same time, PacketCapture Class and UI Class will also be called upon internally. For PacketCapture Class, when a packet has been captured, ReceivePacket function will be summoned to analyze the packet. By analyzing IP header, ReceivePacket function reads and stores information such as its version, protocol, source and destination of the packet, and after determining the Flag type, redundancy check will be performed, after which Node Class is generated to store the information. Subsequently, in CheckStrangeIP, if it becomes suspicious that the packet originating from People's Republic of China or North Korea is harmful, the contents will be displayed on the screen to determine the possibility of DDoS attack - the core function of this program. The threshold value can be calculated in Realtime by analyzing all the received packets every 5 seconds using IsFlooding Class. Should the incoming packet has been judged to have exceeded the value, the program will show the user a Warning Window with ShowMessageListener() function of Tray Class (Huh et al, 2015).

For UI Class, total packet send/receive volume of the computer in use, protocol packets mainly used (e.g. TCP, UDP and ICMP) for DDoS attacks, and received volume of the packets judged to be the harmful ones will be generated/displayed with the graphs. In ConnectURL Class, based on the IPs entered on the UI window by the user, locations of suspicious IPs will be displayed on the Google map using IP location tracking service provided by the Korea Internet Security Agency (KISA).

Figure 2 is a user interface of this detection system. This user interface is coded with Java language to carry light amount of resources so that when it interworks with other programs in real-time, the system's operation will not

Figure 1. Graphic based NIDS' UML for AMI

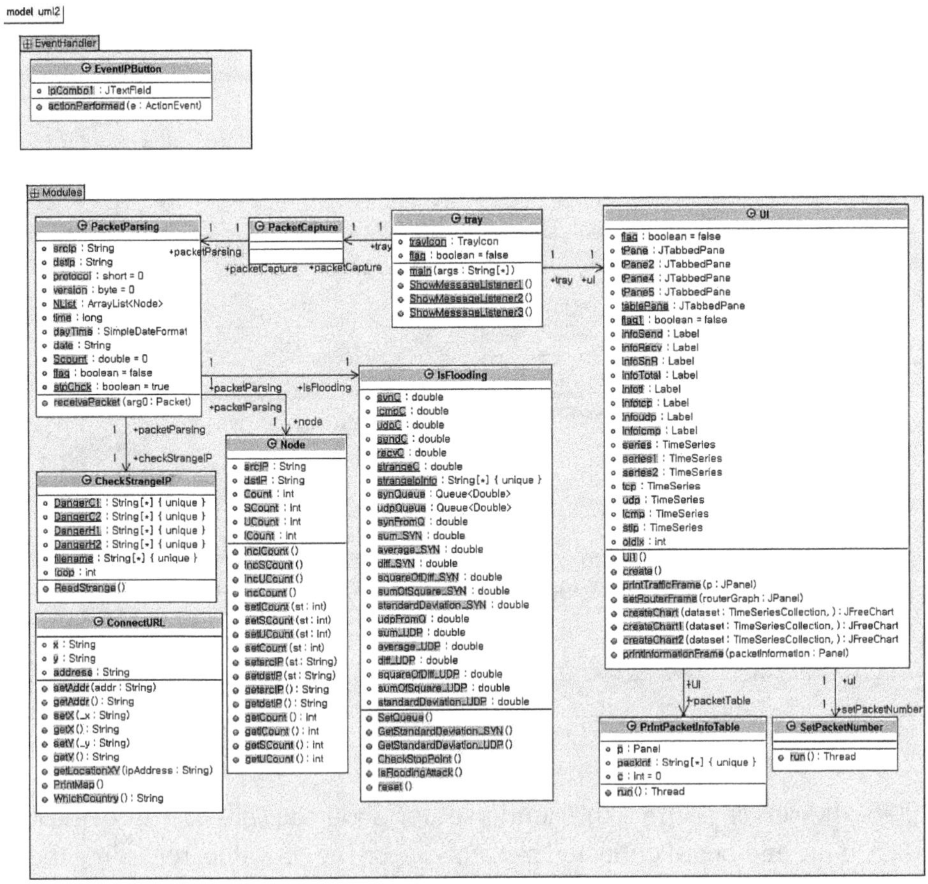

be effected. Additionally, the interface is divided into 8 domains to assist the users with less background knowledge in computer networks. Description of each domain is as follows:

Domain 1, the interface shows the Send/Receive packet traffic volumes with a graphic mode. To be more specific, Domain 1 generates the volume of all incoming/outgoing packets through servers.

In Domain 2, the graphic chart of the strange IP packet traffic volumes will be indicated. It detects TCP, UDP or ICMP Flooding attack packets during the DDoS attack identifying the patterns and outputs their extent of seriousness with graphs. By analyzing each packet pattern and considering the volume of the packets collected in real-time, the program updates the threshold value periodically. After setting the calculated threshold value as

Figure 2. The user interface of the graphic based NIDS for AMI

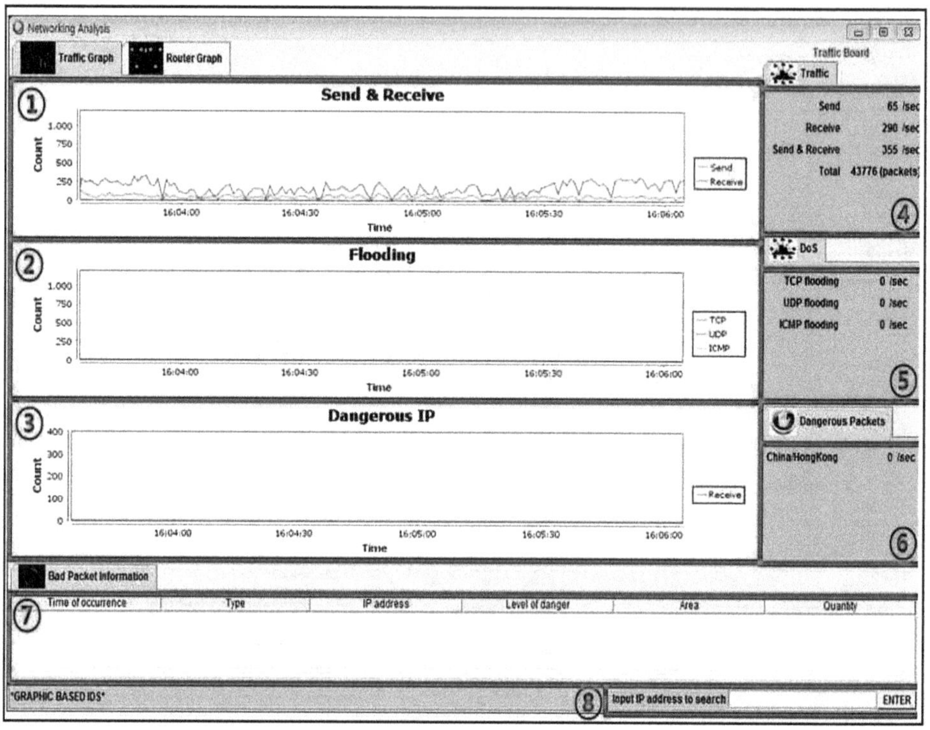

100%, the server notifies the administrator about the possibility of DDoS attack if the number of collected packets exceed the set value, regarding them as abnormal traffics.

Domain 3 sends out a warning to the user when there is a packet transmitted from the suspicious and dangerous countries like People's Republic of China or North Korea. As seen in current DDoS attack events against Republic of Korea, there have been many cases of attacks carried out by North Korean through China, or other countries. Our system was built to block these malicious accesses.

Meanwhile, Domain 4 and 5 denote traffic volumes numerically and the former shows the packet volumes delivered with TCP, UDP and ICMP, and the latter, the traffics in Domain 3 will be indicated. As for Domain 6, the interface has been designed graphically for easier recognition by the server operator to check the detailed information concerning abnormal packets, which includes time of incident, packet type, IP address, risk level, origin and packet numbers. The interface has been designed in Domain 7 to allow the server administrator to check abnormal traffics in detail. Lastly, in Domain 8, it is possible to produce and output a Google map by tracking down initial source router of received packet using the IP address location search button.

3. PERFORMANCE EVALUATION

The proposed program is able to run on various platforms and implemented with Java language which offers strong network communications. Since the system aims at real-time base operation, the program is designed with an adequate structural logic which is to offer speedy response time and execution time. The purpose of the program is to show total network traffic volume in real-time and notify the user in the event of various DDoS attacks (e.g. SYN, UDP and ICMP flooding).

The Wireshark, which could be regarded as a similar product, shows packet structure that is being captured on the network interface card. It's not easy to analyze and access the situation unless the user is an expert and warning function is not provided.

According to the scenario, it seems possible to solve emergency situations using remote function. Once an attack attempt is made, and sense it with real-time surveillance program to notify the programmer standing by, he/she will be able to help initiating the defensive actions and minimize the damages.

Table 1 is a chart that shows DDoS attack status in Republic of Korea in recent years and the test was conducted with the code (i.e. partial code in the section marked) used for the DDoS attack against NH Nonghyup Bank in March, 2013.

Graphic Based Network Intrusion Detection System for Smart Grid was coded with Jpcap on the basis of Java and aimed to lookout all the packets passing through user's network interface card. It was constructed in such a way that the user can grasp network situation at a glance easily, check information regarding the control and other details, and gives warnings using threshold value enabling easy detection of DDoS attack possibilities.

The ultimate objective of the system proposed in this book is to lower the entry barriers to the network analysis and management as well as providing a framework easily understandable to network administrators who do not have adequate expert knowledge yet. Since the program has been constructed based on GUI, the users unfamiliar with Wireshark or other network analysis programs can easily approach it and the program will contribute to lowering

Table 1. DDoS attack status in Republic of Korea in recent years

Date	Events	Description & Damages	Base of Supposition
07/07/ 2009	DDos attack	*435 servers of 61 countries 270,000 + zombie PCs were used -35 sites of Korean & American major organizations	*Head of NIS inspection by National assembly(29/10/09) -After tracking down IP address used for terrorism, it was identified that the address belong to North Korean Communication Ministry
04/03/ 2011	DDos Attack	*746 servers of 70 countries 100,000+ zombie PCs were used -40 sites of Korean major organizations including Cheonwade, National Assembly and the press	*National Police announcement(06/04/11) -As a result of analysis, method of the attack was the same as the one committed on July, 2009 DDos attack
12/04/ 2011	Nonghyup computer network hacking	*27 servers of 13 countries were used -Destroyed 273 PCs of Nonghyup causing network trouble	*Public Prosecutor's Office announcement(03/05/11) -attack origin IP address in the notebook used was the same as the one used by North Korean Reconnaissance General Bureau
09/06/ 2012	Joongang Ilbo hacking	*via 2 domestic & 17 overseas servers(10 countries) was used	*National Police announcement(16/01/13) -Chosun Communication Company(subsidiary of Communication Ministry of North Korea) made the connection with IP bandwidth rented from Chinese company
20/03/ 2013	Destruction of broadcasting, financial companies' computer equipments, etc.	*49 via servers of domestic and overseas employing 76 kinds of malicious codes -Destruction of broadcasting, financial companies' computer equipments and deletion of homepages run by anti- North Korean organizations	*Civil, administration and military joint respond team announcement(10/04/13) -After analyzing IP address and the hacking method, many evidences which support assumption that the act was initiated by North Korea(similar to July 7th DDos attack event)
25/06/ 2013	DDos attacks to Gov't offices and destruction of press companies' communication equipments	*82 kinds of malicious codes, zombie PCs were used -69 cases of DDos attacks and destruction of server units at Cheonwade, Integrated Gov't Computation Center, Newspaper companies	*Civil, administration and military joint respond team announcement(16/07/13) -identified IP address used by North Korea and the method was the same as the March 20th cyber-terror committed by them

of the entry barriers for the network security operation in small network/ server management by individuals. Other contributions would be prevention of unintended errors and protection of user information. This will be achieved by displaying the harmful packets easily missed due to the mistakes and lack of security consciousness of the network administrator clearly on the monitor.

Existing commercial products have adopted a method that applies signatures to make computer systems to avoid invasions or infections. Thus, arranged the test that the existing anti-virus software are unable to detect an emergence of the new Botnets to establish the relevance of the test (Shin & Yang, 2009) and performed comparative analysis for the products listed below: -Product A: Alyac by Eastsoft (EST Security, n.d.) -Product B: Norton anti-virus by Symantec (n.d.) -Product C: Smart Defense by AhnLab (n.d.) -Product D: AD-Spider Dajaba by Digital on net (AdSpider, n.d.) and, -Product E: Online vaccine v6 by nProtect (n.d.).

In the test results, as seen in Table 2, existing anti-virus product lines executed deletions of known Botnets after user's confirmation process but for the unknown Botnets, they continued the detections while the process was being executed. Meanwhile, in the case of Product C and Product E, deletion was executed after detecting some part of unknown malicious codes and this indicates that the deletion process was restrictively applied to the groups having the same types of malicious codes. In general, since the traffic

Table 2. The comparative analysis with advanced research

	Known Botnets(K/M)		Unknown Botnets(U/M)	
	Result	Behavior after infection	Result	Behavior after infection
Proposed system	○	Inform user with alarms	○	Inform user with alarms
Compared system	○	Unintended access detection	▣	Unintended access detection
Product A	○	Delete after detection	▣	Continual detection
Product B	○	Delete after detection	△	Delete after detection
Product C	○	Delete after detection	▣	Continual detection
Product D	○	Delete after detection	▣	continual detection
Product E	○	Delete after detection	△	Delete after detection
○: detection, △ :partial detection, ▣: undetection				

attacks are ones that continuously attack networks, our proposed Graphic based Network Intrusion Detection System has an advantage over others because even if smaller number (i.e. less than threshold value) of packets arrive and there exists abnormalities more than once, the system can detect them. During the test, after constructing the virtual network environment having four clients and initiating DDoS attacks, able to observe that all of the attacks were detected.

4. CONCLUSION

Smart Grids and related facilities are no longer safe from the cyber terrorism. Once a new vulnerability is exposed, the control network that uses the same system can be exposed the same threat and the security threats like malicious codes can be spread all over the world instantaneously. Therefore, a systematic security model for the major infrastructures should be developed urgently. However, even if a suitable model was developed and to be introduced into the ROK, a proper introduction procedure suitable to domestic environmental should be established first and the profile of the framework need to be reviewed in relation to the scale of control system, assessment of risk factors, priorities, etc., especially for the control systems of high-security facilities like Smart Grid plants or Power Grids (Huh et al, 2015).

REFERENCES

AdSpider. (n.d.). Retrieved from: http://www.ad-spider.com/spyware

AhnLab. (n.d.). Retrieved from: http://kr.ahnlab.com/b2b/securityinfo/html/renewasecreport

Jun-Ho Huh, Donghoon Lee, Kyungryong Seo, Implementation of Graphic Based Network Intrusion Detection System for Server Operation, International Journal of Security and Its Applications

nProtect. (n.d.). Retrieved from: http://www.nprotect.com/v6/service

Security, E. S. T. (n.d.). Retrieved from: http://alyac.altools.co.kr/Main/Default.aspx

Shin, D. J., & Yang, H. S. (2009). Design and implementation of an intrusion detection system based on outflow traffic analysis. *Journal of Korea Contents Association*, *9*(4), 131–141. doi:10.5392/JKCA.2009.9.4.131

Symantec. (n.d.). Retrieved from: https://www.symantec.com/

Vol.9, No.2 (2015), SERSC, pp.37-48.

Chapter 5
Bluetooth-Tracing RSSI Sampling Method as Basic Technology of Indoor Localization for Smart Grid

ABSTRACT

In recent years, Smart Grid have become the center of interest for IT companies and construction companies and various types of Smart Grids have been made currently available on the market. Yet, equipment is costly and it is not easy to convert existing equipment for Smart Grid application as they may require additional resources which could also inflict much costs. The extra costs involving the remodeling of existing housing structure and installment of new equipment can be avoided by using advanced wireless technologies. As an example, this book proposed an indoor localization system that adopts Bluetooth technology and uses RSSI (Received Signal Strength Indication) values for localization. Researchers have configured a system where the central control device will recognize all other devices or equipment in the system, communicate with each other, and respond to the commands or the information provided. However, despite the efforts of many researchers, existing RSSI-based indoor localization systems do not show a satisfactory level of accuracy such that we have devised a system that traces the trend in the RSSI samples.

DOI: 10.4018/978-1-5225-2776-3.ch005

1. INTRODUCTION

Like many other IT-advanced countries, Smart Grids are attracting more and more customers in recent years, with devices and equipment associated with Smart Grids continually increasing presenting, more convenient or novel functions. Contrary to such an environment, the cost of constructing or remodeling existing households to adapt Smart functions is increasing as these works (i.e., installing new equipment, setting up new systems and adding new functions) often involve time-consuming and costly operations. Such being the case, we have devised a system that can recognize the conditions of present household appliances, analyze the data and place appropriate commands, and initiate the movements according to them. Here, the 2 most important factors were 'Cost Saving' and 'Accuracy'. The former is related to reducing or avoiding additional device installments and the latter is associated with developing a better algorithm.

As an example of achieving these goals, we have designed an Indoor Localization System that displays a higher level of accuracy than GPS and other similar positioning systems by incorporating Bluetooth technology and RSSI-based algorithm without costly structural changes.

Our system will pinpoint the target objects inside the household by triangulating the signals from the reference terminals (e.g., beacons or other signal-generating devices). Figure 1 shows Bluetooth BLE-based location tracking. Bluetooth Low Energy (BLE) technology has been applied for the beacons considering energy efficiency.

With Bluetooth technology, the intensity of signals and their unique identifiers can be distinguished and the position of a certain beacon can be identified through the relevant identifier. Additionally, the distance between the beacon and a fixed point can be estimated using RSSI so that by accumulating and analyzing such information from all of the beacons, the user's position can be approximated. Since most Smart Phones are embedded with Bluetooth function nowadays, one can avoid purchasing another one to use it as a measuring tool. There are some other products that have a similar function as Bluetooth but they often need to be calibrated with the smooth filter or field data such that they may not provide accurate data. To deal with this problem, we proposed an RSSI sampling method with a trend tracing function.

Figure 1. Bluetooth BLE-based location tracking

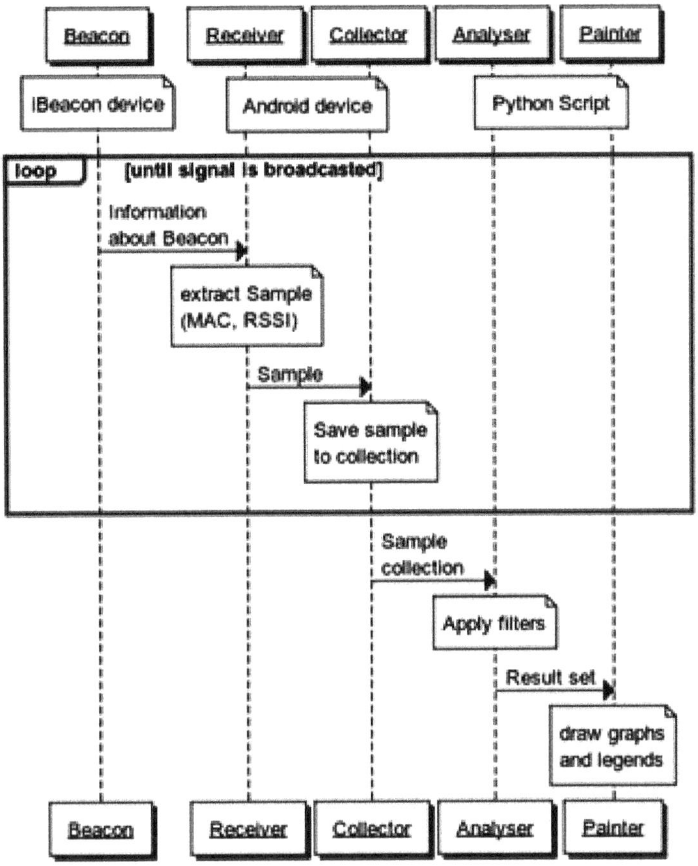

2. RELATED RESEARCH

2.1 iBeacon

The iBeacon is a wireless beacon protocol that has been proposed by Apple, Inc. It is built on the principle of BLE technology and designed to broadcast a beacon's information on a regular cycle to let the device to receive them and make decisions. The information to be broadcasted is listed in the Table 1.

Basically, iBeacon was designed and used for product arrangement guidance for offline shopping so that it include a function of measuring rough distances to a beacon using each beacon's RSSI. However, the measurements are classified into only 3 categories (i.e., Close, Away, Far), and only the

Table 1. Information to be broadcasting

Condition	Information
iBeacon Protocol Prefix	Indicating that the iBeacon protocol is in use
iBeacon UUID	Unique identifier of the installed beacon
Major Code	Major value (e.g., main building) of the installed beacon
Minor Code	Minor value (e.g., 1st lobby) of the installed beacon

adjacent information of the relevant beacon is provided after establishing one-to-one relationship. Currently, many hardware companies manufacture and sell the beacons that follow iBeacon protocol.

2.2 Estimote

Estimote, Inc., is a company that supports hardware sales and carries out application service development for iBeacon. They provide the Software Development Kit (SDK), which facilitates information collection processes following the iBeacon standards, for iOS and Android and recently, they've announced a new SDK called 'the Indoor-SDK' that traces indoor position with iBeacons. However, it only supports iOS and provides basic equipment and SDK, not the application services. Additionally, it only traces and offers 2-dimensional information of indoor space.

2.3 Market Situation

A lot of companies are currently developing and selling the indoor positioning system, however most of their products are based on the different types of technologies and even if they would use similar ones, the markets are still divided by using different types of standards so that the wider use of the solutions is limited, preventing market expansion. Also, there are no domestic companies specializing in this kind of service and most of the time, the similar service will only be developed by the SI development companies if there are any demands in the market. Under such environment, both the accumulation of technological know-how and standardization are far from realization and market growth is contracting in both the short and long terms. Therefore, it is essential to let users benefit from this technology through its standardization and introduction.

2.4 Comparison With Other Studies

The idea of using distance measured by RSSI-based method for indoor localization started early and is still ongoing (B. Nissanka & Priyantha, 2000). Researchers try to improve accuracy by enhancing data processing (Ault et al., 2005). However most of them had to recognize the uncertainty of RSSI (Joonyoung et al., 2013). For instance, they applied different types of smooth filters (Dong et al., 2012), complemented the system by using a network of multiple beacons (Botta et al., 2013) and so forth. While these studies prove to be effective for RSSI-based localization, they don't seem effective enough to produce higher-accuracy products (Huh et al, 2016). Thus, we need to further the study for RSSI-based localization technique.

3. DELTA TRACE SAMPLING FOR RSSI-BASED DISTANCE ESTIMATION FOR SMART GRIDS

3.1 RSSI (Received Signal Strength Indication)

The abbreviated term RSSI (Received Signal Strength Indication) refers to the value of the electrical power received by the wireless receivers. Since it describes the signal powers, both the antenna gain and the circuit loss are not considered. In most cases, the unit dBm, which indicates the gain and attenuation, is used.

3.2 Decibel

Equation (1) shows decibel. Decibel is a dimensionless relative system unit that indicates the ratio between a specific reference volume of A and measured volume of B. Below is a formula to calculate the decibel.

$$dB = 10 * \log_{10}\left(\frac{B}{A}\right) \tag{1}$$

3.3 dBM

Although dB is the dimensionless unit, a reference value of A for the target medium should be set to be used in the engineering field. Thus, 0dBm is defined by setting the reception power that will incur 0dB as 1mW.

3.4 RSSI Model

Equation (2) represents the RSSI model (Kang et al., 2007).

- D: The distance(m) between signal transmission equipment and measuring equipment.
- $P_{r(D_0)} P_r\left(D\right)$: Measured RSSI value on D
- $P_{r(D_0)} P_r\left(D_0\right) P_{r(D_0)}$: Measured RSSI value (dBm) when D=1
- N: The signal attenuation constant. In an ideal condition, the number would be 2 but it can be amplified or attenuated depending on the radio environments.

$$\frac{P_r\left(D_0\right)}{P_r\left(D_0\right)} = \left(\frac{D}{D_0}\right) \tag{2}$$

3.5 RSSI-Based Distance Estimation

Relation model between the distance and RSSI can be obtained from.

- OFFSET: $P_{r(D_0)} P_r\left(D_0\right)$
- RSSI: $P_{r(D_0)} P_r\left(D\right)$

$$\mathrm{D}\left[m\right] = 10\left(\frac{OFFSET - RSSI}{10 * N}\right) \tag{3}$$

Calibration of OFFSET and N: Measure the RSSI value from 1m distance and designate the value as OFFSET. Next, measure the RSSI value again from a different distance and substitute it to the formula derived from above.

- **OFFSET:** RSSI value measured from 1m distance
- **RSSI:** The RSSI value currently measured
- **D:** The distance between the place of RSSI measurement and the equipment

$$N = \frac{OFFSET - RSSI}{10 * \log_{10}(D)} \tag{4}$$

3.6 RSSI Sampling

RSSI values show considerable fluctuations even in a static condition. This phenomenon is usually caused by multiple reasons such as signal interferences with other signals, unstable power level of transmitting equipment, current fluctuations during data transmission and the radio wave strength in surrounding environments. Thus, if the sample errors are not calibrated, exact calibration for the OFFSET or N will also be impossible, causing the problem of not being able to obtain an exact sample that will be used for the distance measurement.

However, since the smooth filter works targeting entire samples, it is easily affected by the changes in the RSSI values, which appear from the momentary signal distortions or other electromagnetic disturbances. For this reason, the margin of error will be larger so that the reliability of measurements themselves cannot be secured easily. Nevertheless, if it's possible to secure enough reliability, the smooth filter can be quite suitable for deducting a trend value among the measurements obtained. Thus, a new approach is needed to secure the reliability for the measured values. Since we will not be dealing with the smooth filters and relevant studies in the development research area, the range average calculation, which is the simplest form of smooth filtering, will be used.

3.7 Delta-Based RSSI Sampling

The reason for the low reliability of measured RSSI values is that each measurement results in a different outcome even if the smooth filtering method has been applied because the range of fluctuation is too wide even in a static condition. Moreover, most of the results obtained from the smooth filtering showed that better accuracy can be achieved when more data samples are available. However, the process speed will be poor as it requires a longer time to measure the samples.

In this regard, we suggest using another sampling method in which only the effective samples will be collected comparing the deltas of measurement values from a small number of samples.

The basic idea of the delta-based sampling is that when the current measurement variations are excessively larger when compared to the recent trends, it will be most likely that the current value is an irregular sample extraneous to recent variation in the measurements.

Following this idea, an algorithm that identifies the effective samples by distinguishing them by through checking the ratio (i.e., proportion between the delta of current sample value and that of recent samples) that falls below a certain level, is described next.

Smooth filter will be used to determine sample trend and for our algorithm, the range average will be used - the simplest smooth filtering method.

- **SAMPLE_LIST:** List of time-sequenced samples to be measured (size(SAMPLE_LIST) >=2)
- **SAMPLE:** Currently measured sample
- **THRESHOLD:** Variation tolerance compared to the tendency (0 < THRESHOLD)

```
var DELTA_LIST = new LIST;
        for (var index = 0; index < size(ALL_LIST) - 2;
index++)
        var delta = ALL_LIST[index + 1] - ALL_LIST[index];
        DELTA_LIST.add(delta);
        end for
        var averageDelta = average(DELTA_LIST);
        var currentDelta = SAMPLE - ALL_LIST.last();
        var deltaRatio = abs(currentDelta / averageDelta);
        var effective;
        if (deltaRatio < THRESHOLD)
        effective = TRUE;
        else
        effective = FALSE;
        return effective;
```

The merit of this algorithm is that we will be able to read off whether the next measurements would be included in the overall changing trend by determining the fluctuation tendency of the values obtained from a small-sized sample group of 20 or so.

Accordingly, even for the real-time measurements, it will be possible to seek out valid samples relatively exactly using recent available samples.

On the other hand, in cases wherein the size of the targeted SAMPLE_LIST is too large, or measuring interval for the samples is too long, one should pay particular attention to setting the size or range for these factors at the time of an actual implementation because there's a possibility of overlooking the fact that the trend may have changed for a short period of time and returned to normal afterwards.

Figure 2 shows the sampling/analysis mechanism. The process and the core algorithm are as follows:

Set A with the sequential RSSI values is,

$$S = \{S_1, S_2, \ldots, S_n\} \tag{5}$$

and, Set D, which consists of deltas between neighboring elements in the Set A, is defined as,

$$D = \{d_1 = 0, d_2 = S_2 - S_1, \ldots, d_n = S_n - S_{n-1}\} \tag{6}$$

Function T(n), which estimates the trend of elements in the Set D, is an arithmetic mean of the Set D.

Figure 2. Sampling/analysis mechanism

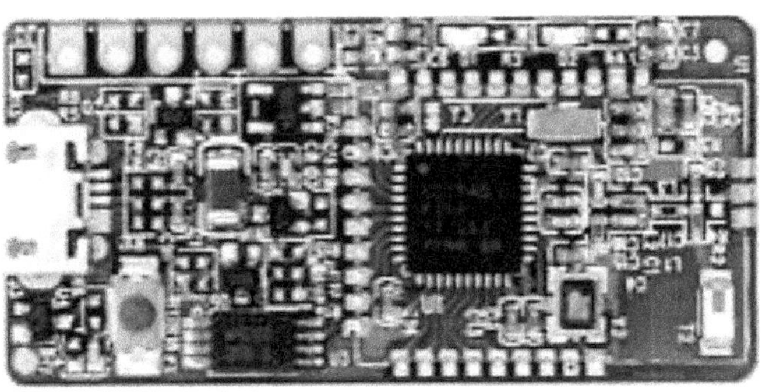

$$T(n) = \frac{1}{n}\sum_{k-1}^{n} d_k \tag{7}$$

Then, when S_k has been provided, a Function $R(S_k)$ that estimates the trend of S_k will be defined as,

$$R(S_k) = \frac{d_k}{T(k-1)} \tag{8}$$

When a random value R_{max} is given, a Function $V(S_k)$ which determines the validity of S_k will be defined as,

$$V(S_k) = \begin{cases} False <= R(S_k) < 0 \\ True <= 0 \leq R(S_k) \leq R_{max} \\ False (= R(S_k)) R_{max} \end{cases} \tag{9}$$

∵ Since the directions will be opposite when R_(S_k) is negative, S_k does not coincide with the trend.

∵When $0 \leq R(S_k) \leq R_{max}$, the directions of d_k and T(k) will be the same so that S_k coincides with the trend.

∵When $R_{max} \leq R(S_k)$, d_k and T(k) will bear the same direction but their variations differ much so that S_k does not coincide with the trend.

If n becomes large enough, the accuracy of the trend estimation T(n) will be decreased. Thus, the improved T(n) is defined as,

$$T'_n = \begin{cases} \frac{1}{n}\sum_{k-1}^{n} d_k <= n \leq T_{count} \\ \frac{1}{T_{count}} \sum_{k=n-T_{count}}^{n} d_k (= n) T_{count} \end{cases} \tag{10}$$

Here, T_{count} sets the random values that represent the number of elements used to estimate the trend. T'(n) is similar to the average-based smoothing filters.

4. DELTA TRACE SAMPLING FOR RSSI-BASED DISTANCE ESTIMATION FOR SMART GRID

For the Android platform, we've test-implemented the above-mentioned algorithm by following its rules to perform sampling, distance measurement and calibration of N and OFFSET using a BLE device. For the embodied test system, a function of reading that the BLE device has moved more than a certain distance from the designated position has been added to check whether the variation measuring function is working properly.

Figure 3 shows TI CC2540 Module. A communication module manufactured with TI CC2540 (small BLE SoC) was used as a transmission device for the implementation.

Additionally, in order to measure the device's RSSI periodically, a firmware was compiled and installed on the ROM to broadcast information in accordance with the iBeacon standard.

Figure 3 shows a conceptual diagram for implementation. For the measurement, an application that registers/measures broadcasting information

Figure 3. TI CC2540 module and conceptual diagram for implementation

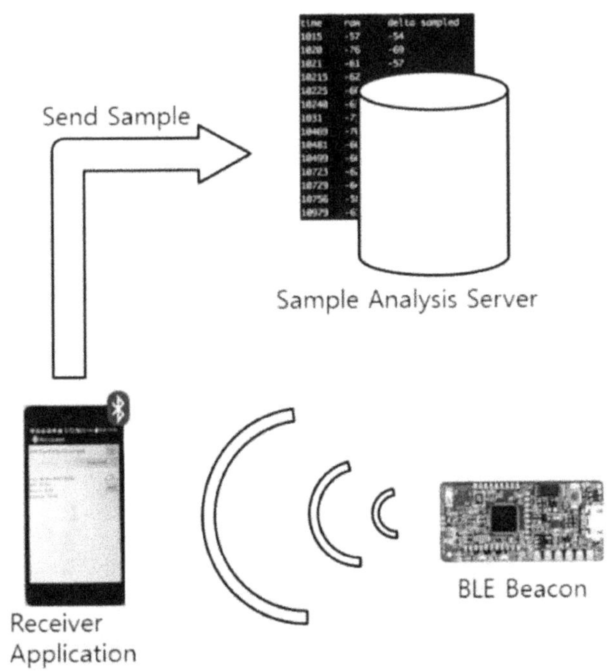

and RSSI by using an android device (phone) and that sends them to the server was developed. Both the server that collects/ stores the transmitted sample measured with the device and the Delta Sampling algorithm, which is to process stored data, were implemented using the Python Script.

5. PERFORMANCE EVALUATION

5.1 Experiments

Figure 4 shows actual embodiment. For the measurement, a test was conducted by setting the range size and threshold as 20 and 7.5 (test-1), respectively following the small-scale preliminary test. During the test, collection of the RSSI samples every 100ms was conducted shifting the distance the range from 1m to 3m over 70 seconds. The test was repeated several times and the results were collected for the calculation. For the measurement, a test was conducted by setting the range size and threshold as 5 and 7, respectively following the small-scale preliminary test. During the test, RSSI samples were collected every 375ms. The test was repeated several times and the results were collected for the calculation.

We have two types of tests. In test-2, we collected 100 RSSI samples while the device was fixed at three meters from beacon. In the test-3, the

Figure 4. Actual embodiment

RSSI samples were conducted shifting the distance the range from 1m to 3m over 30 seconds.

5.2 Analysis

As shown in Figure 5, we were able to confirm that the range of fluctuation was more stable than the case of raw samples, which had shown a wide fluctuation range. From this result, we learned that finding the valid values with our algorithm is not only effective in correct sampling but also effective in extracting more precise data during the smoothing process. The result shows better performance levels than other available methods.

The Figure 5 shows the comparison of measurements collected with the raw sampling and Delta sampling methods. It can be observed that the range of fluctuation in the measurements collected with the Delta sampling method is more stable than that of the raw sampling method. The minimum and maximum values and their standard deviation obtained from the raw sampling were -90, -54 and approx. 5.95 respectively while they were -81, -57

Figure 5. Comparison of raw sample and Delta based sample (Test-1)

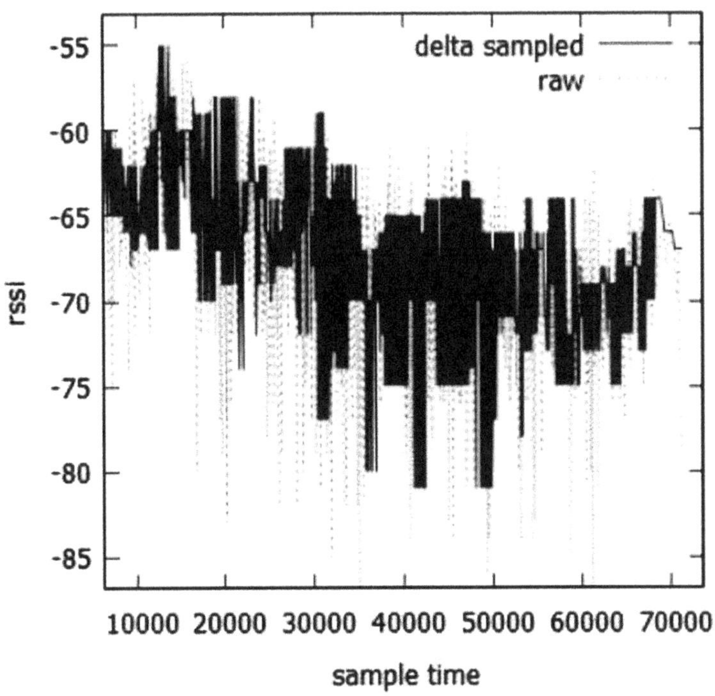

and approx. 5.02 each from the Delta sampling. Thus, one can assume that the delta sampling collects a little more stable RSSI measurements compared to the raw sampling method.

Furthermore, if it's possible to find a suitable threshold depending on the usage environment through repetitive experiments, a higher degree of precision could be expected. Therefore, as a future work, we shall pursue experimental studies for an algorithm that is suitable for the trend estimation of raw data as well as finding an appropriate threshold for each different usage environment.

As shown in Figure 6 and Figure 7, we were able to confirm that the range of fluctuation was more stable than the case of raw samples which had shown a wide fluctuation range. Figure 6 and Figure 7 show the comparison of measurements collected with the raw sampling and Delta sampling methods. It can be observed that the range of fluctuation in the measurements collected with the Delta sampling method is more stable than that of with the raw sampling method.

Figure 8 and Figure 9 show comparison of average smoothing filter, delta based trend trace sampling, and a combination of both methods. The graph is smoother than raw RSSI samples when using the average smoothing filter.

Figure 6. Comparison of raw sample and Delta based sample in the (Test-2)

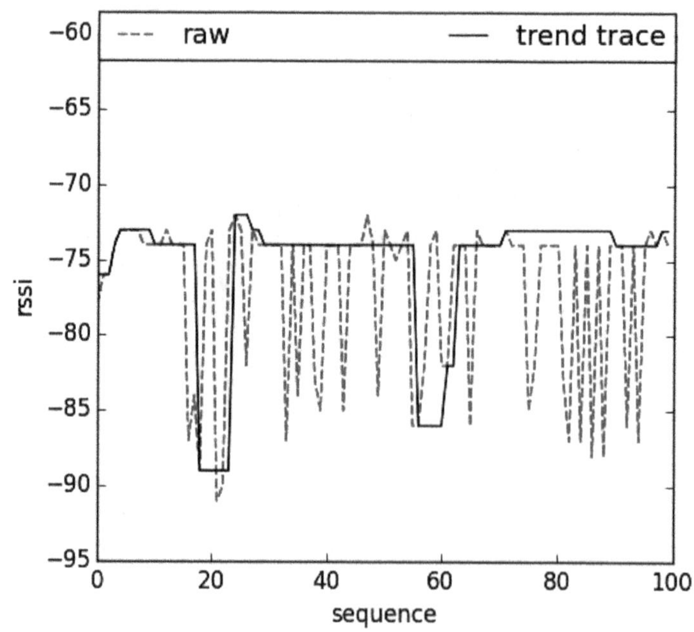

Figure 7. Comparison of raw sample and Delta based sample in the (Test-3)

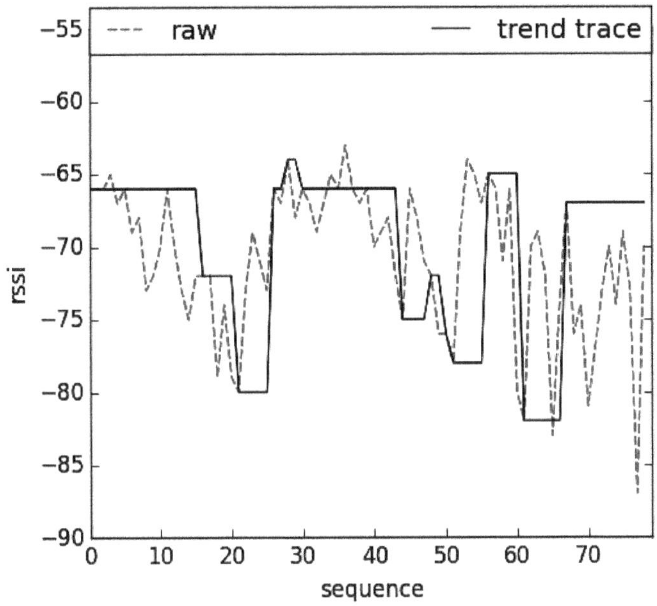

Figure 8. Comparison of average filter, trend tracing and MIX in the (Test-1)

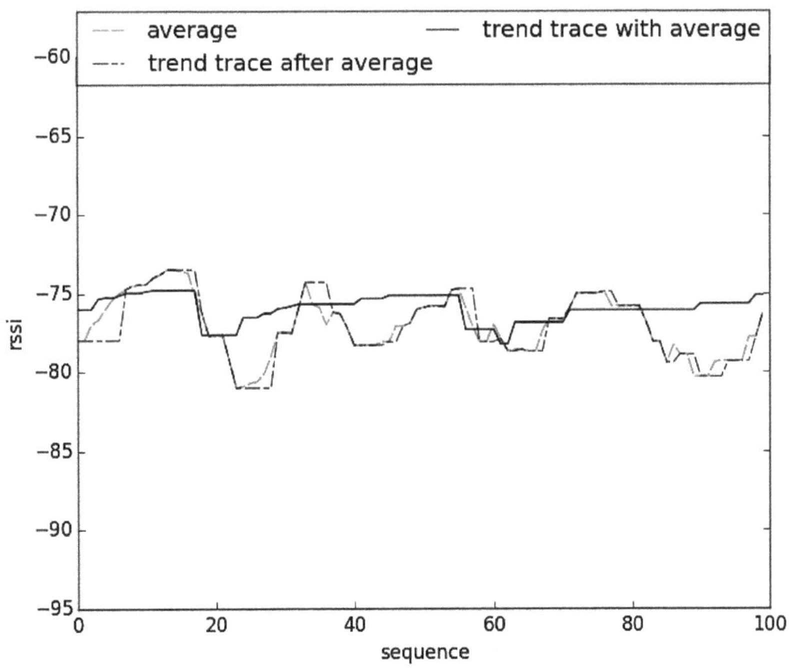

Figure 9. Comparison of average filter, trend tracing and MIX in the (Test-2)

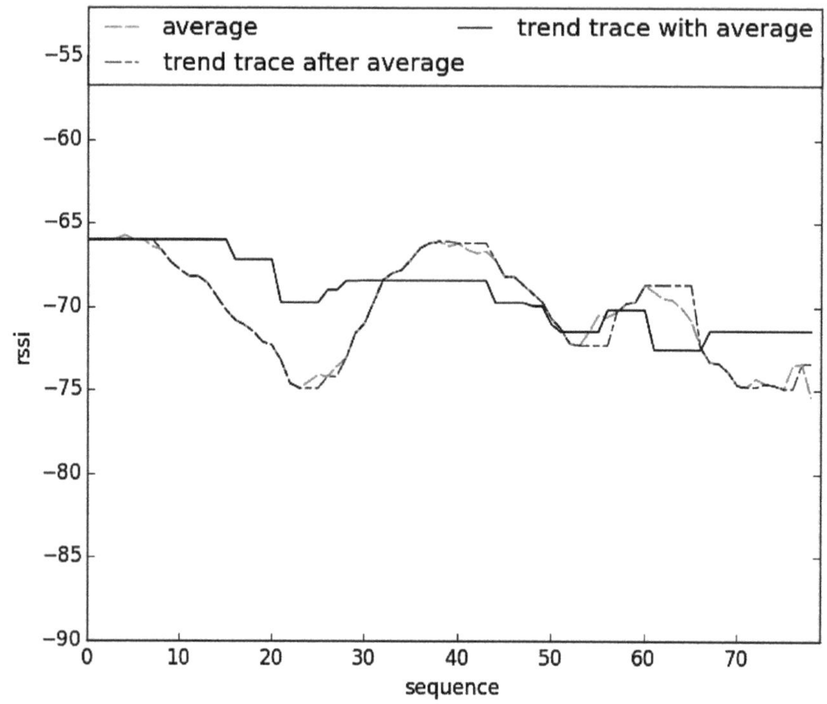

However, the graph spiking, even though the values are decreasing, because it was exposed to an error of the RSSI samples. And it followed the trend. In other case that used delta based trend tracing that combined with average smoothing, the graph is stable and less affected by the error.

In a last case that used delta-based trend tracing after average smoothing, it was not effective. From this result, we learned that finding the valid values with our algorithm is not only effective in correct sampling but also effective in extracting more precise data during the smoothing process. The result shows better performance levels than other available methods when it was mixed with others.

6. CONCLUSION

Our experiment results have shown that the proposed algorithm was effective and efficient in extracting usable and valid samples in an -RSSI data set but there were a few limitations. That is, to trace the trend, researchers used the range average algorithm but it was clear that the measurements tended to depend on the sample sizes and the number of valid samples varied in each sampling rate or sampling environment. Moreover, a raid position shifting affects the measurement results so that the trend estimation becomes inaccurate. Another notable problem was that it was anticipated that when the threshold value in the experiment was too high, the noise would have more impacts on the result and when it was kept low, low accuracy could be expected.

Therefore, researchers plan to find a more sophisticated algorithm to improve the accuracy of the system in our future research to deal with these problems. While integrating the Delta-based RSSI sampling method with other complementary methods may improve the accuracy, proving its effectiveness in Smart Grids will require much work. Additionally, it is important to mention about the BLE-exclusive devices. For example, researchers suggest that the beacons with CR2045-model batteries should be used for the experiments and in actual Smart Grids for their low power consumption nature and a long service life (1.8 ~ 28.7 months). In addition to this research, our future study will consider these 2 factors more seriously.

Lastly, our method can be applied to many other fields (e.g., theft prevention, warehouse management, and etc.) as a base technology and we expect that we will be able to enhance it to deal with a more complex environment (Huh et al, 2016).

REFERENCES

Ault, A., Zhong, X., & Coyle, E. J. (2005). K-nearest-neighbor analysis of received signal strength distance estimation across environments. In *Proceedings of the First Workshop on Wireless Network Measurements.*

Botta, M., & Simek, M. (2013). Adaptive distance estimation based on RSSI in 802.15 4 network. *Radio engineering, 22*(4), 1162- 1168.

Dong, Q., & Dargie, W. (2012). Evaluation of the reliability of RSSI for indoor localization. In *Proceedings of the IEEE International Conference, Clermont Ferrand.*

Huh, J. H., Bu, Y., & Seo, K. (2016). Bluetooth-tracing RSSI sampling method as basic technology of indoor localization for smart homes. *International Journal of Smart Home, 10*(10), 9–22. doi:10.14257/ijsh.2016.10.10.02

Jung, J., Kang, D., & Bae, C. (2013). Distance Estimation of Smart Device using Bluetooth. In *Proceedings of the Eighth International Conference on Systems and Networks Communications*, Italy (pp. 13-18).

Kang, J., Kim, D., & Kim, Y. (2007). RSS self-calibration protocol for WSN localization. In *Proceedings of the IEEE 2nd International Symposium on Wireless Pervasive Computing* (pp. 181-184).

Priyantha, N. B., Chakraborty, A., & Balakrishnan, H. (2000). The Cricket Location-Support System. In *Proceedings of the 6th annual international conference on Mobile computing and networking* (pp. 32-43).

Chapter 6
Comparison With Other Systems

ABSTRACT

This chapter compares the system explained throughout this book with other systems expressed in other research. 'The Trend on the Smart Grid Systems in the Republic of Korea' shows a normal system in Smart Grid Test Bed in Jeju-island, and the author distinguishes this approach with that of a new proposed method using RUDP (Reliable User Datagram Protocol).

1. INTRODUCTION

The preceding research Figure 1 is 'The Trend on the Smart Grid Systems in the Republic of Korea', which describes a typical system selected at the Smart Grid Test Bed in Jeju-island.

This study in Figure 2 distinguishes itself from the preceding research by proposing the PLC communication system and presenting the benefits of using RUDP in the Transmission Layer.

The preceding research (Yao, 2011) has been written as a graduate of Purdue University and the difference with this book is that it uses a different simulation tool. To be specific, while a DDoS detection system was simulated in the application of the precedent research, the same was tested by directly implementing it in the Application Layer used in this study.

DOI: 10.4018/978-1-5225-2776-3.ch006

Figure 1. A trend on the smart grid in Republic of Korea

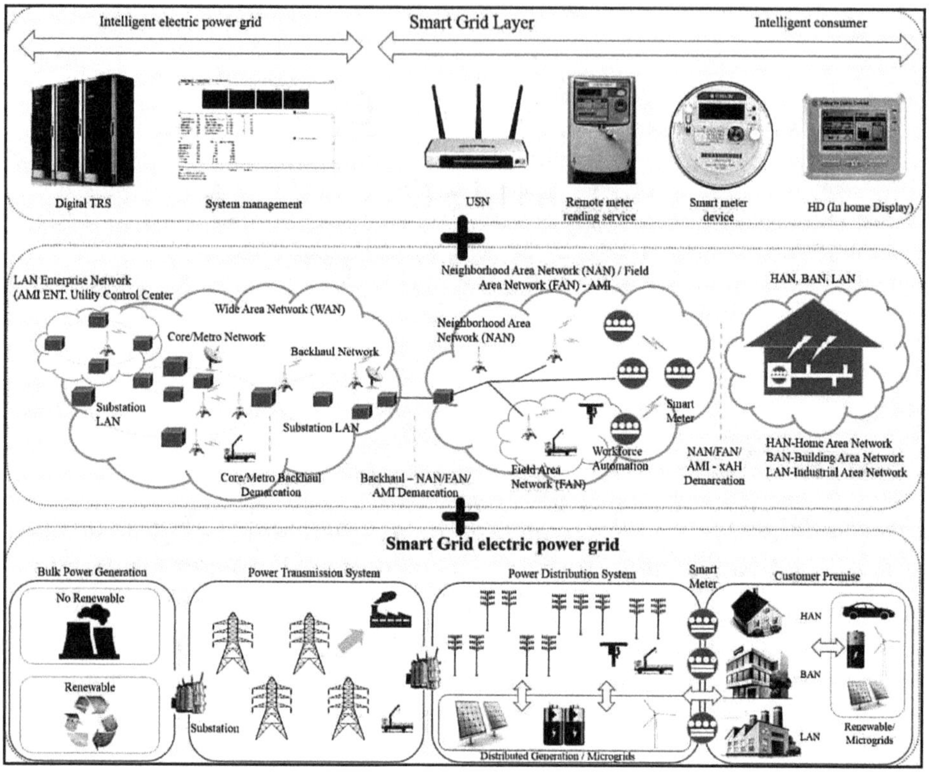

Figure 2. Smart grid test bed (proposed method)

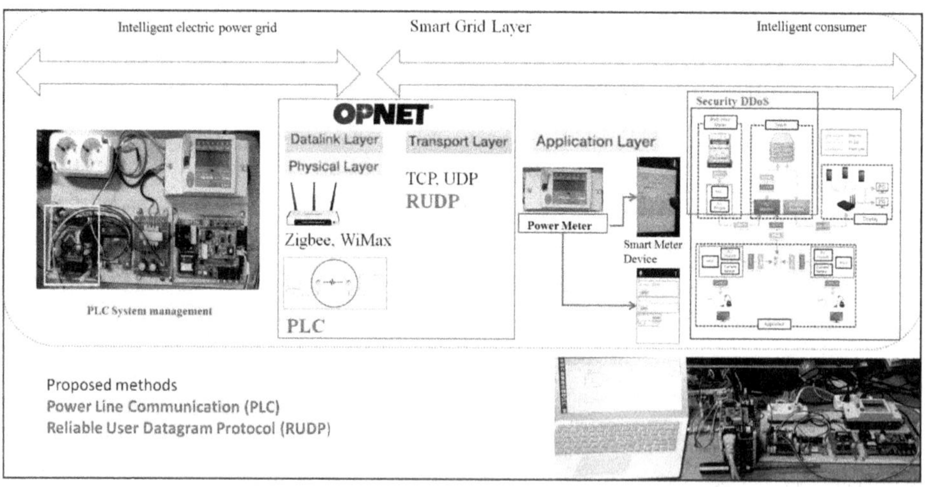

The preceding research is a Paper published on Journal and has different test bed software and the range of implementation Figures 3-9 (Soetedjo, Lomi & Nakhoda, 2015).

Figure 3 describes the architecture of the power distribution network of proposed Smart Grid test bed. The network is comprised of three separate power resources such as PV generator, wind turbine and utility grid, two power demanders (Home-1 & Home-2) and five power switches. These are linked with 12V DC-bus. The power switch assumes the vital role of managing the flow of electricity and controlled by the micro-controller which acquires commands sent from the Smart Grid Center through wireless communications.

Figure 4 shows the architecture of information network within the test bed. The architecture includes five nodes and a single center linked through Xbee communications. The Xbee module is mounted on an embedded system utilizing the Arduino micro-controller. The Smart Grid Center uses the SCADA software which runs on PC. Embedded Xbee module-equipped system interfaces with PC to establish a connection with each node. Followings are the detailed description of the components being discussed in this section.

PV simulator including the simulation software and the adjustable power supply is described in Figure 5. This power supply controlled by the Arduino micro-controller correlates with the output of the model. For a simple implementation model, the micro-controller sets the position of servo

Figure 3. Architecture of power network

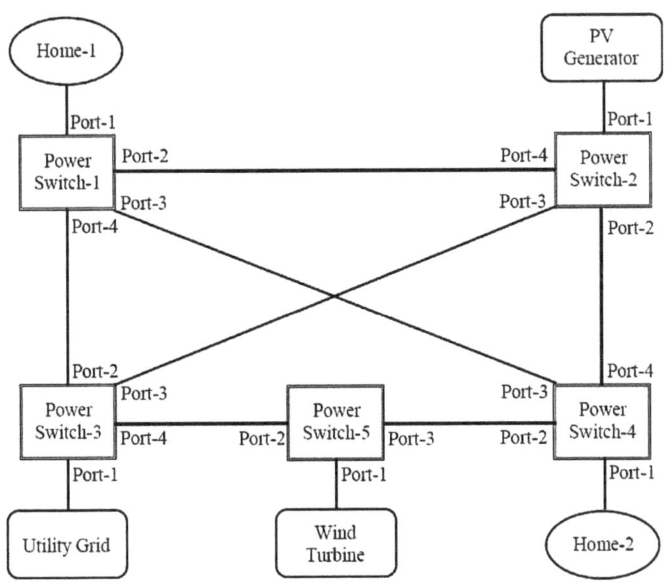

Figure 4. Architecture of information network

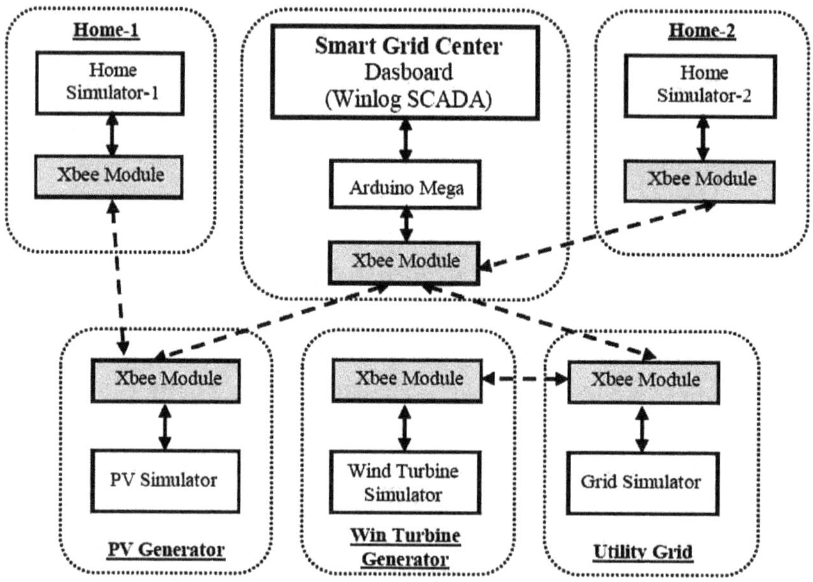

Figure 5. Block diagram of PV simulator

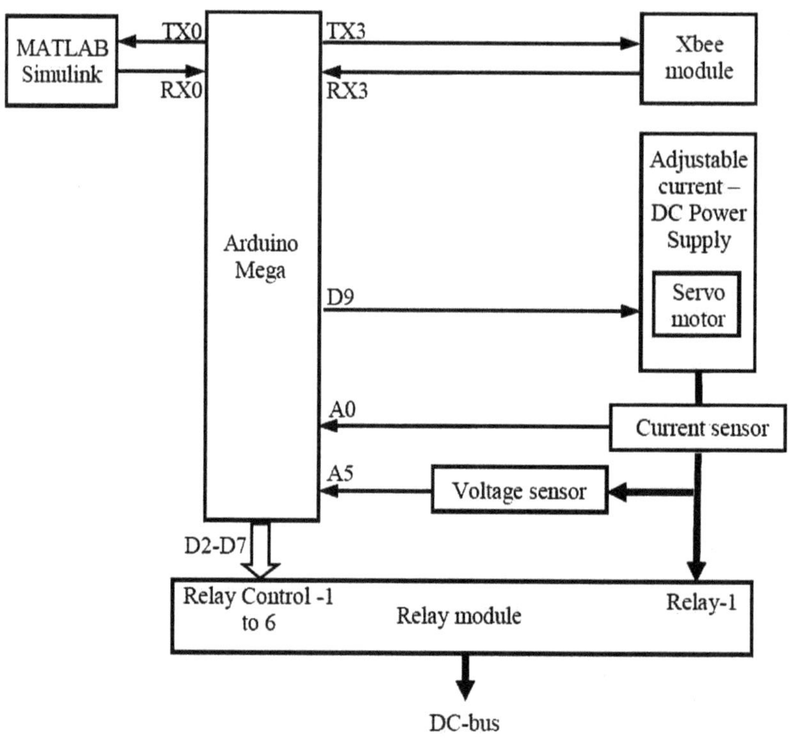

motor mounted on the potentiometer of current-controlled power supply, and the current and voltage level measured by each exclusive sensor will be checked by the Arduino micro-controller who sends the data to the Smart Grid Center through Xbee module. Simultaneously, it controls the power switch(s) following the commands given by the center.

Designing a wind turbine simulator is similar to that of PV simulator but the difference can be found in software simulations. That is, the Simulink models the wind turbine generator, not the PV generator. Hardware interfacing is analogous to the PV simulator.

To simulate demanders and the load levels in a Smart Grid, the home simulator will be used. The major element is the load unit consisting of four lamps Figure 6. The lad level can be adjusted or controlled (on or off) by the load relay controlled by the micro-controller. The load level or condition

Figure 6. Block diagram of home simulator

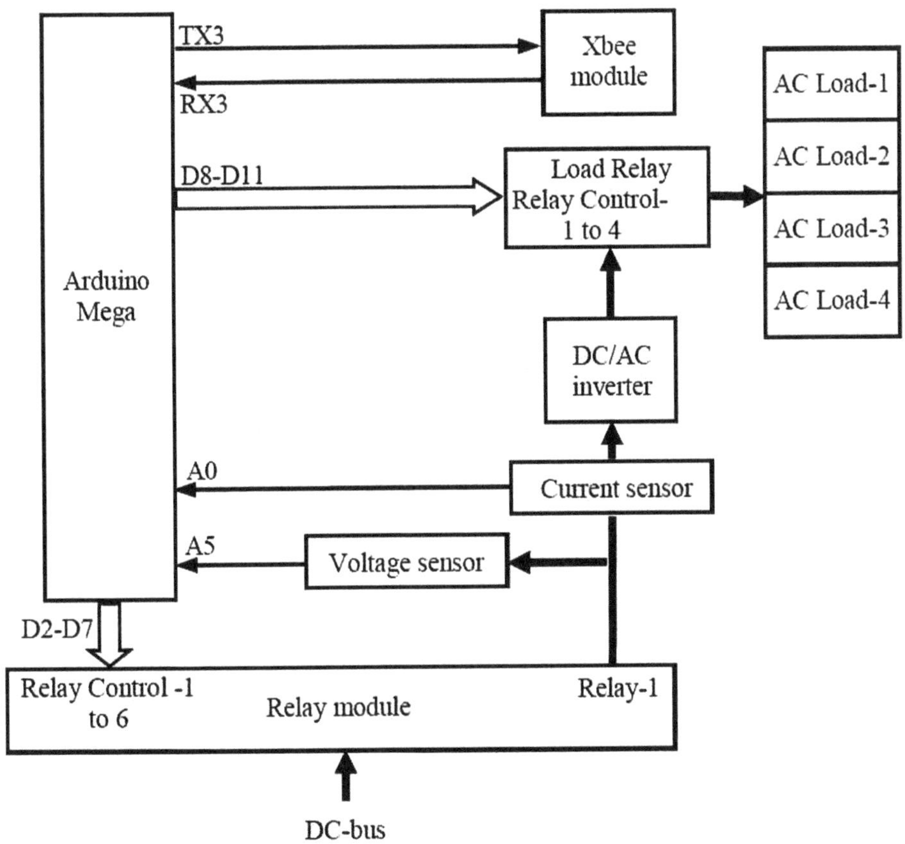

Figure 7. PV simulator hardware

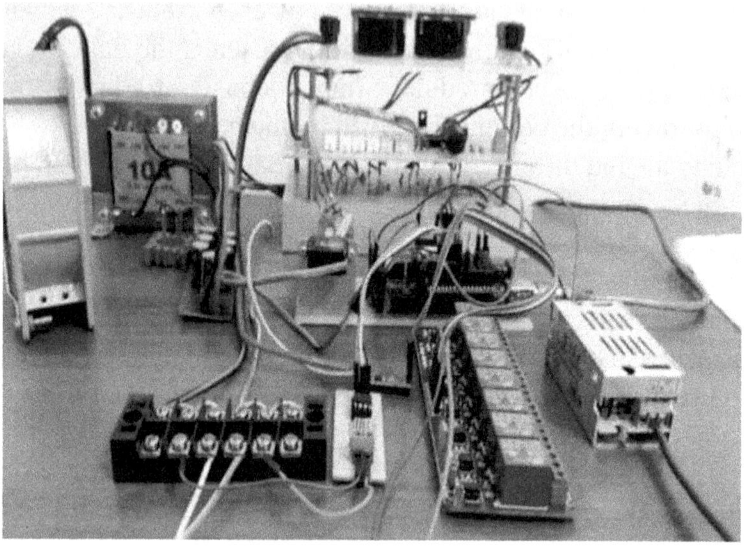

can be defined by the user and relevant simulations are performed using the SCADA software in the Smart Grid Center. Thus, configuration by the user will be convenient. Moreover, it is possible to use such scheme for the direct lad control algorithm which is considered to be one of the critical features in Smart Grid application.

The hardware test bed simulator consisting of adjustable power supply with servo motor, the micro-controller, the Xbee module, and the Power Switch is illustrated in Figure 7. Each component or node's hardware module has been designed separately. The electrical connections have been made with the DC-bus cables whereas the information connections are made with wireless means.

REFERENCES

Soetedjo, A., Lomi, A., & Nakhoda, Y. I. (2015). Smart grid testbed using scada software and xbee wireless communication. *International Journal of Advanced Computer Science and Applications*, *6*(8), 86–92. doi:10.14569/IJACSA.2015.060811

Yao, W. M. (2013). *Increasing scalability in network simulation and testbed experiments* (Ph.D thesis). Purdue University, West Lafayette, IN.

Chapter 7
Design and Implementation of Monitoring System Using PLC for ICT-Integrated Fish Farm

ABSTRACT

The key to a Smart Grid is to manage electric power within the grid efficiently. There are many advantages in building a smart grid but it is useful in using it as an ancillary power grid for the business that can be damaged extensively due to blackout or interruption of power caused by hurricanes at power plants, substations and other related infrastructures. Fish farms in the republic of Korea use a considerable power and suffer a serious loss from a mass stranding caused by power interruptions. This chapter introduces a case study which aims to design a power line-based communication for the fish farms/vertical farms as the cost of constructing an internet network at the farm with UTP cables and other communication equipment is very high. This section includes a designing technique for the entire system and its necessary applications.

1. INTRODUCTION

To achieve a "Creative Economy" in the Republic of Korea (ROK), economic democratization must be realized first. People will be able to put their best efforts with hopes only when a fair market order is established. It is a very important goal for the government to implement a development policy for

DOI: 10.4018/978-1-5225-2776-3.ch007

the small and medium-sized companies to let them succeed in their business and to coexist with larger firms. For this, the Korean government is trying to eradicate unfair trading practices of the past and the present so that company workers and owners in every industry can demonstrate their best ability. Contrary to such efforts, in the aquaculture industry, there are still many troubling cases and the power interruption is one of major problems. Power interruptions usually result from natural disasters but human-related incidents cannot be ignored.

Thus, this book proposes a crisis-response system and an Android application for the small and medium-sized fish farms, using the Power Line Communication and Information & Communication Technology (ICT) technologies.

The power management is a major parameter for the land-based and water recirculating fish farms. The mass mortality caused by the natural disasters (e.g., typhoons, abnormal temperatures or red tides, etc.) are the major problems and the power interruption at the farming site has been considered to be the worst situation. The proposed system has been designed to cope with power interruptions. For the system model, the use of Reliable User Datagram Protocol (RUDP) on a transport layer which is one of the seven OSI Layers, is proposed to compensate transmission losses caused by noise powers typical in the Power Line Communication models.

2. VERTICAL FARM

"Vertical Farm" is an alternative farming system invented by Prof. Dickson Despommier, Columbia University in 1999 to solve the problems of shortages of food and farmlands. This system allows farmers to grow their products regardless of weather conditions by controlling lights, water, temperature and humidity. It's often referred to as a "Plant farm". For the agricultural sector, the test bed experiments were completed in Japan, US and Europe, where some of their organizations have entered the business stages whereas the livestock sector is actively studying the system with some pharmaceutical companies (Zaballos et al., 2009; Gao et al, 2008; Zaballos et al., 2007; Ropitault et al., 2013).

Figure 1 shows target of Vertical Farm. This book is focused on selecting the suitable sites. In theory, Prof. Despommier's system can crop thirty times a year in the city area and grow plants as much as one needs no matter what the

Figure 1. Target of vertical farm

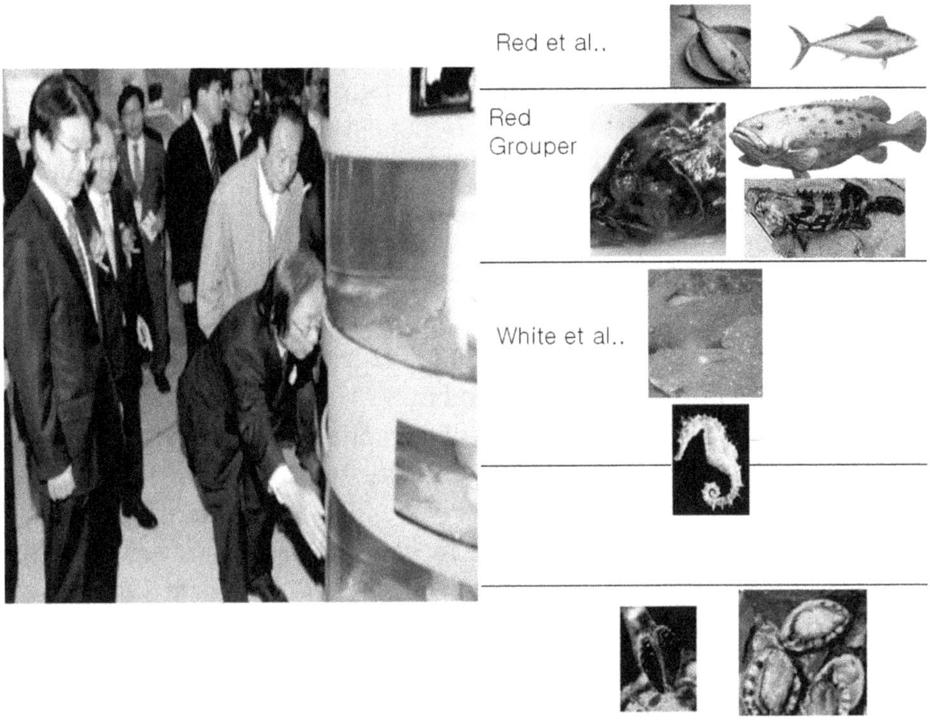

weather conditions (e.g., droughts, floods and typhoons, etc) (Despommier, 1999; 2009; 2010; 2014).

Left part shows the test bed experiment completed in the Republic of Korea for flatfish and sea cucumber and the right is a diagram of the polyculture system involving both future target species and currently developed species.

This "Future Farming" in the time of abnormal climate changes uses LED lights instead of sunlight and sprinkles nutrient-mixed sprays on the roots instead of soaking them in the water. According to the simulations, it is possible to grow foodstuffs that are enough to feed fifty thousand people with the thirty-story farm.

3. DESIGN AND IMPLEMENTATION OF MONITORING SYSTEM USING PLC FOR ICT-INTEGRATED FISH FARM

There have been many cases of mass mortality of fish due to the power interruptions caused by natural disasters, human errors or acts of sabotage by the employees hired by rival companies, discouraging the small and medium-sized fish farmers. Thus, in this book, a system that can monitor the power status of fish farms to cope with such situations is being proposed. The system utilizes both PLC and ICT technologies to respond to crisis.

Electric power is a critical parameter for both the land-based and the water recycling fish farms. The mass mortality of fish due to natural disasters was the highest risk in the aquaculture industry. The proposed system allows immediate response for the power interruptions and consistent power monitoring by the employees. To compensate the signal transmission loss often caused by the noise power typical in the PLC-utilized models, Reliable User Datagram Protocol is proposed for the transmission layer. Also, for the Android application, a graphic user interface was designed to manage overall fish farm activities, photoperiod, seawater control, Smart aquafarm, access control, location check, withdrawal period, and secondary battery check. Figure 2 shows target of OSI layer system design for ICT-integrated Fish Farm monitoring.

When operating a fish farm, human resource management is deeply associated with monetary aspects. Financial losses can result from accidents or wrong-doings by the employees when they exhibit some problems due to their situations and behaviors. Such losses are the risk factors for the small business companies but they can be efficiently reduced with the ICT technology. Figure 3 shows system design for ICT-integrated Fish Farm monitoring.

3.1 Employee Monitoring

By using the fish farm theft-prevention tags, the management can monitor working situations of employees. The farm owner can have the advantage of easily and precisely checking individual employee's location, working situation and monthly attendance. This will not only prevent financial losses but also can secure employees' safety by informing the dangers to them when accidents or problems should occur. On the other hand, employees do not have to rely on the time cards to check their attendance records, which are the basis of their wages, to avoid unnecessary dispute.

Figure 2. Target of OSI layer system design for ICT-integrated fish farm monitoring

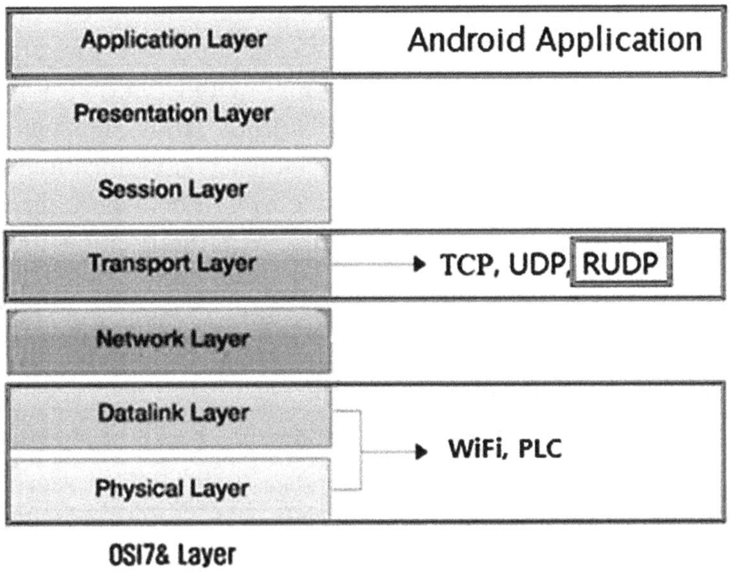

Figure 3. System design for ICT-integrated fish farm monitoring

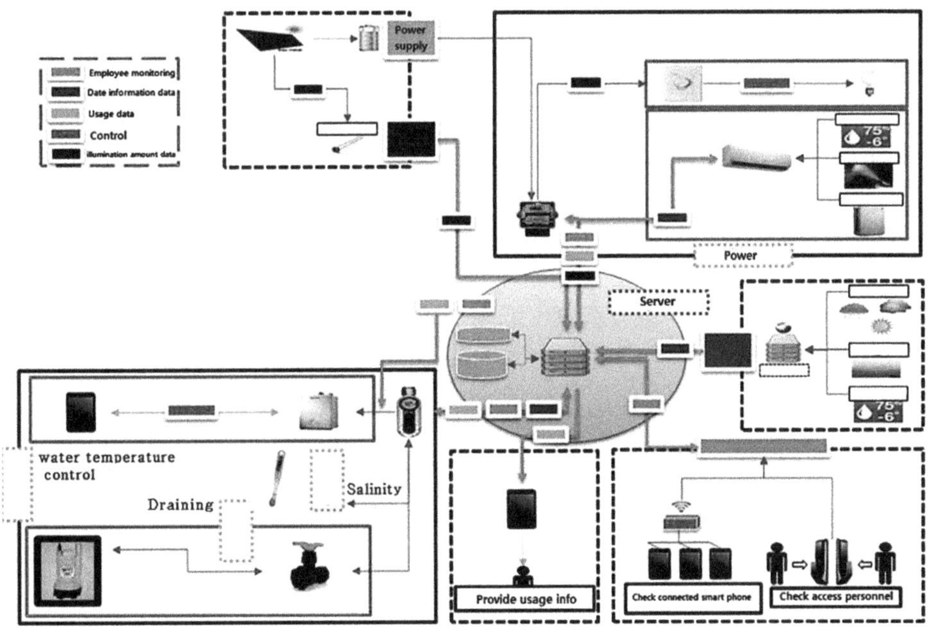

The technology being described was designed based on a method that was used for a door-lock system using wireless AP, and Reliable User Datagram Protocol was selected to improve communication efficiency and reliability. The system application has been programmed with C and C++ languages followed by the implementation with Java Android. This system has the merit of not requiring any additional circuits to establish communications between devices or equipment as existing power line(s) will be used. Since electric power is available for most of fish farms despite of their remote locations in the agriculture and fishing villages, farmers will be able to establish a PLC network with low costs using the existing facilities provided by the Korean Electric Power Corporation (KEPCO). The core algorithm of the employee monitoring system is described in Figure 4.

By using the employee monitoring algorithm in addition to positioning service, it is possible to prevent employee's negligence of duties and thefts of -value fishes. Moreover, the system can save employees in times of

Figure 4. The core algorithm of the employee monitoring system

A_p : Address accessed in the past

A_n: Current address being accessed

$f(x) = (\sum Number(A_p) - \sum Number(A_n))$

//Comparison of the numbers of the past and current addresses approached

if $f(x) = 0$ //If the numbers are identical

if $\sum Time(A_n) \geqq 1Hour$ //If the current approach time exceeds 1 hour

Then Warning Message transmission to the User's phone

Not $Return(f(x))$ //Return to f(x) comparison

Not $(= f(x) \neq 0)$ // If the numbers are not identical

if $\sum Number(A_n) \geqq 2(= g(x))$ //Assume that the number of current //approaching address is 2 or more

Then $Return(g(x))$ //Return to g(x) comparison

Not

if $\sum Time(A_n) \geqq 0.5Hour$

//If the current approach time exceeds 0,5 hour

Then Visitation request message transmission to the User's phone

Not $Return(f(x))$ //Return to f(x) comparison

emergencies in an early stage by checking the whereabouts of them periodically and informing them of fore-coming dangers. Figure 5 show the method of employee monitoring.

Most of fish farms hire people living close by the fishing villages so that the major portion of the employees is comprised of older generations as many younger ones are leaving the area. Such situation brings about dangerous situations (e.g., exhaustion fainting by heat waves, etc) especially in summer for these employees but with the system equipped with the proposed algorithm and design, their lives can be saved by checking their movements. The system will recognize the danger when a employee does not move for long hours.

Fish farms ship out fishes and shells when their growth have reached the scheduled levels so that the seasonal, not continuous, transactions will be carried out. The theft or the accidents in such periods will bring a devastating damage.

The theft prevention systems used at the existing fish farms require much investment and more expenses are needed for the installment by the professional companies. Compare to this, the Power Line Communication-based system will require moderate expenses in an early stage of installment and can be operated continuously efficiently. After designating the people who can access the farm, tags will be attached to them to check their access times, access statuses and periods. The access by one without the tag will be notified to the owner or the management through the application embedded in the smart phones, preventing thefts.

Figure 5. A method of monitoring employees

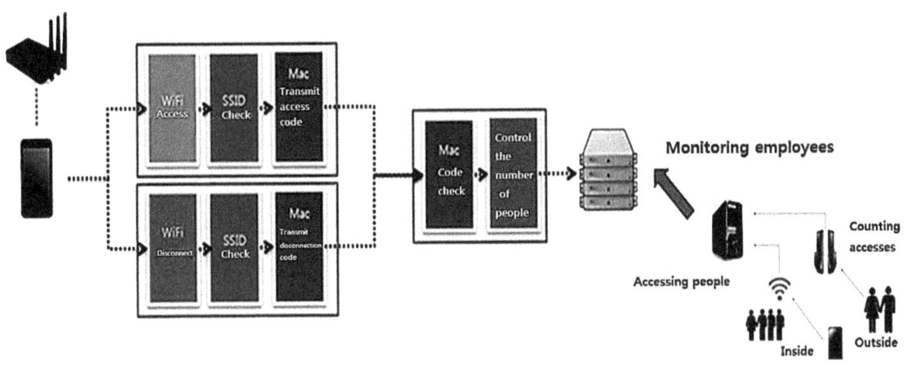

3.2 Farming Environment Monitoring

Various factors must be balanced within an appropriate environment when culturing marine products. For the ocean cage-fishing farms, seawater temperatures, weathers, currents and tides should be considered and for the land-based farms, water temperatures, pH and salinity are the important factors. Especially the latter is often influenced by the temperature changes caused by the nearby factories who use seawater as a coolant or by the changes in salinity within the tanks resulting from salty substances carried from the seaside, leading to mass mortality of fishes and shells. Since these changes cannot be visually recognized, it is difficult for the small and medium-sized fish farms. Even a little change could present much damage and if they accumulate, a greater damage can be expected.

However, it is possible to check these changes with the proposed application immediately and by setting the checking categories to issue an alarm, farmers will be able to prevent damages and reduce manpower waste. Meanwhile, by creating a database for the fish-oriented diseases and intelligently operating the system, relevant information can be sent to nearby marine disease control centers, the Ministry of maritime Affairs and Fisheries, and National Fisheries Research and Development Institute with speedy feedbacks to prevent damages beforehand.

3.3 Management of Photoperiod

Spawning and growth of fish can be controlled by using the water temperature and photoperiod. It is possible for the fishes to spawn and grow faster with the "long-day treatment (lengthening of photoperiod with artificial lights)", especially for the flatfish for which the photoperiod is more effective than temperature. If the photoperiod can be controlled with automatic light adjustments through the application, the number of fishes will be multiplied much faster.

3.4 Aspects of Energy and Environmental Conversation

3.4.1 Auxiliary Batteries and Solar Cells

Currently, typhoons during the summer are the major concern of fish farmers as the power interruptions by them can be fatal to their farms. Even a short interruption can result in entire mortality within the farm so that they prepare for the emergency situations when typhoon alerts are issued. To deal with such situations, the auxiliary batteries, which can be charged at dawn and used in the daytime during which the same batteries can be re-charged through solar pipes, can be considered. In this way, farmers can save the power within the batteries twenty-four hours a day to get ready for unexpected or notified power interruptions, reducing damages.

3.4.2 Controlling Withdrawal Period

Use of antibiotics are inescapable when growing the fishes and shells. Since antibiotics often remain in the body of marine organisms for a certain period of time, fishes and shells must be shipped out after going through withdrawal time. However, for the reason that there are so many kinds of antibiotics and they are being used frequently, the management and workers do not know their exact quantities for use and recommended/required withdrawal times. Thus, the application includes a Guide to Medicines for Marine Products (Korea National Fisheries Research and Development Institute, 2014) in their database to notify the ending day of withdrawal period by alarm once the kind and quantity of antibiotic being used are entered into the system. This function will provide a greater safety image for the marine products.

The two most important elements for aquaculture, dissolved oxygen and temperature, need to be continuously observed and managed but in many cases, employees at the site were unable to cope with emergency situations. Figure 6 shows design of Android application.

3.4.3 Design of Medication Intake Notification Service DataBase

The database has been prepared with the initial basic data included in the application and the database constructed by the user, which stores the data when in use. The initial basic data includes ‘medicine_table’ which contains the information of medicines, ‘ingredient_table’ that contains the ingredients of relevant medicines, ‘ingredients_info_table which is a general information of the medicines, and ‘sideeffect_table’ that has the information concerning the side effects of the medicines.

The list of medicines which are being administered to a patient will be stored at ‘medication_table’ as data. As for ‘Medicine_table’, design elements

Figure 6. Design of fish farm Android application

are unique_id (unique number of the medicine), name, effect, side_effect, attention (cautions) (Figure 7).

The database has been prepared with the initial basic data included in the application and the database constructed by the user, which stores the data when in use. The initial basic data includes 'medicine_table' which contains the information of medicines, 'ingredient_table' that contains the ingredients of relevant medicines, 'ingredients_info_table which is a general information of the medicines, and 'sideeffect_table' that has the information concerning the side effects of the medicines.

The list of medicines which are being administered to a patient will be stored at 'medication_table' as data. As for 'Medicine_table', design elements are unique_id (unique number of the medicine), name, effect, side_effect, attention (cautions) Figure 8.

'ingredients_table' contains the data of a certain medicine and comprised of 'medicine_id' and 'ingredient_id', both of which are unique numbers (Figure 9).

'ingredient_info_table' has the information about the ingredient and comprised of 'unique_id', 'effect', 'side_effect' and 'attention' (Figure 10).

Figure 7. Ingredients_table diagram

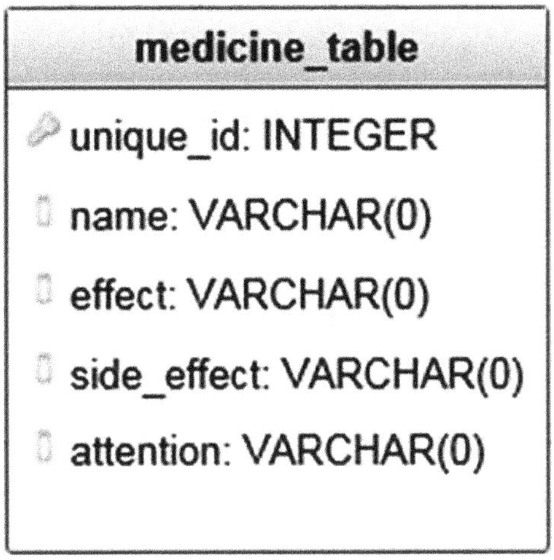

Figure 8. Medicine_table diagram

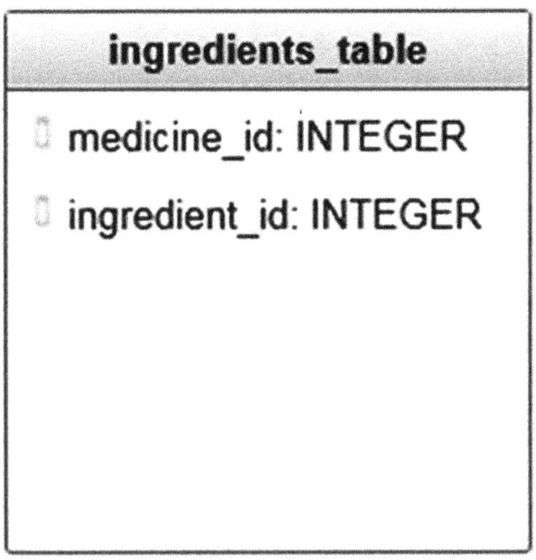

'side_effect_table' contains the information concerning the possible side effects when two or more medicines are mixed and comprised of 'ingredient_a', 'ingredient_b' and 'side_effect' (Figure 11).

Whole DataBase design is shown in Figure 12.

'medication_table' that contains the information of current medication has been designed with the items such as 'index' and 'unique_id' (Figure 13).

Figure 9. Ingredients_table diagram

Figure 10. Ingrediant_info_table diagram

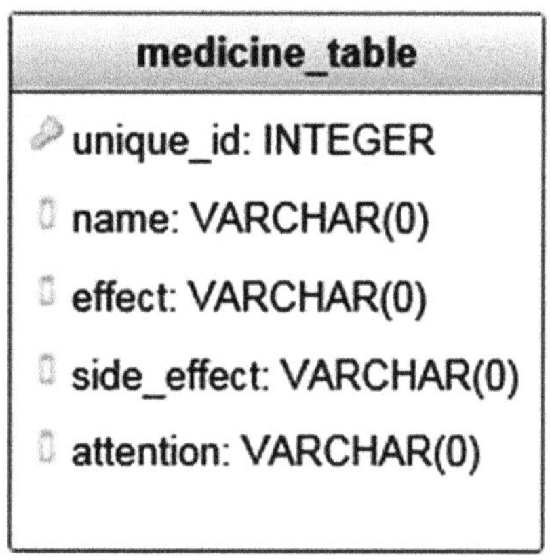

Figure 11. Ingrediant_info_table diagram

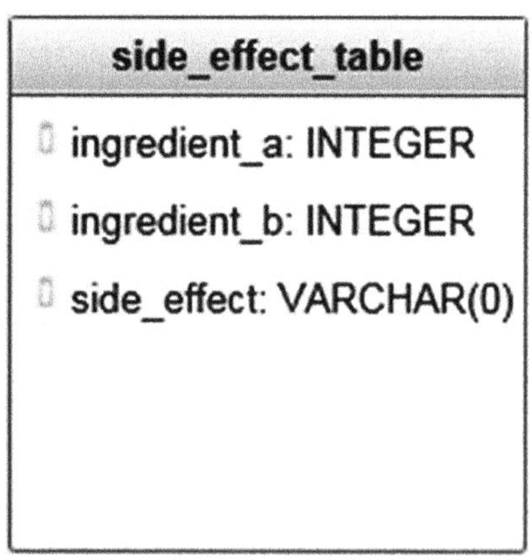

Figure 12. Whole DataBase design of mobile medication-hour notification system

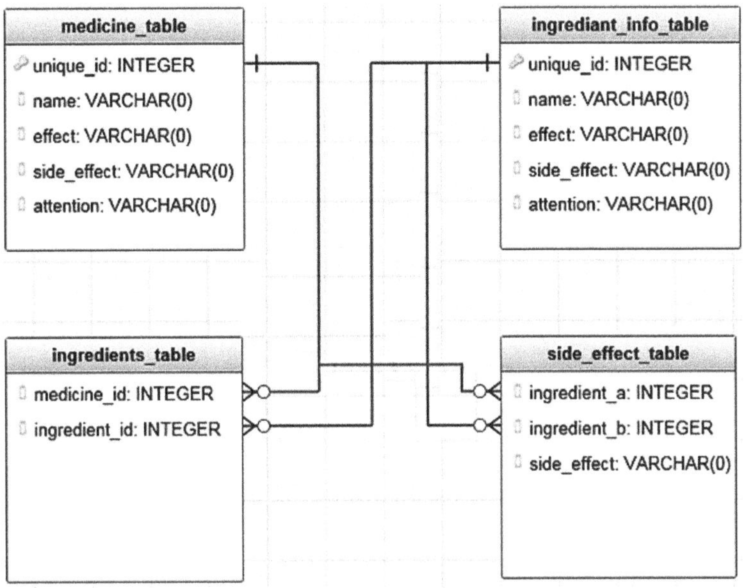

3.4.4 Implementation of Medication Intake Notification System

The data for the basic medicine intake notification system has been constructed as Figure 14 and developed with Java Android. The 'Medicine' section to contain the information of the medicine and 'Ingredient' section to contain the ingredients of the medicine were constructed so that the user can manage his/her medication records using the names or unique numbers of the medicines. In this model, the data can be generated and managed automatically by using the ingredient list stored at the DB. All the information concerning current medications will be managed with the 'MedicineStoreService' class and the user can obtain all the current list of medicines in use. The data can be either added or deleted.

If the user is to take two or more medicines, the system checks all the relevant ingredients of each medicine through 'SideEffect- ManageService' designed as below in Figure 15 and warns possible side effects to the user to prevent misuse and abuse.

To notify the medicine intake time to the user, the 'AlarmService' class has been implemented (Figure 16). If there is any additional medications, add an alarm(s) to notify the intake time(s) at intervals of hours or days, depending

Figure 13. Medication_table diagram

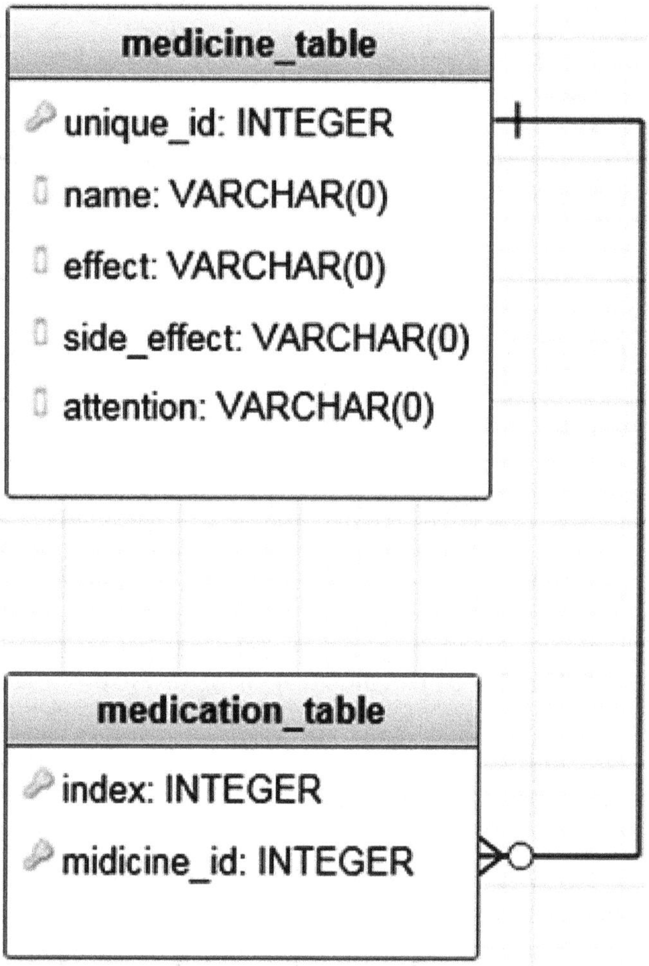

on the type of the medicine. Notification alarm(s) will be cleared as soon as the current medication(s) are removed from the list.

3.4.5 DataBase Processing Mechanism

The 'Notification' function of Android is being used to alert the hour of medicine administration, and in this distance, administration period, amount to be administered, and cautions will be indicated as well. 'Notification' was designed to send out notifications every pre-set period through 'Alarm

Figure 14. Data object modeling and servicing

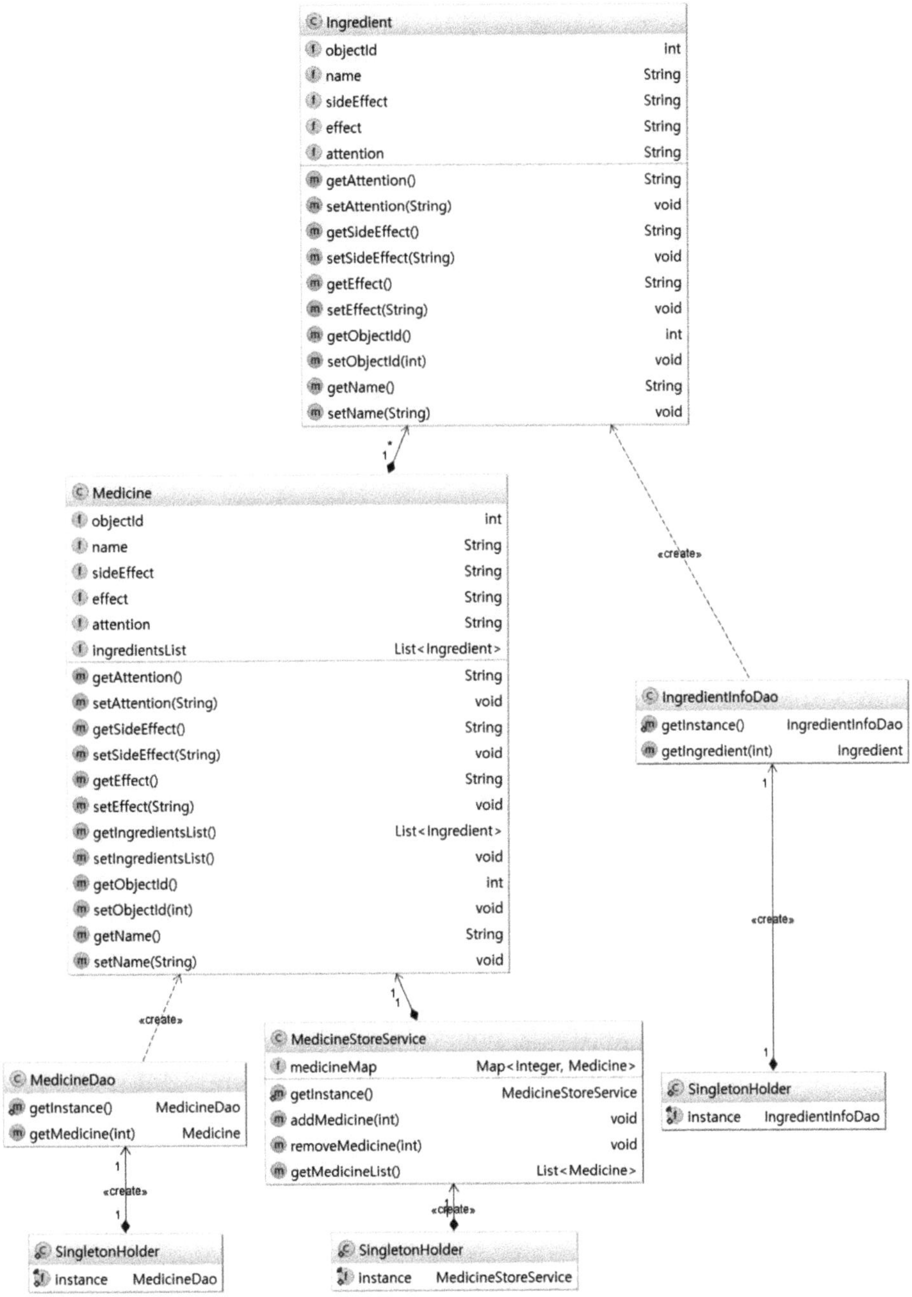

Figure 15. Side effect management service using the ingredient information

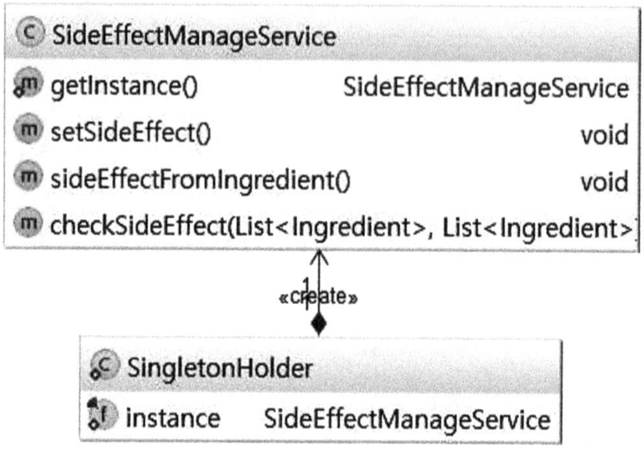

Figure 16. Alarm management service object

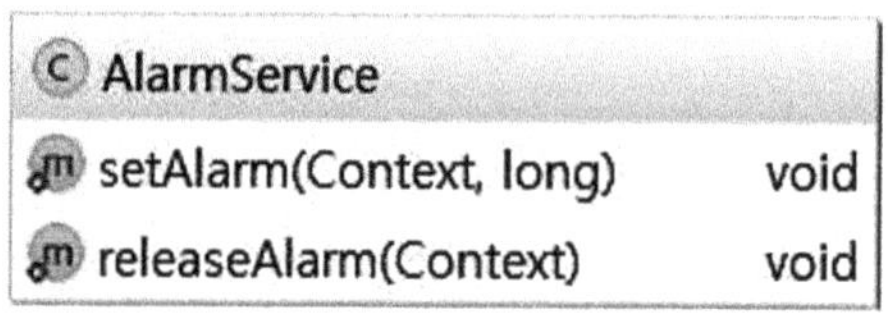

Manager'. 'Alarm Manager' receives the name(s) of medicine(s) and periods or specific dates and hours from the database in use.

The structure of this database is comprised as Table 1 'objectId' is used as a primary key to distinguish individual databases; 'name' refers to the medicine name; 'period_hour' allows to set the medicine administration periods indicating hours; 'period_day' sets this cycle on daily basis; and 'noti_hour' is a basic time setting to notify the hour of medicine administration once the daily administration alerts have been set. With this function, one can avoid receiving the notifications after 12 midnight.

Meanwhile, a list of medicine ingredients will be stored as shown in Table 2.

Table 1. Period table

Designation	Attribute
objectId	Int
ingredient_name	String

Table 2. Medicine ingredient list

Designation	Attribute
objectId	Int
name	String
period_hour	Int
period_day	Int
noti_hour	Int

The table that stores the ingredients of medicine is comprised of 'objectID' which refers to the 'objectID' of a medicine and 'ingredientID' which refers to 'ingredientID' Table 3 while Table 4 is comprised of 'objectID' which refers to the 'objectID' of an ingredient, 'Hamful_object' that identifies harmful ingredients, and 'side_effect' in which side effects will be documented. Possible side effects can be informed to the user using this table.

4. CONCLUSION

The electrical power is a critical parameter for both the land-based and the water recycling fish farms. Especially, the power interruptions caused by typhoons present grave consequences to the farms leading to a mass mortality. The proposed algorithm and application will assist the farmers to deal with such situations and allow an efficient and continuous monitoring.

In this book, the use of Reliable User Datagram Protocol has been proposed for the transport layer to compensate the transmission loss usually caused by the typical noises of the PLC-based communication models. Also, for the Android application a graphic user interface has been developed to manage fish farm activities, photoperiod, and to control seawater condition, Smart aquafarm, access, as well as checking the location of employees, withdrawal

Table 3. Medicine ingredient ID list

Designation	Attribute
objected	Int
ingredient_Id	Int

Table 4. Harmful ingredient list

Designation	Attribute
objectId	Int
Harmful_object	Int
side_effect	String

period and secondary batteries. This research will be disclosed on a journal after registering and applying the technology for a patent.

Author expects that the small and medium-sized fish farms will easily find this system acceptable as it can reduce a small portion of repetitive damages every year. Furthermore, since the application will control the use of antibiotics, which is the major concern of our public, the fear will be allayed and trust of marine products will be increased.

REFERENCES

Akinnikawe, A., & Butler-Purry, K. L. (2009). *Investigation of broadband over power line channel capacity of shipboard power system cables for ship communication networks. In Proceedings of the IEEE Power and Energy Society General Meeting*, Calgary (pp. 1–9). doi:10.1109/PES.2009.5275639

Antoniali, M., Tonello, A. M., Lenardon, M., & Qualizza, A. (2011). *Measurements and analysis of PLC channels in a cruise ship. In Proceedings of the 2011 IEEE international symposium on power line communications and its applications* (pp. 102–107). Udine. doi:10.1109/ISPLC.2011.5764372

Barmada, S., Bellanti, L., Raugi, M., & Tucci, M. (2010). Analysis of power-line communication channels in ships. *IEEE Transactions on Vehicular Technology*, *59*(7), 3161–3170. doi:10.1109/TVT.2010.2052474

Bova, T., & Krivoruchka, T. (1999). Reliable UDP protocol (IETF internet draft).

Despommier, D. (1999). *Reducing the impact of agriculture on ecosystem functions and services*. USA: Department of Environmental Health Sciences, Mailman School of Public Health, Columbia University.

Despommier, D. (2009). The rise of vertical farms. *Scientific American*, *301*(5), 80–87. doi:10.1038/scientificamerican1109-80

Despommier, D. (2010). The vertical farm: Controlled environment agriculture carried out in tall buildings would create greater food safety and security for large urban populations. *J. Cons. Prot. Food Saf.*, *6*(2), 233–236. doi:10.1007/s00003-010-0654-3

Despommier, D. (2014). *Vertical farms in horticulture. In Encyclopedia of Food and Agricultural Ethics* (pp. 1791–1799). Netherlands: Springer.

Dickson, D. (2013). Farming up the city: The rise of urban vertical farms. *Trends Biotechnol. Cell*, *31*(7), 388–389. doi:10.1016/j.tibtech.2013.03.008

Ding, Q., Ma, D., Li, D., & Zhao, L. (2010). Design and implementation of a sensors node oriented water quality monitoring in aquaculture. *Sens Lett*, *8*(1), 70–74. doi:10.1166/sl.2010.1204

Economist. (2003). The promise of a blue revolution. The economist Special report: Fish farming. Retrieved from http://www.economist.com/printedition/2003-08-09

Halal, W. E. (2008). *Technology's promise: expert knowledge on the transformation of business and society*. UK: Palgrave Macmillan. doi:10.1057/9780230582538

Huh, J. H. (2016). *Design and android application for monitoring system using PLC for ICT-integrated fish farm, advanced multimedia and ubiquitous engineering. In Advanced Multimedia and Ubiquitous Engineering, LNEE* (Vol. 393, pp. 617–625). Singapore: Springer.

Huh, J. H., Je, S. M., & Seo, K. (2016). *Communications-based technology for smart grid test bed using OPNET simulations, ICISA 2016. LNEE* (Vol. 376, pp. 227–233). Singapore: Springer.

Huh, J.-H., Kim, N., & Seo, K. (2016). Design and implementation of mobile medication-hour notification system with push service function. *Int. J. Appl. Eng. Res.*, *11*(2), 1225–1231.

Huh, J. H., Koh, T., & Seo, K. (2016). Design of a shipboard outside communication network and the test bed using PLC: for the workers' safety management during ship-building process. In *Proceedings of the 10th international conference on ubiquitous information management and communication, ACM IMCOM '16* (Vol. 43, pp. 1–6). ACM.

Huh, J. H., Lee, D. G., & Seo, K. (2015). Design and implementation of the basic technology for realtime smart metering system using power line communication for smart grid. In *Advances in Computer Science and Ubiquitous Computing, LNEE* (Vol. 373, pp. 663–669). Singapore: Springer. doi:10.1007/978-981-10-0281-6_94

Huh, J.-H., Otgonchimeg, S., & Seo, K. (2016). Advanced metering infrastructure design and test bed experiment using intelligent agents: Focusing on the PLC network base technology for smart grid system. *The Journal of Supercomputing, 72*(5), 1862–1877. doi:10.1007/s11227-016-1672-4

Huh, J. H., & Seo, K. (2015). *RUDP design and implementation using OPNET simulation. In Computer Science and Its Applications, LNEE* (Vol. 330, pp. 913–919). Heidelberg: Springer.

Huh, J. H., & Seo, K. (2015). Hybrid advanced metering infrastructure design for micro grid using the game theory model. *Int. J. Softw. Eng. Appl., 9*(9), 257–268.

Hwang, S. I., Kim, O. Y., & Lee, S. Y. (2014). A case study on the ICT-based smart aquaculture system by applying u-farms. *J. Kor. Inst. Comm. Inform. Sci., 39*(2), 173–181.

Jin, Y. H. (2009). *A study on an application model of RFID technology for manufacturing and distribution: a case of clean cheju's supporting system for high quality of u-aquaculture* [MS. Thesis]. The Graduate School of Business Administration, Univ. of Seoul. (In Korean)

Kim, O. Y. (2011). *Design of farming fishery traceability management system using RFID* [Ph.D. Dissertation]. Dept. Comput. Eng., Univ. of Changwon. (In Korean)

Thammadi, A. (2011). *Reliable user datagram protocol (RUDP)* [Master's Thesis]. Kansas State University.

Yang, Y.-M., & Fang-Tsou, C.-T. (2014). System aquaculture cloud management. *J Inform Technol Appl, 8*(2), 1–3.

Chapter 8
Simulation of Power Line Communication Using OPNET for Vertical Fish Farm

ABSTRACT

OPNET is a network simulation tool which can simulate various elements in a network. It is able to analyze traffics and even can simulate security events recently. System designers can identify vulnerabilities or find an efficient algorithm/protocol that could reduce costs if simulations are performed prior to actual implementation/construction of a large-scale system. Thus, this chapter includes the method of simulating a PLC-system, designing a PLC-based vertical fish farm and an expandable simulation test bed with which students or researchers will be able simulate their implementation methods. With these methods, it will be possible to perform large-scale simulations.

1. INTRODUCTION

In recent years, the systems related to the Vertical Farm and Internet of Things (IoT) are being studied widely. Although there are some useful technological means (e.g., RS-232, Zigbee and Wifi) to establish these systems, we are concentrating our book on the Power Line Communication technology. Communications utilizing existing power line are generally called the Power Line Communication and this technology uses high-frequency signals to transmit or store data through power lines, during which frequency signals are

DOI: 10.4018/978-1-5225-2776-3.ch008

separated with the dedicated power line modem(s) as low or high frequency signals before reaching to a terminal device. Two different speed levels are available for the Power Line Communications: high-speed and low-speed. The former uses frequency bands between 0.5 and 30MHz reaching speeds of 14Mbps to 1Gbps (as of 2015) while the latter relies on the bands between 10 and 450KHz achieving a maximum speed of 9.6Kbps normally used for electric equipment or devices. In fact, the Power Line Communication media are originally designed for power transmissions.

It's been assumed that the channel characteristics of Power Line Communications are unsuitable for a data communication purpose due to rather heavy noises and serious signal attenuation. One way to improve reliability of the Power Line Communication is to enhance its physical layer associated with channel estimation and selection, filtering design, power distribution or others.

Meanwhile, China is one of the major producers of aquaculture products and their production volume is the largest among these countries. In this country, the aquafarms distant from the cities are attempting to surmount high costs of internet installation by constructing the internet network through existing power lines. Since the problem of silent node often appears in the long-distance Power Line Communications, this book targeted the fishing villages where the internet service is not available.

The Power Line Communication technology is being applied to the home-network construction or factory automation nowadays but due to the improvements made in transmission speed and stability achieved with recent digital power line technology, its applicability is still growing and getting much attentions (Tsuzuki, Areni & Lehnert, 2012). One of the main reasons for such interest is because of the feasibility of low-cost implementation using existing power line distribution and smart meter networks. Meanwhile, the situation-specific simulations and modeling of Power Line Communication-based solution will be needed to implement Power Line Communication system in a noise-filled channel environment to improve both the stability and reliability. Although many studies have been carried out for modeling Power Line Communication's physical characteristics with a variety of simulation tools so far, they many more are yet to come (Huh & Seo, 2014; Gao et al., 2008; Sabolic, 2003).

The Figure 1 shows a structure where data is sent through power line by installing two Power Line Communication adapters in the wall outlet. Using the power line installed at the time of construction, the electricity is supplied with the low frequency band of 60Hz and the data will be transmitted with

Figure 1. The mechanism of PLC and PLC Slient node problem

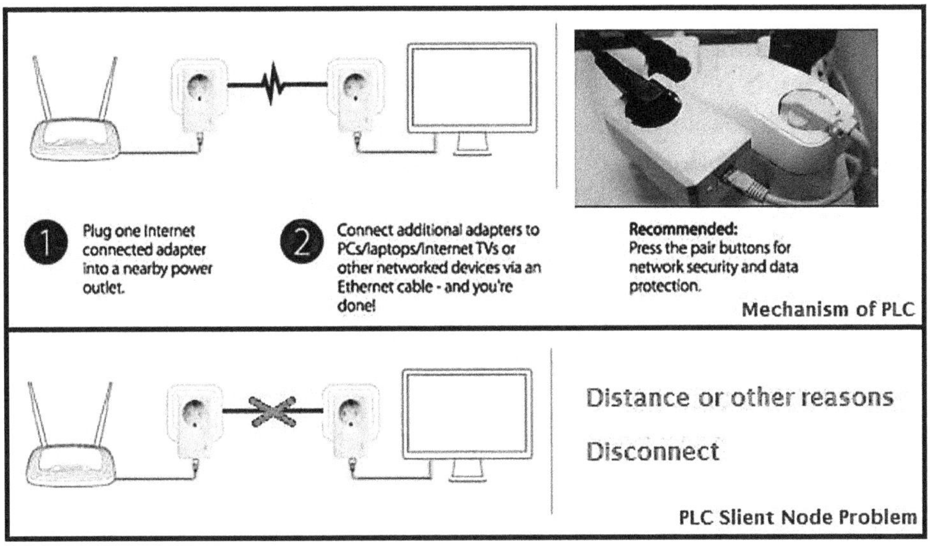

the remaining high frequency bands available. New product lines of adapters with the maximum speed of 1Gb/bps have been released recently.

The Power Line Communication technology is increasingly accepted in the fields of building automation, security control and others as it enables convenient and cost-effective communication system. One such system is the Automatic Meter Reading (AMR) system where meter readings can be sent over to a power company through the power lines available in most of the industrial and public sectors. Not requiring separate or additional infrastructures is a very strong advantage and quite suitable for establishing the low-cost AMR network.

For the simulations, the scenario in which the Power Line Communication-applied environment has been created and was applied to the households divided into A and B blocks. For both blocks, twenty homes standing 20m apart against each other have been arranged equally. Only the difference was that a router was installed at a point of 200m distance in the B-block but none for the A-block to test the performance of the Power Line Communication network. Finally, a performance analysis was performed for each of TCP, UDP application to measure their performances. The problem of communication cut-off, which represents the silent node problem, appeared at the point of 200m distance and we simulated this phenomenon by using OPNET Modeler

14.5 PL8. Simulation environment for performance evaluation, hardware and software are shown in Table 1.

The silent node problem was simulated based on the previous successful implementation of the Power Line Communication using OPNET Modeler 14.5 PL8 (PlanetLab, n.d.). The C++ language was used partially for the coding. The simulations were carried out under the two different environments; with or without routers. The simulated transmission distance was 400m for each environment.

3. TECHNOLOGY TRENDS

3.1 Power Line Communication (PLC)

The current Power Line Communication can speed up to 1Mbps-1Gbps and has reached commercialization for the home and the multimedia networks. Its biggest problem is data loss due to the noises generated by the power lines but they can be compensated by using transformers. This technology has been applied to light and home appliance control or telephone networks and is now receiving the spotlight as a core technology for IoT system construction. The major advantage of using Power Line Communication for IoT construction process is that no additional communication network construction is necessary to exchange information between devices or equipment since communications can be established with Power Line(s) only, without any other communication circuits. For instance, in the Republic of Korea,

Table 1. Simulation environment

Working Environment Computer Specifications	
CPU	3.0+ GHz using x86, EM64T, x86 AMD, or AMD64 (dual-core)
RAM	2GB
System File Space	3GB
Working File Space	100MB
Display Resolution	1024x768
Development Environment	
OS	Windows 7 Professional (32 bit)
Compiler	Microsoft Visual Studio 6.0
M&S Tool	OPNET Modeler 14.5 PL8

Logistics, Energy Korea Co., Ltd. (KOSDAK listed company) developed the attachable Power Line Communication modems and the external Power Line Communication modems, both for watt-hour meters. Other instance is that the Korea Electrotechnology Research Institute has succeeded in developing the Power Line Communication-based broadcast circuit for the first time in the Republic of Korea.

It's being expected that use of the Power Line Communication technology will be increased in other application fields but there always has been a problem of reliability in communication process.

3.2 Meter Unit Model of AMR Simulation Modeling

The "Silent Node" is thought to cause incomplete communications between a metering unit (meter) and the Data Concentrator (DC). That is, a signal carrying the metering data will not be delivered to its final destination due to signal attenuation, noise or varying impedances in a Power Line Communication network. Here, the "node" represents a metering unit and its performance will depend on the network environments to work properly or to be silent.

The state transition probability of the meter unit under specific conditions (e.g., Good and Bad) can be represented with the Two-state Markov model as below (Huh & Seo, 2014; Gao et al., 2008; Sabolic, 2003)

"Good" and "Bad" in this model refer to the possible conditions under which a meter is being placed. While the meter unit in the "Good" condition can communicate with DC or other relay meter units, the meter in the "Bad" condition will be in silent node (Huh & Seo, 2017; Huh & Seo, 2014; Q. Gao et al., 2008; D. Sabolic, 2003).

As for Pg and Pb, they indicate the possibilities of the meters currently placed under both conditions. Pgg is the possibility that the meter under the "Good" condition would still remain at the same environment whereas Pgb

Figure 2. Two-state transition Markov model for a meter unit

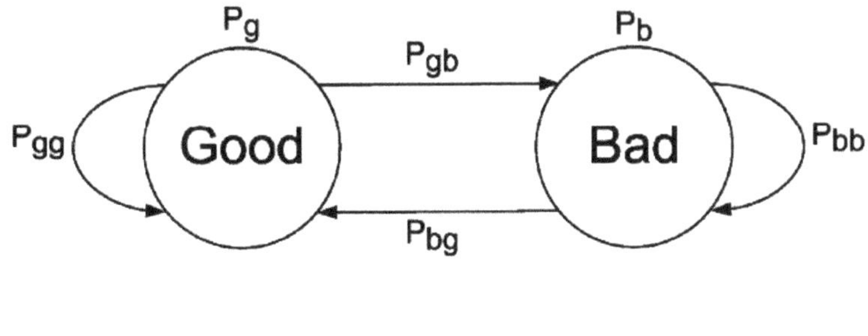

shows the possibility that the meter would move to the "bad" condition. Similar possible reverse movements of the meter in the "Bad" condition are represented as Pb and Pbg. Following Gao (2008), this two-state transition can be quantified using the equations (1) - (3) so that the local balance can be formulated as (4).

$$P_g + P_b = 1 \tag{1}$$

$$P_{gg} + P_{gb} = 1 \tag{2}$$

$$P_{bb} + P_{bg} = 1 \tag{3}$$

$$P_g P_{gb} = P_b P_{bg} \tag{4}$$

3.3 Ship-PLC and Three-Phase Three-Line Method

The role of an APF is to compensate the reactive power and eliminate the harmonic components due to nonlinear loads at the same time. To be specific, using the APF forces the utility current to take a sinusoidal waveform and be in phase with relevant utility voltage to allow more stable and balanced power flow. Thus, the main aim of the proposed APF here is the utility current.

Being different from other APF control methods, it is important to understand that the concept involved in using the APF is to command the utility current not the output current of converter. The positive sequence components of three-phase voltages in a three-phase four-line system can be postulated as

$$v_{an}\left(t\right) = V_s \sin\left(wt\right) \tag{5}$$

$$v_{bn}\left(t\right) = V_s \sin\left(wt - 120^0\right) \tag{6}$$

$$v_{cn}\left(t\right)=V_{s}\sin\left(wt+120^{0}\right) \tag{7}$$

The Zig-Zag transformer can compensate for the zero-components of load current as it is linked between the APF and the load. Only the positive- and negative-sequence components are contained by the three-phase load currents for the APF and the load currents can be expressed with the Fourier as below:

$$i_{La}\left(t\right)=\sum_{n-1}^{\infty}I_{L_{an}}\sin\left(wt+\varnothing_{an}\right) \tag{8}$$

$$i_{Lb}\left(t\right)=\sum_{n-1}^{\infty}I_{L_{bn}}\sin\left(wt+\varnothing_{bn}\right) \tag{9}$$

$$i_{Lc}\left(t\right)=\sum_{n-1}^{\infty}I_{L_{cn}}\sin\left(wt+\varnothing_{cn}\right) \tag{10}$$

With compensation of reactive power and suppression of harmonic power by the AFP, one can expect sinusoidal and balanced three-phase utility currents which will contain only the basic positive-sequence components expressed as below where I_1 denotes the amplitude of the utility currents.

$$i_{sa}\left(t\right)=I_{1}\sin\left(wt\right) \tag{11}$$

$$i_{sb}\left(t\right)=I_{1}\sin\left(wt-120^{0}\right) \tag{12}$$

$$i_{sc}\left(t\right)=I_{1}\sin\left(wt+120^{0}\right) \tag{13}$$

The fundamental idea of such method is to drive the three-phase utility currents as described above (11)-(13) and become sinusoidal and balanced ignoring the load characteristic such that their waveforms and phases can be presumed, leaving the amplitude as the only unknown parameter. The amplitude can be estimated by power balancing among the involved factors (i.e., AFP, utility and load). Assuming the role of an energy buffer, the DC capacitor on the power converter's DC side generates a stable DC voltage for the converter in the steady-state condition. Should the active power supplied from the utility is insufficient to meet the demanded active power of load, the difference must be compensated by injecting an additional power to the DC capacitor or receiving it from the capacitor. This difference in active power will lead to the voltage variation of the DC capacitor but it is important that the power factor maintains the conformity after the compensation by the APF. Thus, the difference of active power can be represented with below equations.

$$\Delta_p = P_s - P_L = \frac{3}{2} V_s I_1 - P_L \tag{14}$$

In (14), PL denotes the demanded active power of load so that if $\triangle$p is negative, it indicates that the supplied active power is smaller than the active power demanded, causing reduced voltage of the DC capacitor, and vice versa when $\triangle$p is positive. For this reason, $\triangle$p must be adjusted to be zero in a steady state. According to (14), it is evident that the amplitudes of both the utility voltages and demanded active power cannot be controlled and the difference of active power needs to be controlled by adjusting the amplitude of the utility currents only. By widening the amplitude of utility currents, the difference of active power can be increased in case the DC capacitor voltage stays below the steady-state value. Contrawise, decreasing the amplitude will reduce the difference if the same capacitor's voltage is staying over the steady-state value. It is clear that the amplitude of utility currents is determined by the amplitude of the DC capacitor voltage using the control circuit. Also, since the current sensors used in the proposed APF detects the three-phase utility currents only, matching adjustment of current sensors (i.e., sensors that detect the load currents and APF output currents) in the conventional APF can be avoided. This streamlines the design of the control circuit and the number of current sensors can be reduced accordingly.

3.4 Implementation of Power Line Communication Using OPNET for Vertical Fish Farm

As the scenarios, five equipment had been deploy in the home network. five units of Power Line Communication equipment models and a switch were deployed for the Power Line Communication scenario and five units of wireless equipment models and a router for the Wireless scenario in order to construct them on the same level. For the traffics, five demands had been connected and G.729 traffic generation was observed.

Figure 3 shows a Power Line Communication Simulation Using OPNET for Vertical Fish Farm that consists of Power Line Communication Router nodes model and node model. The simulation was to check the changes according to the speed of the Power Line Communication link and we placed the minimum number of elements to analyze the results. And to ensure that all traffics do not have the same processing priority, QoS was not used.

Figure 3. Power line communication simulation using OPNET for vertical fish farm

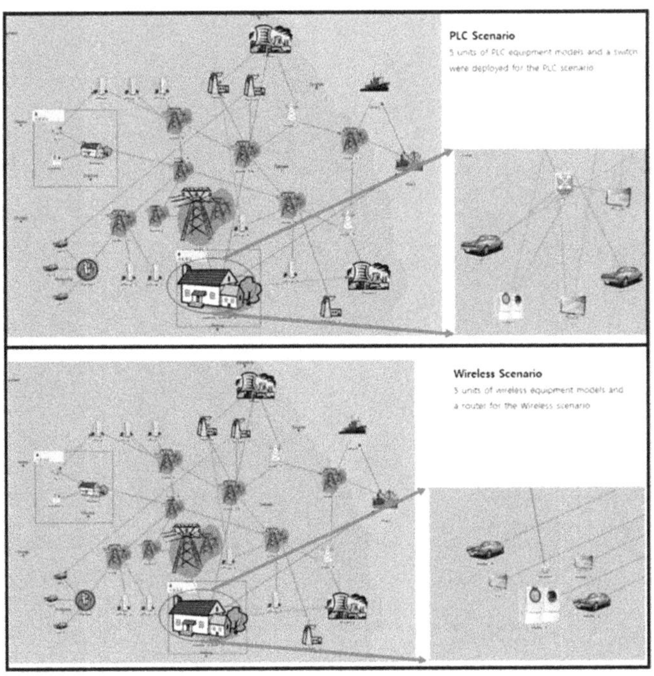

3.5 PLC Slient Node Problem Using OPNET for Vertical Fish Farm

Overall scenario is shown in Figure 4 where the topologies of A and B blocks are described. In the A-block, there are 20 homes standing 20m apart and no communication-support equipment has been installed. As for the B-clock, the same arrangement has been made but a router exits at a point of 200m distance.

The A-block construction is as shown in Figure 5. The "Silent Node" scenario was executed here and the result showed that communication was not successful beyond 200m.

On the other hand, in the B-block (Figure 6), communication had been carried out normally as there was a router. It is evident that the Power Line Communication-based network needs some kind of supporting communication equipment to reach remotely located destinations to overcome some of the problems involved in the Power Line Communication networks. Similar

Figure 4. Overall scenarios

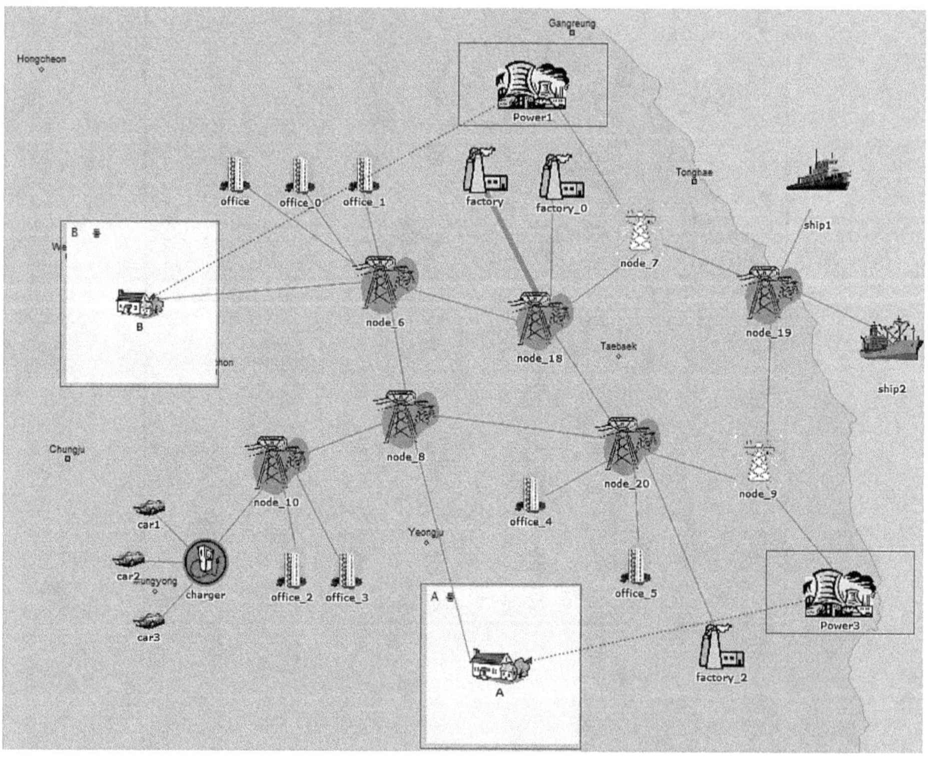

Figure 5. A-block construction

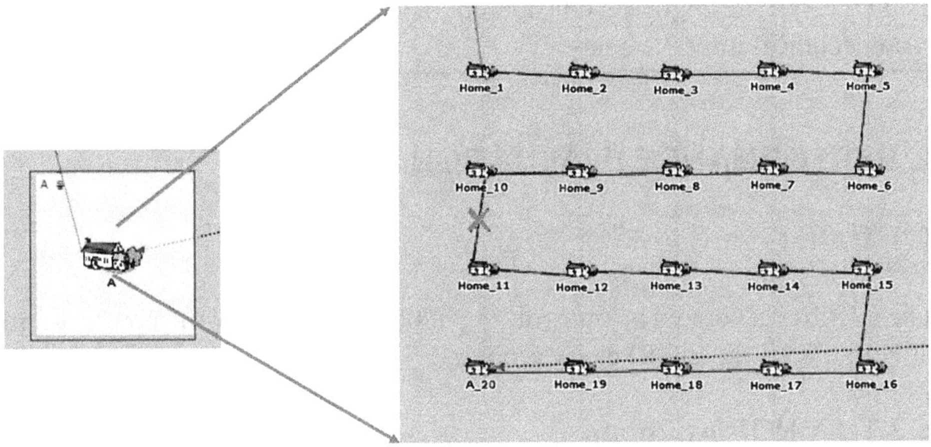

Figure 6. B-block construction

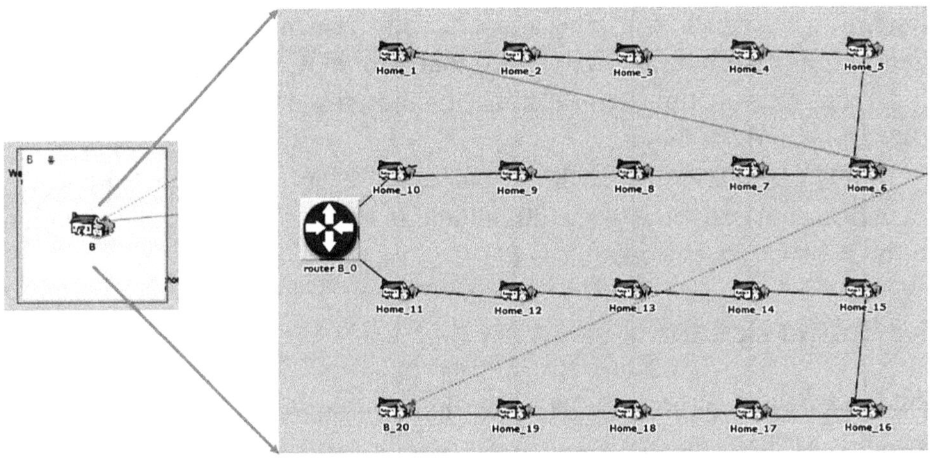

network has been applied to a deluxe apartment complex construction site in Shanghai where the Power Line Communication Smart Grid system is being used for the communications. Not limited to this site, world's leading construction companies and IT companies are forming consortiums to adapt the Power Line Communication-based smart grid system in other rapidly growing housing markets (i.e., China and other developing countries). That is, the silent node problem is one of the weaknesses in a long-distance Power Line Communication and serious enough to cause communication disruptions. Their recent goal is to offer IoT technology to the customers using the system. Zigbee is considered to be another possibility but it needs more technical improvements.

4. PERFORMANCE EVALUATION

4.1 Traffic

The traffic is generated by transmitting 300bit packets from the power origin of both blocks (A and B) every second.

4.2 TCP, UDP Analysis

Not a single communication was successful in the control group where the silent node problem appeared. It is the "Silent Node" problem in AMR systems, when a metering data cannot directly transmit messages to the Data Concentrator (DC) or receive valid signals from DC.

It required more time as the distance increased and the traffic did not reach 20th home in the A-block.

In this case also, further the distance, more time was required and the traffic was not delivered to the 20th home in the A-block. Yet, the situation in the B-block was successful.

4.3 Traffic Scenario

The scenario is one that the "Traffic" has been introduced in the Baseline scenario. Minimal elements were input to analyze the results. Traffic scenario configuration information is shown in Table 2. Arrangement of the nodes, connections of the links, and the traffics were randomly determined.

4.4 Results of Analysis

Traffic T/R volumes, delays and jitters were analyzed comparatively.

Table 2. Traffic scenario configuration information

Arrangement of the Node	5 units of PLC equipment models and a switch were deployed for the PLC scenario and 5 units of wireless equipment models and a router for the Wireless scenario in order to construct them on the same level.
Traffic	5 demand connection (G.729 traffic)
Demand Connection (G.729 Traffic)	11Mbps
Execution time	1 hour
Performance Indicators	Traffic transmission rate, end-to-end transmission delay, Jitter

4.4.1 Traffic T/R Rate

The average rate was 30kbps equaling to the generated traffic volume and there were no messages losses when traffic T/R rates were compared. Both networks had generated the traffics within 11Mbps and 100% volume was received without any loss.

The loss could have occurred in case of wireless network but owing to the WLAN's CSMA-CA (collision avoidance) and ARQ functions, the loss did not occur.

4.4.2 End-to-End Delays

The delay for Power Line Communication-based network was 0.0008sec and for the wireless network, it was 0.0013sec showing around 1.7fold differences. Since Wireless type uses CSMA-CA method which considers times taken for Backoff, collision avoidance and retransmission, more delays are longer than the Power Line Communication type. Figure 8 shows comparison between Power Line Communication and wireless type End-to-End delays (sec).

4.4.3 Jitters

The average jitter for Power Line Communication type was 0.00005sec and 0.00045sec for Wireless type showing around nine-fold differences. Such difference seems to have been caused by the same reason mentioned above - the characteristic of the CSMA-CA. Figure 9 shows comparison between Power Line Communication and wireless type Jitters (sec).

Figure 7. Comparison between PLC and wireless type T/R rates

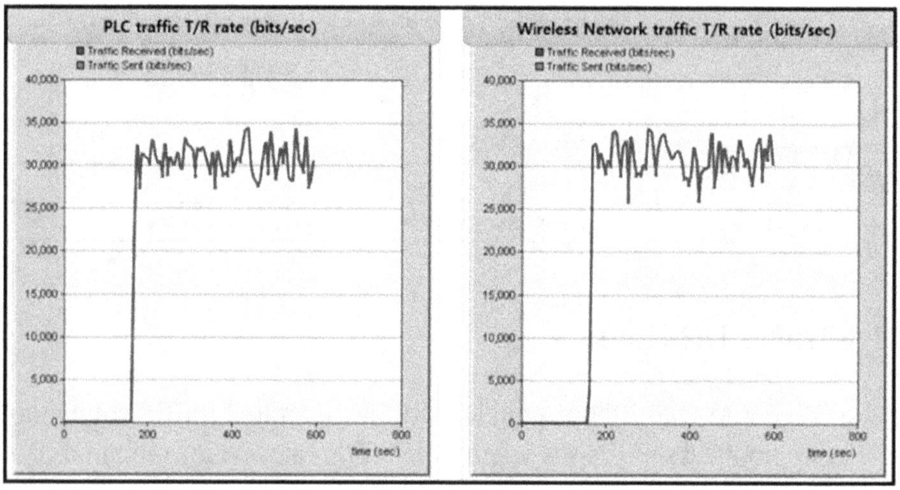

Figure 8. Comparison between PLC and wireless type End-to-End delays

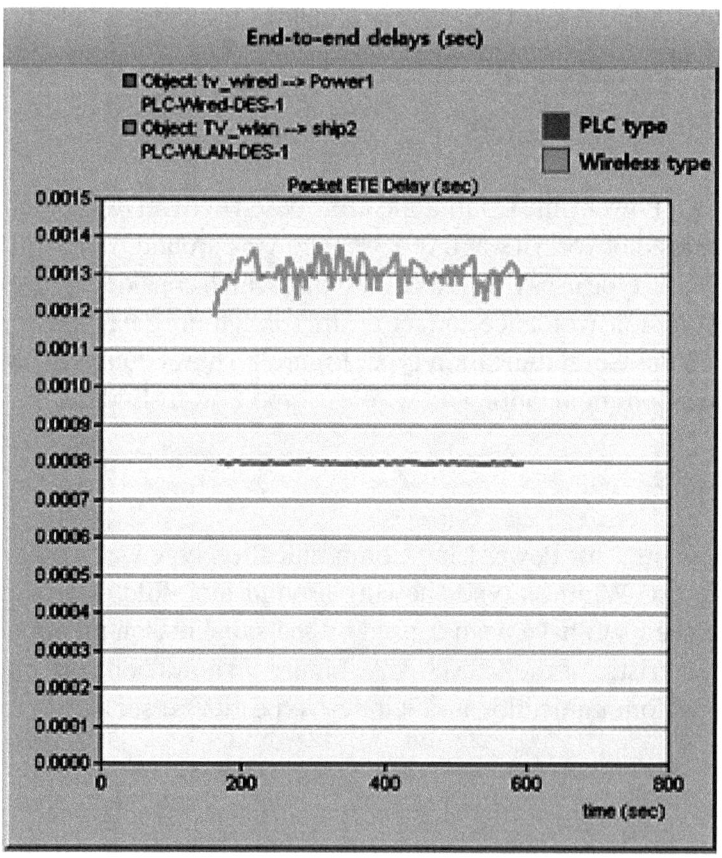

Figure 9. Comparison between PLC and wireless type jitters

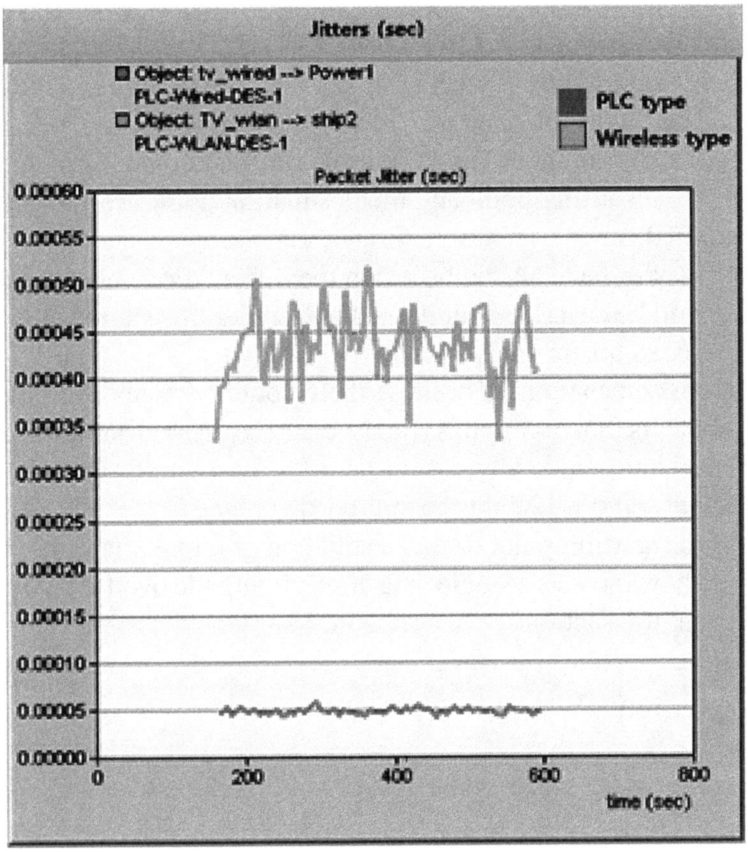

5. IMPLEMENTATION OF PART OF IDS USING JAVA

5.1 Classification of Protocols and Services Using Port Numbers

The items needed to show traffics graphically on a screen are starting point and destination addresses, the type of protocol used by the captured packet, and the relevant services used. That is, with your own client as the center, indicate the address of opponent's starting point or destination with a dot and then connect them with a line. After doing so, write the relevant protocol and service under the line. For example, if the captured packet uses the TCP protocol and comes in or out through port No. 80, mark it as 'HTTP'. And to do that, assign some exceptions for the port numbers corresponding to respective protocols and collect related information. Box 1 is a simple implementation of the relevant contents. It can be referred when representing a certain traffic graphically.

Check the captured pack whether it is a packet belonging to the relevant protocol or not with an operator 'instanceof'. In this case, the compare corresponding port number and output the relevant contents, including the addresses of the starting point and the destination point. Results are shown in the Figure 10.

The traffic will be analyzed based on the information collected here and the process will be described similar to the form in the picture below. Figure 11 shows a jncap traffic analysis tool.

A problem comes up here because there could be many overlappings in captured contents. Since countless points could be printed out when one tries to represent them graphically, it will be difficult to understand the picture and inefficient. Figure 12 shows overlapped contents.

Thus, if the starting point or the destination point are the same, it would be better to show them as a single data after solving the overlapping problem in the relevant transaction.

Box 1.

```
import jpcap.*;
import jpcap.packet.*;
public class TestMain {
NetworkInterface[] list = null;
JpcapCaptor captor;
//TestMain constructor - open by selecting device No. 0
public TestMain() throws Exception{
list = JpcapCaptor.getDeviceList();
//1000ms
captor = JpcapCaptor.openDevice(list[0], 65535, false, 1000);
captor.loopPacket(-1, new CheckedPacket());
captor.close();
}
public static void main(String arg[]) throws Exception{
new TestMain();
}
}
class CheckedPacket implements PacketReceiver{
public void receivePacket(Packet pack){
//classify with port No. when Packet is not NULL
if(pack==null)
{}
else if(pack instanceof TCPPacket&&(((TCPPacket)pack).src_port==21
|| ((TCPPacket)pack).dst_port==21))
{
```

continued on following page

Box 1. Continued

```
System.out.println("FTP");
TCPPacket tc=(TCPPacket)pack;
System.out.println(tc.src_ip+" --> "+tc.dst_ip);
System.out.println("=====================================");
}
else if(pack instanceof TCPPacket&&(((TCPPacket)pack).src_port==20
|| ((TCPPacket)pack).dst_port==20))
{
System.out.println("FTP Data");
TCPPacket tc=(TCPPacket)pack;
System.out.println(tc.src_ip+" --> "+tc.dst_ip);
System.out.println("=====================================");
}
else if(pack instanceof TCPPacket &&(((TCPPacket)pack).src_port==80
|| ((TCPPacket)pack).dst_port==80)){
System.out.println("HTTP");
TCPPacket tc=(TCPPacket)pack;
System.out.println(tc.src_ip+" --> "+tc.dst_ip);
System.out.println("=====================================");
}
else if(pack instanceof TCPPacket &&(((TCPPacket)pack).src_port==110
|| ((TCPPacket)pack).dst_port==110)){
System.out.println("POP3");
TCPPacket tc=(TCPPacket)pack;
System.out.println(tc.src_ip+" --> "+tc.dst_ip);
System.out.println("=====================================");
}
else if(pack instanceof TCPPacket &&(((TCPPacket)pack).src_port==53
|| ((TCPPacket)pack).dst_port==53)){
System.out.println("DNS");
TCPPacket tc=(TCPPacket)pack;
System.out.println(tc.src_ip+" --> "+tc.dst_ip);
System.out.println("=====================================");
}
else if(pack instanceof UDPPacket &&(((UDPPacket)pack).src_port==53
|| ((UDPPacket)pack).dst_port==53)){
System.out.println("DNS");
UDPPacket up=(UDPPacket)pack;
System.out.println(up.src_ip+" --> "+up.dst_ip);
System.out.println("=====================================");
}
else if(pack instanceof UDPPacket &&(((UDPPacket)pack).src_port==67
|| ((UDPPacket)pack).dst_port==67)){
System.out.println("DHCP");
UDPPacket up=(UDPPacket)pack;
System.out.println(up.src_ip+" --> "+up.dst_ip);
System.out.println("=====================================");
}
else if(pack instanceof UDPPacket &&(((UDPPacket)pack).src_
port==69
|| ((UDPPacket)pack).dst_port==69)){
System.out.println("TFTP");
UDPPacket up=(UDPPacket)pack;
System.out.println(up.src_ip+" --> "+up.dst_ip);
System.out.println("=====================================");
}
}
}
```

Figure 10. Resulting screen

```
HTTP
/192.168.52.101 --> /216.34.181.96
========================================
HTTP
/192.168.52.101 --> /216.34.181.96
========================================
DNS
/192.168.52.101 --> /192.168.52.254
========================================
DNS
/192.168.52.254 --> /192.168.52.101
========================================
HTTP
/192.168.52.101 --> /216.34.181.96
```

Figure 11. A jncap traffic analysis tool
(Huh, 2017)

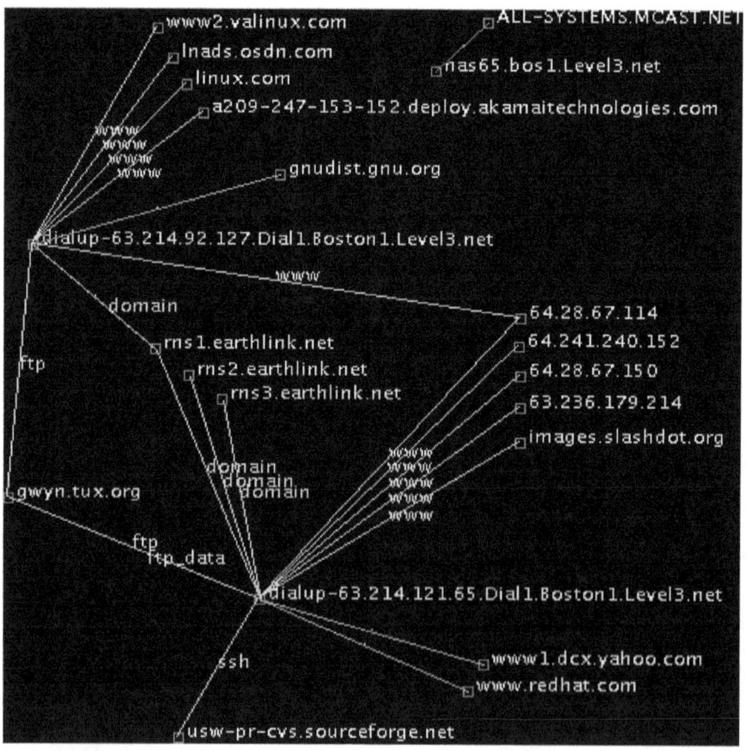

Figure 12. Overlapped contents
(Huh, 2017)

```
HTTP
/192.168.52.101 --> /216.34.181.96
========================================
HTTP
/192.168.52.101 --> /216.34.181.96
========================================
HTTP
/192.168.52.101 --> /216.34.181.96
========================================
HTTP
/192.168.52.101 --> /216.34.181.96
========================================
HTTP
/192.168.52.101 --> /216.34.181.96
========================================
DNS
/192.168.52.101 --> /192.168.52.254
========================================
DNS
/192.168.52.254 --> /192.168.52.101
========================================
HTTP
/192.168.52.101 --> /216.34.181.96
========================================
HTTP
/216.34.181.96 --> /192.168.52.101
========================================
```

5.2 Printing Out Captured Contents in the Text Area

We thought that it would be helpful if the captured contents are printed out in a text form prior to picturizing them thoughtlessly. The Port number-based classification program prepared in advance was linked with the GUI. Then, just by pushing the 'start' button, the packets will be captured and the contents will be printed out in a text area.

In order to output the relevant address in the text area when a packet had been captured, 'append' was used. Also, to output a string-type 'rs' that has an address value with 'append', InnerClass was created. We did not use InnerClass at the beginning and it was unclear how to hand over the value but while using the threads to prevent kinks in a packet capturing logic, we hit on a new way to output values. That is, after creating 'InnerClass' and outputting the values with 'append'. Additionally, EventListener, which corresponds to Start and Exit buttons, was created. Get the network device by pushing the Start button and open it, followed by execution of 'looppacket' with the thread. When all of these are implemented, the UI similar to the one described in Figure 13 will emerge.

Box 2.

```
import java.awt.*;
import java.awt.event.*;
import java.io.*;
import java.net.*;
import javax.swing.*;
import jpcap.*;
import jpcap.NetworkInterface;
import jpcap.packet.*;
public class TestGui extends JFrame{
JPanel p1 = new JPanel();
JPanel p2 = new JPanel();
JButton btn = new JButton("Start");
JButton btn1 = new JButton("Exit");
JTextArea txt = new JTextArea();
JScrollPane jp = new JScrollPane(txt);
NetworkInterface[] list = null;
JpcapCaptor captor;
public TestGui() throws Exception{
Container c = this.getContentPane();
//set 2 panels as a border layout
c.setLayout(new BorderLayout());
//set the background color of the panel - the black area will be used to output
the picture
//at this point, output to text area
p1.setBackground(Color.black);
p2.setBackground(Color.black);
//add a button on the panel2 setting it as FlowLayout
p2.setLayout(new FlowLayout());
p2.add(btn);
p2.add(btn1);
//mount ActionListener on the button
btn.addActionListener(new ButtonListener());
btn1.addActionListener(new ButtonListener());
//add the panel and TexArea to the frame
c.add("Center", txt);
```

continued on following page

Box 2. Continued

```
txt.setLineWrap(true);
txt.setBorder(BorderFactory.createTitledBorder("Packets"));
c.add("East", p2);
setSize(600, 400);
setVisible(true);
setDefaultCloseOperation(JFrame.EXIT_ON_CLOSE);
}
//perform capturing sequence with thread
java.lang.Runnable rnner=new Runnable(){
public void run(){
captor.loopPacket(-1, new CheckedPacket());
}
};
Thread t;
public void startCapThread(final JpcapCaptor captor){
t = new Thread(rnner);
t.start();
}
//set actions with InnerClass
public class ButtonListener implements ActionListener{
public void actionPerformed(ActionEvent e){
Object obj = e.getSource();
if(obj == btn){
try {
list = JpcapCaptor.getDeviceList();
captor = JpcapCaptor.openDevice(list[0], 65535, false, 1000);
txt.append("Starting Capture\n");
txt.append(" \n");
//initiate packet capturing
startCapThread(captor);
System.out.println("Start");
} catch (IOException e1) {
e1.printStackTrace();
}
}else if(obj == btn1){
System.exit(0);
}
}
}
//set action options when capturing packets
class CheckedPacket implements PacketReceiver{
public String rs;
public void receivePacket(Packet pack){
//classify with port No. when Packet is not NULL
if(pack==null){
}
else if(pack instanceof TCPPacket&&(((TCPPacket)pack).src_port==21|| ((TCPPacket)
pack).dst_port==21))
{
TCPPacket tc=(TCPPacket)pack;
rs = tc.src_ip+" --> "+tc.dst_ip;
txt.append("FTP\n");
txt.append(rs + "\n");
txt.append("==============================\n");
}
else if(pack instanceof TCPPacket&&(((TCPPacket)pack).src_port==20|| ((TCPPacket)
pack).dst_port==20))
```

continued on following page

Box 2. Continued

```
{
TCPPacket tc=(TCPPacket)pack;
rs = tc.src_ip+" --> "+tc.dst_ip;
txt.append("FTP DATA\n");
txt.append(rs + "\n");
txt.append("==============================\n");
}
else if(pack instanceof TCPPacket &&(((TCPPacket)pack).src_port==80 ||
((TCPPacket)pack).dst_port==80)){
TCPPacket tc=(TCPPacket)pack;
rs = tc.src_ip+" --> "+tc.dst_ip;
txt.append("HTTP\n");
txt.append(rs + "\n");
txt.append("==============================\n");
}
else if(pack instanceof TCPPacket &&(((TCPPacket)pack).src_port==110 ||
((TCPPacket)pack).dst_port==110)){
TCPPacket tc=(TCPPacket)pack;
rs = tc.src_ip+" --> "+tc.dst_ip;
txt.append("POP3\n");
txt.append(rs + "\n");
txt.append("==============================\n");
}
else if(pack instanceof TCPPacket &&(((TCPPacket)pack).src_port==53 ||
((TCPPacket)pack).dst_port==53)){
TCPPacket tc=(TCPPacket)pack;
rs = tc.src_ip+" --> "+tc.dst_ip;
txt.append("DNS\n");
txt.append(rs + "\n");
txt.append("==============================\n");
}
else if(pack instanceof UDPPacket &&(((UDPPacket)pack).src_port==53 ||
((UDPPacket)pack).dst_port==53)){
UDPPacket up=(UDPPacket)pack;
rs = up.src_ip+" --> "+up.dst_ip;
txt.append("DNS\n");
txt.append(rs + "\n");
txt.append("==============================\n");
}
else if(pack instanceof UDPPacket &&(((UDPPacket)pack).src_port==67 ||
((UDPPacket)pack).dst_port==67)){
UDPPacket up=(UDPPacket)pack;
rs = up.src_ip+" --> "+up.dst_ip;
txt.append("DHCP\n");
txt.append(rs + "\n");
txt.append("==============================\n");
}
else if(pack instanceof UDPPacket &&(((UDPPacket)pack).src_port==69 ||
((UDPPacket)pack).dst_port==69)){
UDPPacket up=(UDPPacket)pack;
rs = up.src_ip+" --> "+up.dst_ip;
txt.append("TFTP\n");
txt.append(rs + "\n");
txt.append("==============================\n");
}
}
}
public static void main(String arg[]) throws Exception{
new TestGui();
}
}
```

Figure 13. Output screen
(Huh, 2017)

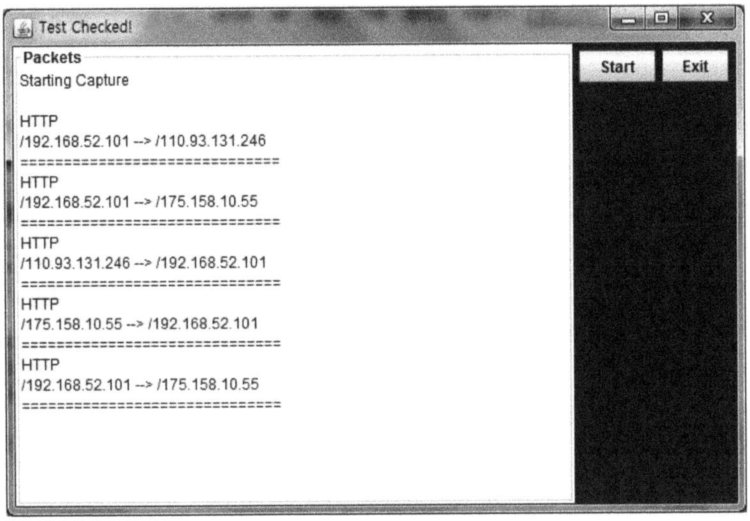

6. CONCLUSION

In this chapter, the Power Line Communication Router node and Power Line Communication node-based modeling had been performed and on the basis of this model, designing and implementation of the Smart grid Home Network System were achieved. For a simulation environment, OPNET Modeler 14.5 PL8 was used, and to ensure all the traffics would not have the same processing priority, QoS was not considered.

In this chapter, the results obtained from designing and performing the implementation scenario for the Power Line Communication-based Vertical Fish Farm were compared and analyzed with that of IEEE 802.11 WLAN MAC(the foundation technology at Jeju Smart grid Test bed)-based Smart grid home network. OPNET 14.5 PL8, OSI 7 layer, Power Line Communication router nodes and Power Line Communication nodes had been used for the designing and implementation simulations of both systems.

Additionally, QoS was not considered in order to guarantee that all the traffics would not have the same processing priority.

And, OPNET Modeler 14.5 PL8 was used for the performance analysis focusing on the silent node problem. The tests conducted for each case (i.e., TCP/UDP bases) revealed different results. That is, the traffic did not reach the node located beyond 200m distance in A-block while all the nodes in B-block received the traffic as router had been installed at the point of 200m.

I anticipate that our chapter of the silent node problem will be fundamental to the Vertical Fish Farm. Also, we are scheduled to conduct a larger scale simulations in future book where protocols and security related issues will be dealt with and then the book will be presented to construction companies as a system proposal.

This chapter is a continuing study where some descriptions of additional experiments have been introduced into the existing study published earlier. Even though it was not possible to carry out an ideal simulation in prior study because of the fact that we were unable to generate the silent node problem itself successfully, we plan to proceed with a large-scale research in our future work based on this book.

REFERENCES

Battré, D., Hovestadt, M., Lohrmann, B., Stanik, A., & Warneke, D. (2010). Detecting Bottlenecks in Parallel Dag-Based Data Flow Programs. In *Proceeding of 2010 IEEE Workshop on Many-Task Computing on Grids and Supercomputers*. doi:10.1109/MTAGS.2010.5699429

Behm, A., Borkar, V. R., Carey, M. J., Grover, R., Li, C., Onose, N., & Tsotras, V. J. et al. (2011). Asterix: Towards a Scalable, Semi-Structured Data Platform for Evolving-World Models. *Distributed and Parallel Databases*, *29*(3), 185–216. doi:10.1007/s10619-011-7082-y

Beyer, K. S., Ercegovac, V., Gemulla, R., Balmin, A., Eltabakh, M. Y., & Kanne, C. C. et al. (2011). Jaql: a Scripting Language for Large Scale Semi-Structured Data Analysis. In *Proceeding of International conference on Very Large data Bases* (Vol. 4, pp. 1272-1283).

Boden, C., Karnstedt, M., Fernandez, M., & Markl, V. (2013). Large-Scale Social Media Analytics on Stratosphere. *In Proceedings of the 22nd International Conference on World Wide Web (pp. 257-260). ACM.*

Borkar, V., Carey, M., Grover, R., Onose, N., & Vernica, R. (2011). Hyracks: A Flexible and Extensible Foundation for Data-Intensive Computing. In Proceeding of International Council for Open and Distance Education (pp. 1151-1162).

Gao, Q., Yu, J. Y., Chong, P. H. J., So, P. L., & Gunawan, E. (2008). Solutions for the Silent Node Problem in an Automatic Meter Reading System Using Power-Line Communications. *IEEE Transactions on Power Delivery*, *23*(1), 150–156. doi:10.1109/TPWRD.2007.910990

Huh, J. H. (2017). Designing of Image Processing Curriculum Considering Network Security. Asia-Pacific Journal of Multimedia Services Convergent with Art, Humanities, and Sociology, 7(6), 861-869.

Huh, J. H., Kim, H. B., & Seo, K. R. (2016). A Preliminary Analysis Model of Big Data for Prevention of Bioaccumulation of Heavy Metal-Based Pollutants: Focusing on the Atmospheric Data Analyses. *Advanced Science and Technology Letters, 129*, 159–164. doi:10.14257/astl.2016.129.32

Huh, J. H., Koh, T. H., Kim, N. J., & Seo, K. R. (2015). Design and Test Bed Experiments of Smart Grid-Based PLC Slient Node Problem Using OPNET. In *Proceedings of the 11th International Conference on Multimedia Information Technology and Applications* (pp. 317-322).

Huh, J. H., & Seo, K. (2017). Simulation of Power Line Communication Silent Node Problem Using OPNET. Journal of Korea Multimedia Society, 20(3), 491-499.

Huh, J. H., & Seo, K. R. (2014). PLC-Based Smart Grid Home Network System Design and Implementation Using OPNET Simulation. *Journal of Multimedia and Information Systems*, *1*(2), 113–120.

Huh, J. H., & Seo, K. R. (2015). Design and Implementation of the Basic Technology for Solitary Senior Citizens Lonely Death Monitoring System Using PLC. *Journal of Korea Multimedia Society*, *18*(6), 742–752. doi:10.9717/kmms.2015.18.6.742

Panchadcharam, S., Taylor, G. A., Pisica, I., & Irving, M. R. (2012). Modeling and Analysis of Noise in Power Line Communication for Smart Metering. In *Proceedings of the Power and Energy Society General Meeting* (pp. 22-26). IEEE. doi:10.1109/PESGM.2012.6345488

Sabolic, D. (2003). Influence of the Transmission Medium Quality on the Automatic Meter Reading System Capacity. *IEEE Transactions on Power Delivery*, *18*(3), 725–728. doi:10.1109/TPWRD.2003.813609

Sivaneasan, B., So, P. L., & Gunawan, E. (2010). A New Routing Protocol for PLC-Based AMR Systems. *IEEE Transactions on Power Delivery*, *26*(4), 2613–2620. doi:10.1109/TPWRD.2011.2161000

Tsuzuki, S., Areni, I. S., & Yamada, Y. (2012). A feasibility Study of 1Gbps PLC System Assuming a High-Balanced DC Power-Line Channel. In *Proceeding of 2012 16th IEEE International Symposium* (pp. 386-391). doi:10.1109/ISPLC.2012.6201334

Xiao-sheng, L., Liang, Z., Yan, Z., & Dianguo, X. (2011). Performance Analysis of Power Line Communication Network Model Based on Spider Web. In *Proceeding of 8th International Conference on Power Electronics ECCE Asia* (pp. 953-959). doi:10.1109/ICPE.2011.5944653

Conclusion

The Zigbee- and WiMAX-based Smart Grid environments and a PLC-based communication environment have been implemented and proposed in this book. As for the Transmission Layer, RUDP was tested and presented as the most suitable protocol that includes benefits of both TCP and UDP for the PLC networks. The simulation results showed that the functionality of RUDP surpassed that of TCP and UDP.

RUDP mainly allows client-to-client communications. First of all, in a common Client/Server structure, the communications between the Clients gain their reliability by sending data through the server but the speeds are slower than UDP and the server can be overloaded. For these reasons, RUDP, which has the major benefits of TCP (Reliability) and UDP (higher speed), has been devised. TCP requires that one side has to be on standby assuming the role of a server but UDP can immediately initiate communications without doing so. Therefore, as the most suitable protocol for the Smart Grids, RUDP has been proposed.

The simulations demonstrated better speeds and reliability compared to UDP and TCP, respectively. Additionally, an AMI application embedded with the Intelligent Agent was proposed.

The communication rates have been tested in three different environments: the first environment where many devices were linked to the power line and power consumptions consistently changed; the second, where devices were linked similar to the first environment but the consumption rates were steady; and the last, when the intelligent agent was involved. The observation was made for respective conditions. In other words, the situations were that the communication was carried out by using PLC, devices were linked directly with RS232, and the intelligent agent was used in the process, respectively. The PLC communication resulted in about 10% failure even though there were no interferences, and 20-35% failure when interferences were present. Meanwhile, almost 0% failure had resulted when direct linking method was used for the first two environments. Next, the results when the Intelligent Agent has been applied and the communication failure rates in both environments are significantly lower than that of the PLC in average, showing the improvements. From the results of performance evaluation, it was confirmed that the application was effective in reducing the problems often seen in the PLC networks and detecting malicious attacks including DDoS attacks.

It is expected that this research will contribute in devising some new and more efficient means of establishing Smart Grids and developing technologically better protocols and applications.

In this book, experiments were conducted for four out of seven OSI layers for a Smart Grid. The test bed experiments for remaining Network, Session and Presentation Layers using upgraded version of OPNET and other simulators, or by actual implementation, remain as future research tasks to be pursued.

Also, in the end, the Smart Grid system provides consumers with their power use amounts regularly in order to let them monitor and control their usage patterns hourly, instead of just handing over the bills every month. Desirable power usage pattern may change consumer behavior to save power in their daily lives but if the information given to the users is not properly checked and examined they could be used inappropriately. Thus, we plan to simulate proposed protocol and propose a more secure protocol.

Appendix 1

1. PRACTICE 1: EXTENDABLE FUTURE RESEARCH

This study focuses on a Smart Home System that has utilized both the voice and face recognition technologies which have been theorized in the author's previous works. The voice recognition system converts a voice command into the text form on a smart phone by using the speaker- independent speech recognition technology to deliver it to the system which in turn sends out an adequate response after executing the command. Meanwhile, the face recognition is used for the intruder recognition module to recognize an intruder by comparing the his/her face captured by the cam with the faces of the users registered in the system database and send out the response.

The face recognition system determines face matching/mismatching by comparing the relative distance or size of each facial organ such as the eyes, nose and mouth. Basically, the Smart Home System proposed in this study allows wireless communications between the modules through Wi-Fi and Raspberry Pi is used as its central control server. While the system was developed without any special Arduino board, the esp-12f Arduino module was used in each module except the intruder recognition module which uses Raspberry Pi to process images. Since each function has been modularized, an additional function(s) can be added by creating a separate independent module.

1.1 Introduction

Although the modern IT technologies have been continuously developed to provide vast amount of information and convenience to the people, the speed of development is still not satisfactory to the modern people who are swamped with their everyday lives. Modern people are so busy that they have not much time to take care their families. Going to work in the morning and working for a whole day is very common and they just go to sleep when they get home without enjoying their freedom or leisure activities. The technologies to escape from such a dreary life are the ones that derived from the IoT technologies. The IoT systems, which are based on the extended conception of M2M, are the ones that aim to control 'Things' and deliver information between them through networking. Current 'Things' include household appliances like washing machines, microwave ovens, TV's, Boilers and others, even cars and electronic door locks. The number of the IoT technology users are rapidly increasing and the Smart Home System where the IoT technology has been applied entirely by connecting the most of the household appliances is no longer an unique technology but a common commercial system establishing a strong base in the housing market. Nevertheless, the current Smart Home Systems certainly have their technical limits. That is, most of them are still taking on a form of simple physical network where all the connected "Things" are controlled/operated passively. Such a passive IoT technology is not enough to improve the quality of modern people's lives. Therefore, more of active aspects should be added to the existing Smart Home Systems. For instance, the IoT-based systems should be able to notify or alert the family members staying outside about what's happening in their home in the event of home invasion, fire or water leaking automatically. In other words, the information should be delivered to the families actively to allow them to take an adequate measure. Thus, in this practice, an IoT-based Smart Home Solution in which the intellectualized system concept has been added (so called Internet of Everything; IoE) can be developed.

2. EXTENDABLE PRODUCTS

- Android-based service applications that recognize voices
- Smart Home Solutions that can solve various household issues such as lighting & air conditioning controls, intrusion detection, shutting off gas valves or appliances, or monitoring in-house situations
- A show house applied with a certain Smart Home Solution

2.2 Scenarios

2.2.1 Command Delivery

Figure 1 shows the command delivery scenario. When controlling the devices, a user first delivers his/her recognition command to operate service application. The transmission of the recognition command can be achieved with a button or an App registered on the system. The application usually stays in a paused state and starts to operate and sends out a confirmation message that the operation is in progress after receiving the command. Only then the user can command the devices with his/her voice. The voice command can be given with the natural language or the word(s) and will be delivered to the server after being converted to a character string. At the server, the string-form command is analyzed and then transmitted to the target device to take control. The device executes the command and returns the results. The server delivers the results to the application after converting them in a form easily recognizable by the user. Finally, the application displays the results on a screen for the user.

2.2.2 Access of an Authorized User

Figure 2 shows the access scenario of an authorized user. When a user enters home, a camera device recognizes the him/her and delivers the image data to the server where the data is analyzed/compared to the image data already

Figure 1. The command delivery scenario

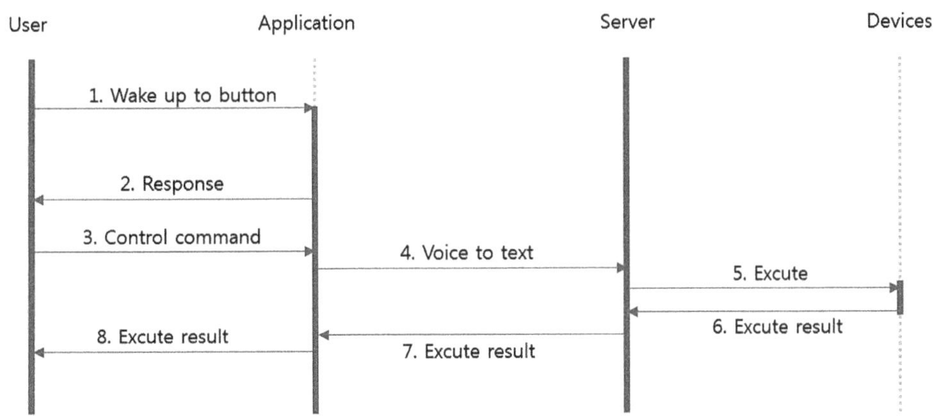

Figure 2. The access scenario of an authorized user

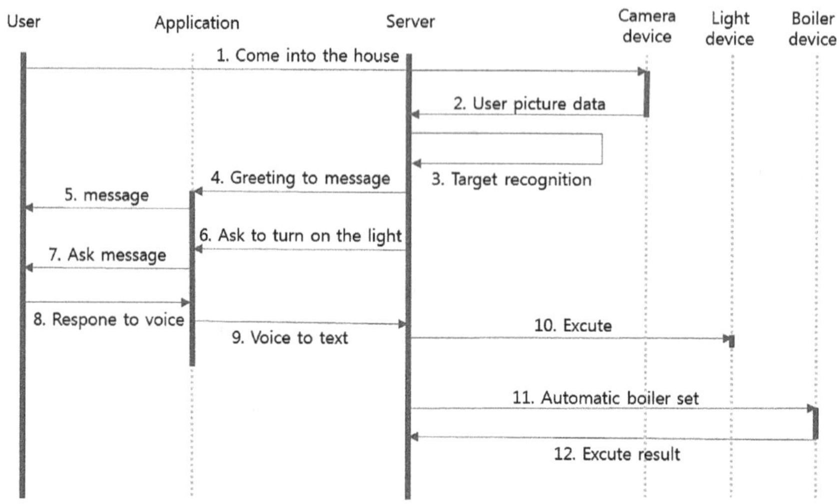

stored in the server. If the user is identified as a family member after analysis, the server sends out the message to the user application that the recognition process has been completed and then asks the application whether to turn on the light if the room light has been turned off. Then the application displays the question received from the server on a screen and the user gives his/her voice reply to the application, which will convert the voice data into a string data to deliver it to the server where the string data will be analyzed and send out a control signal to the lighting apparatus. After that, the server also delivers a pre-programed input value to the boiler system, which in turn changes its value and returns its performance results to the server.

2.2.3 Access of an Unauthorized User

Figure 3 shows the access scenario of an unauthorized user. If the analysis result of the image data received from the camera device confirms that the user is not one of the family members, the server transmits an intrusion alert to every smart phone belonging to the family members. Then, the server autonomously sounds the alarm.

Figure 3. The access scenario of an unauthorized user

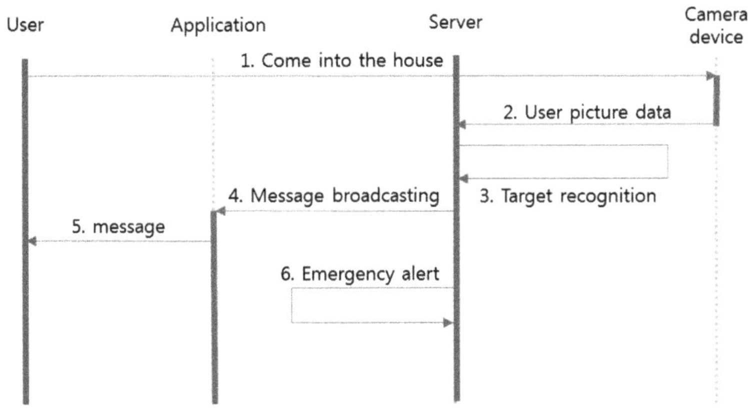

2.3 Voice Recognition

The voice recognition technology aims to develop the dictation technique that takes down vocalized contents, the automatic interpretation technique which translates Korean language into foreign languages, and the intelligent technique that process the voice commands by itself. This task voice recognition technology is used for the development of an intelligent assistance service where the voice commands are processed automatically. Basically, the voice recognition devices print out text information based on the voice signals mostly using either the language model or the acoustic model. The latter model commonly learns phonemes through the deep learning process and use them in the voice recognition algorithm. The former model learns the use patterns of words or vocabularies and draw statistics. The statistical models are generally used for the language models. Figure 4 shows basic structure of the voice recognition devices.

To apply the speaker-independent technology and to establish a natural command system, this task extracts voice commands as texts. Additionally, based on the statistics drawn from a language model, a list of texts that would match the voice commands has been compiled to compare it with the command system stored in the database. If they match, the command will be executed.

Figure 4. Basic structure of the voice recognition devices

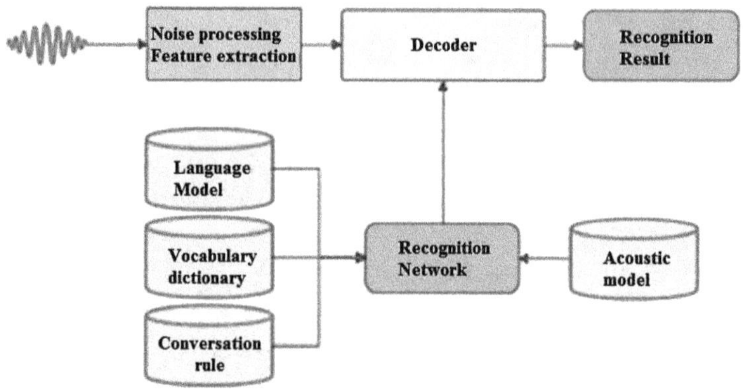

2.4 Face Recognition

The face recognition technology first creates a database of the facial pictures taken in advance and the identification process is proceeded by comparing a facial image taken by the camera with the stored image. The process generally goes through two steps: face detection and face recognition.

2.4.1 Face Detection

This is the technology which detects the position of a person's face in the image. The knowledge-based, Feature-based, and template-matching methods are often used for this technology:

- **Knowledge-Based Method:** The knowledge-based method detects faces based on the researcher's knowledge on the premise that a person's face includes an eyebrow(s), nose, and mouth and that it is common that each structural element has general distance and positional factors. This method easily finds faces using the rules concerning the positions and sizes of facial elements in the images.
- **Feature-Based Method:** This method uses a mixed-type information (i.e., information concerning the facial features such as size and shape, interrelationship, facial colors and textures) to detect faces. This method has the merit of finding a face easily in a short period of time but also has the demerit of being quite sensitive to the noises.

- **Template-Matching Method:** This is one that detects a face by comparing the entered (captured) facial image with the standard template created for all the faces targeted. Although this method is rarely affected by the noises, it is sensitive to the factors such as face rotation angles and size variations due to distances. Also, it is quite difficult to create an average template that contains all the individual information.

2.4.2 Face Recognition

The face recognition algorithm quantifies the features of each face by distinguishing the major elements (i.e., eyes, nose, ears and mouth) after analyzing its facial image. For example, the quantification involves the factors such as the distance between eyes, size of nose, and the ratio between area and height of forehead. By integrating all the quantified values, a “faceprint” can be created so that this faceprint is used to find the right face by comparing it with the faceprint stored in the data base.

2.4.3 Operating Structure of MQTT Protocol

Figure 5 shows a Server-Client communication architecture. A major point in the relationship between the server and the client is that the communications are established by using the MQTT protocol. MQTT is an abbreviation of Message Queuing Telemetry Transport and this protocol is used for the remote controlling and measuring activities. This protocol has been chosen as an IoT standard protocol by the OASIS (Organization for the Advancement of Structured Information Standards) administered by the International Organization for Standardization for the large-volume message delivery since 2013.

Figure 5. A server-client communication architecture

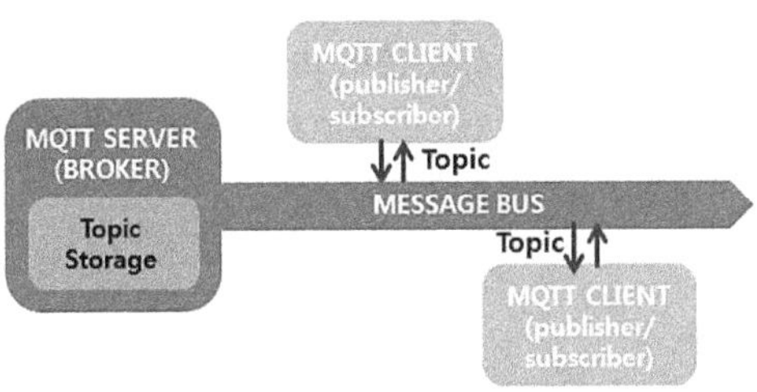

In this project, mosquitto (mqtt server service) will be installed in a Linux environment to construct a MQTT server while a mqtt library provided by the Eclipse Paho project ecwill be installed in an Android environment to create a MQTT client. In terms of connection, when used in an environment where Wi-Fi, 3G, or LTE connections are frequently disconnected, the MQTT protocol has a merit of allowing the broker to push the message sent from the server at the point the connection has been made with the user.

The delivery of control and state messages between the client and server is achieved in the following manner. The MQTT push service often used in Facebook or Messenger is executed in the background and has the advantage of consuming lesser packets and battery power compared to the periodic polling. In the average chatting services, the increased server load can occur as the number of users increases because they have to create more sessions and manage them. However, the MQTT method can distribute the load by letting the messages to be delivered by the MQTT broker after distinguishing the them as publish or subscribe category.

To detail the method of checking the condition of hardware (appliance) by the application, a topic is published after creating the data for each hardware and this topic will be delivered to the MQTT broker through the server. Then the broker delivers the topic with a relevant application so that the user can subscribe his/her desired topic (i.e., the condition of of hardware). The same process is applied when the application generates and delivers the command topics. Figure 6 shows MQTT.

For the MQTT, the system's wireless network has to remain as a substructure. The IEEE 802.15.4 standard suitable for the large scale, low-power, and low-cost systems often used by the common wireless technologies (e.g., Zigbee, Wi-Fi and etc.) is used in the first layer while the standard defined by the 6LoWPAN working group is used in the second layer. The 6LoWPAN-defined networks are the local area networks where IP addresses are assigned to directly connect 3L (low-power, low-cost and low-bandwidth) devices with internet and offer higher stability.

Ipv6/RPL currently subjected to standardization is used in the third layer. As it is expected that around twenty-six billion IoT devices will be connected to internet by year 2020, Ipv6 is more adequate than Ipv4 which can offer only about 4.3 IP addresses. RPL is an Ipv6 routing protocol has been designed to fit the low-power network environment.

Figure 6. MQTT

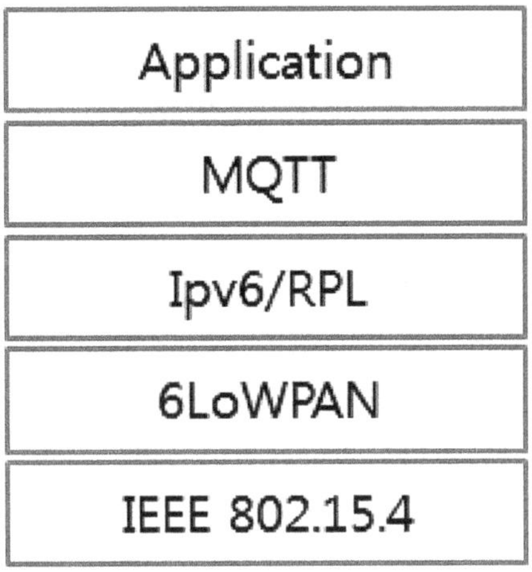

3. PRACTICE 2: EXTENDABLE FUTURE RESEARCH

3.1 Application

The user's voice commands are delivered to the server through STT (Speech to Text, a voice recognition API) after being converted to text form). The voice recognition function starts when a relevant button is clicked and the application delivers the commands in real-time and receives information from the server while working in the background. The devices can be controlled/ monitored by either voice commands or application. Figure 7 shows voice recognition API.

3.2 Application-Server Communications

Users can control hardware through the voice recognition function embedded in the application and receive resulting state on screen. The current state of hardware can be checked this way also. Figure 8 shows application - server communications.

3.4 Server-Hardware Communications

Basically, the communications between the server and hardware(s) are achieved in one direction by using the MQTT protocol but the cameras

Figure 7. Voice recognition API

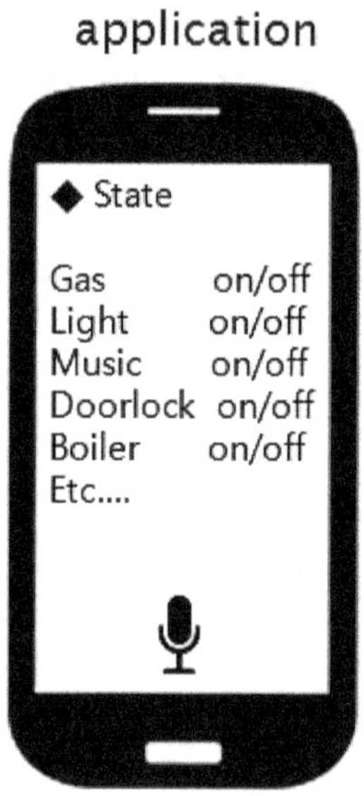

Figure 8. Application-server communications

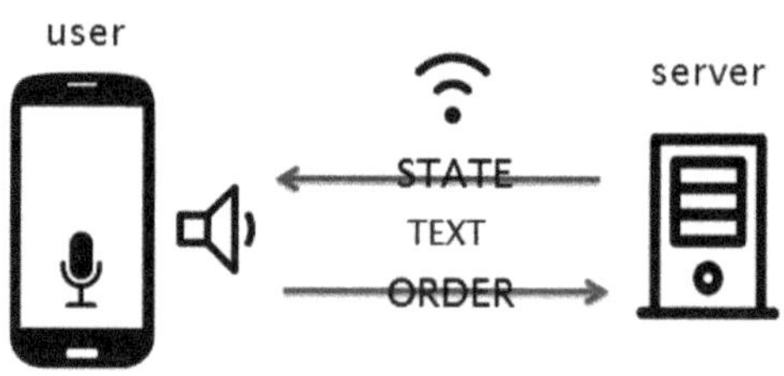

Figure 9. Server operation mechanism

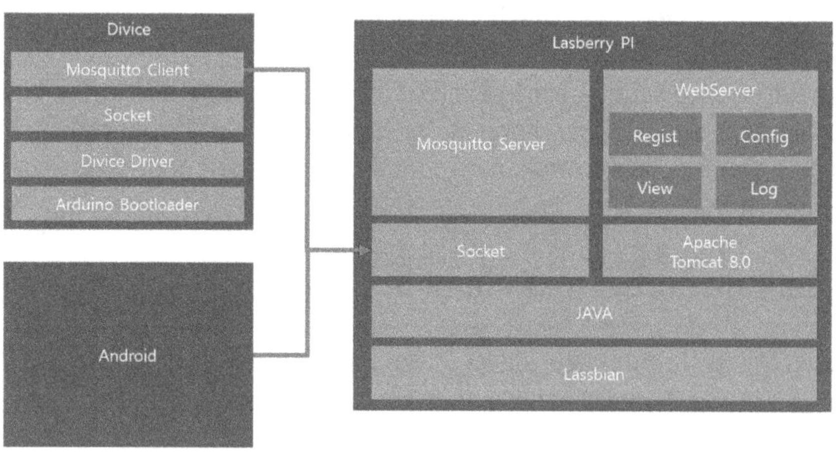

Figure 10. Problem

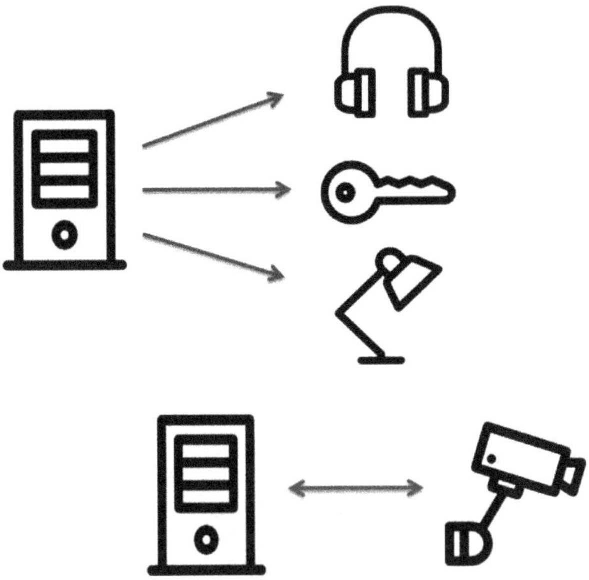

perform bi-directional communications with the server. The images from the system camera are delivered to the server with the RTP protocol, or the MQTT protocol, vice versa.

3.5 Hardware (Appliances)

3.5.1 Camera

The camera which has been programmed with OpenCV to process images will be able to send facial data (family members) signals to the server or receive commands from it.

3.5.2 Lighting

If the camera-captured facial image corresponds with one of the facial images of the family members, the lighting(s) will be turned on or off according to the voice commands.

Figure 11. Camera

Figure 12. Lighting

3.5.3 Door Lock (Electronic)

The door-lock will be locked/unlocked by the user's voice commands through application.

3.5.4 Boiler

Both the temperature control and automatic temperature maintenance functions performed by the user's voice commands through application.

3.5.5 Audio System

Audio system can be turned on or off, or move to next or previous tune conveniently with the voice commands.

Figure 13. Key

Figure 14. Boiler

3.5.6 Curtain

Daylighting can be controlled with the voice commands through application.

3.5.7 Refrigerator

The temperatures of freezer or refrigerator compartments and defrosting function can be controlled with the voice commands.

Figure 15. Headphones

Figure 16. Curtain

Figure 17. Refrigerator

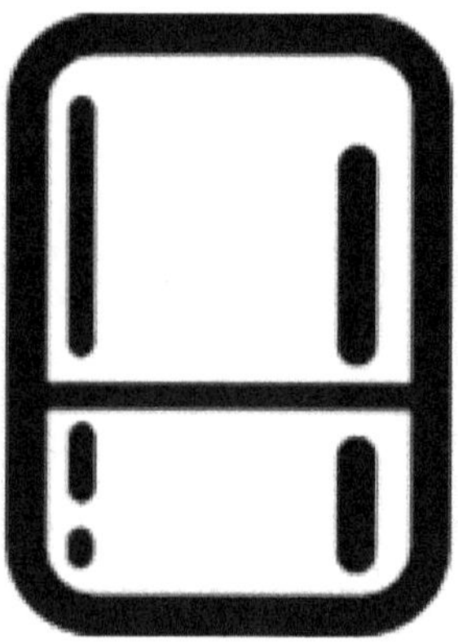

Figure 18. Water supply

3.5.8 Water Supply

Realtime monitoring of water meter and water leaking incidents, or warning of water pipe rupturing due to cold waves can be achieved by the application.

3.5.9 Gas Detector

The trace of gas in the indoor air can be detected in realtime and issue a warning in case of gas leak through the application.

4. PRACTICE 3: EXTENDABLE SCENARIO

We have always dreamt of a smart system like Jarvis the butler in the movie Iron Man and now, with the emergence of an artificial intelligent like AlphaGo and the arrival of the IoT era, such a system is becoming more realistic technically. Thus, we examined the possibility of creating a Jarvis-like system that utilizes user's location information. The most popular traditional method to gain location information is GPS (Global Positioning System). The introduction of GPS system has made a significant contribution in changing many areas of people's lives. That is, by providing the user's location in real-time together with the distance information and the possible routes to save time and money, the quality of life has improved. However, despite of many advantages, most of the GPS systems cannot be used in the spaces where satellite signals are weak or blocked (e.g., inside of building, underground, or tunnel). Thus, the necessity of the indoor positioning systems has emerged and many relevant studies and developments are being conducted

Figure 19. Flame

actively recently. Both the infrastructure-based positioning technology and the terminal-based positioning technology can be used for an indoor positioning system. The infrastructure-based positioning technology involves the WLAN, UWB, RFID and IR technologies while the terminal-based technology uses the inertia sensors embedded in the accelerometers or the gyroscopes. Indoor positions will be estimated in this task by using the BLE beacons in a WLAN network. The BLE technology is one of the promising technologies currently used widely as it consumes very little power. Based on the studies related to the BLE-utilized indoor positioning, user's position will be determined from the moment one arrives his/her home and with this information will be used to control the household equipment or devices. Moreover, by processing the data by relating the position information with the user's device-controlling acts, it will be possible to anticipate a certain function that the user seeks.

4.1 Task and Goal

We attempt to create a system that can estimate the indoor coordinates of the target object (e.g., smart phone, smart watch and etc.) with the RSSI (Received Signal Strength Indicator)-based trilateration technique based on the signals from the BLE (Bluetooth Low Energy) beacons and if this three-dimensional information determines that the target object is positioned in a certain area, the devices within the household will be controlled through the application.

In Figure 20a, the area allowing control of TV set was set near the sofa so that if the target object is positioned within the area, TV will be turned on, or turned off if it moves away from the area. Figure 20b shows that the display will preferentially indicate the cameras positioned within the designated area when there are multiple number of surveillance cameras indoors.

In order to achieve this, the BLE beacons installed in the room first send signals to the target object which in turn will send the RSSIs to the system server, where the position of the target object will be estimated and checked whether it is in the designated area. If it is determined that the target object is in the appropriate area, the system will control the device(s) assigned to the area. The control area for each device can be set with an App installed on the smart phone.

Figure 20. RSSI task & goal (1)

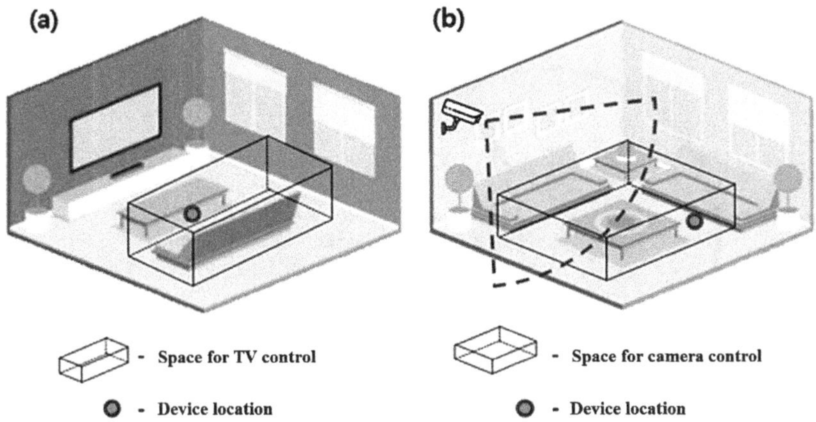

Figure 21. RSSI task & goal (2)

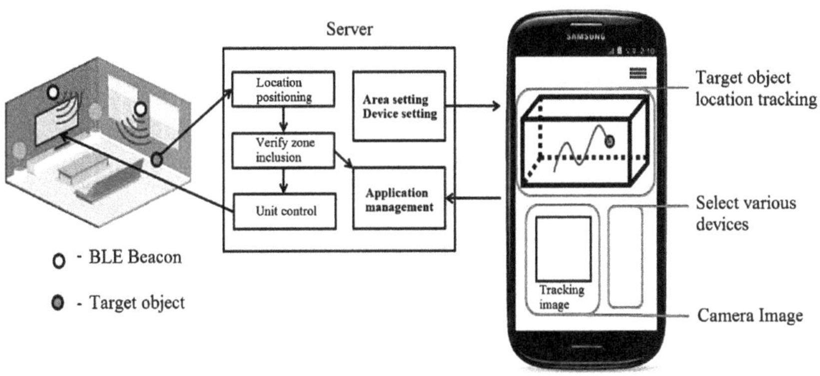

4.2 Current Relevant Technologies

4.2.1 Trilateration

In order to estimate the exact position of B (Position B) on a two-dimensional plane where the distance from the reference point to Position B has been already calculated, three reference points are needed. As described in below Figure 22, if the distances (r1, r2, and r3) from the respective reference points (P1, P2 and P3) to Point B have been determined, the exact position of Point B can be estimated.

Figure 22. Trilateration

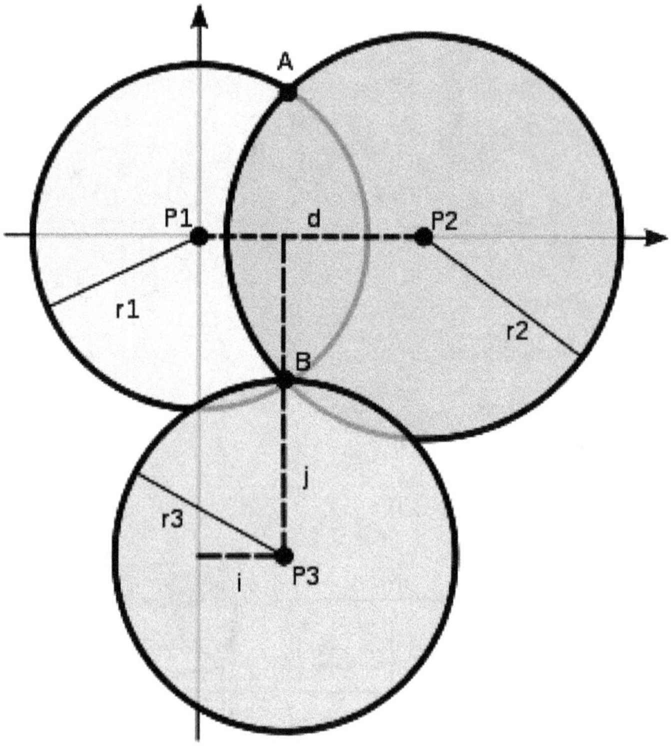

4.2.2 Trilateration in a Space

When there are two reference points in a space, Position B will be located somewhere on a circle. However, if the same points exist on an indoor wall, the shortest distance (radius of a circle) to Position B will be determined.

Using this theory, this task attempts to estimate the position of a "Thing" located indoors by placing two BLE devices as references on each wall forming a rectangular room. The strength of signal from the 'Thing" is then converted to the distance to calculate the shortest distance to each wall. Finally, the position of the "Thing" will be estimated after calculating these distances. Referring to the below graph inserted in the reference paper (Huh & Seo, 2014), there was a margin of error (1~2m) in estimating the expected position of the "Thing" by using RSSIs when the distance from the reference point (beacon node) to the "Thing" was 10m. The leading cause of such an error is the obstacle(s) that may be present in between the reference point and the 'Thing". Currently, many studies are being conducted to reduce the errors

Figure 23. Trilateration in a space

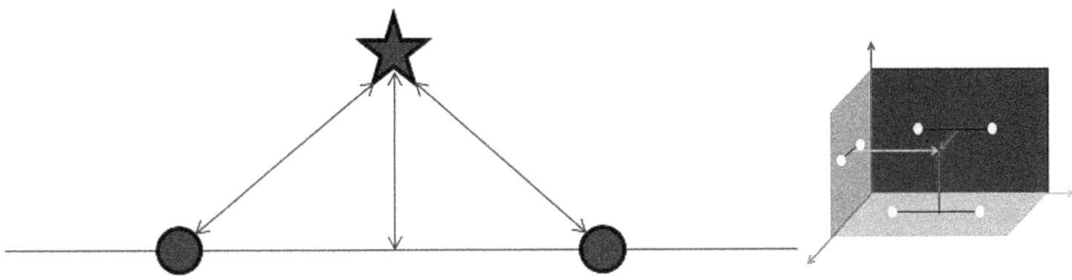

occurring in the RSSI-based positioning systems. The trilateration technique was used in this task to reduce the positioning error.

4.2.3 Measuring Distance Between Reference Point and Target Object

In addition to using the signal strengths, there are methods that use propagation time or transmission time depending on the distance to measure the distance, Time of arrival (ToA) and Time Difference of Arrival (TDoA). Even though these methods provide a relatively precise distance measurement and remain unaffected by the environmental factors, the hardware module that can measure to the nano-seconds is required for the measurement device. On the other hand, although the method that uses RSSI values would provide a relatively large margin of error due to variations in distances or the obstacles, it still is a well-suited technology for the IoT era considering its suitability to wireless communications.

4.3 Features of Beacon

1. A small volume of packets (168 bit = 21 byte)
2. Periodic signals
3. Pairing is not required
4. Low power requirement (3V coin-type battery lasts about 2 years, 200~300ms cycle base)
5. Max. arrival distance of 50m (stable at the distance of 20~30m)
6. Composed of UUID+major+minor+RSSI (only the ID values of the beacons and RSSIs will be transmitted). Figure 24 shows a diagram of beacon signals.

Figure 24. A diagram of beacon signals

The Universally Unique Identifier (UUID) is a 16-byte identifier also called as a unique identification number. It is defined with RFC 4122 specification and generated by using a special algorithm. Also, it is made of the 15-digits hexadecimal numbers. A probability of UUIDs becoming the same is as low as 1.6 billion to one.

- **Major Identifier**: This is used to distinguish the Beacons that have the same UUID (1~65535)
- **Minor Identifier**: This identifier is used to identify the Beacons that have the same UUID or Major Identifier

If a Minor Identifier is to be used, a Major Identifier must be provided as well.

4.4 RSSI (Received Signal Strength Indicator)

RSSI is a common name indicating the strength of received wireless/RF signal that includes noises RRSIs can be measured and confirmed. RSSIs remain in the values ranging from -99 to -35. To put it simply, they are the only and the most basic indicators used when measuring the distance between a certain Beacon and a smart phone, for example. Since the strength of radio signal is inversely proportional to the squared distance value and the RSSI value is proportional to the log of arbitrary definition, the distance information can be provided quite easily. However, the RSSI values are could be affected by several environmental factors. To mention a few, there are the direction of the radio antenna, obstacle(s) existing in between Beacons, and signal jamming.

4.4.1 Method of Using the Database

The database-based algorithm generates a RSSI database by using the signal strength vectors or the probability distribution of signal strengths between Beacons in respective areas of deployment. They are unique following their locations and deployed areas. The values are stored at the central base station and the node who is unaware of its own position attempts to collect at least more than one RSSIs from the surrounding available Beacons to create its own RSSI database and transmit it to the central base station.

4.4.2 Distance Prediction Method

The distance prediction method is one that can minimize the required database. In this case, the distance can be calculated from the source RSSI data by using a signal propagation model. Distance calculation can be made with the difference in the RSSI values taken at two points, considering a possibility of path loss which refers the attenuation in the transmission signal strength during propagation. Since such an attenuation is inversely proportional to the square distance, the signal propagation distance can be calculated with the path loss. It shows that the path loss increases as the distance increases. The path loss can be expressed as below:

Path Loss = (Tx power) - (Rx power)

$$PL = 20\log\left(\frac{d1}{d2}\right)$$

Formula 1 shows that the path loss is the difference calculated between the transmitted signal and the received signal while Formula 2 indicates that the path loss depends on the distance.

$$distance = 10^{(T_x - R_x)/20}$$

The distance-based formula can be expressed as below:

Distance = ...

By substituting the RSSI values into Tx and Rx, the distance can be found.

- **Hata Model:** A typical Distance-Loss model

$$P_{(d)=P_{(d_0)}-10n\log\left(d/d_0\right)+X}$$

In this normal distribution, $P(d)$ and $P(d_0)$ are the received electric power at the distances of d and d_0, n is an attenuation constant, and X is a random variable. Here, d_0 is an arbitrary reference point used to calibrate the n value (a loss factor). The margin of error in measuring the distance based on the calculated RSSIs can be approximated with the Gaussian. When the distance between the Beacon and the receiver has been measured in a three-dimensional space [assuming that the values for the Beacon's spatial position (XiYiZi, i = 1,2,3,...N) have been already entered], the position of the receiver can be estimated by trilateration or multi-positioning techniques based on the calculated distance.

$$\left(x-x_i\right)^2+\left(y-y_i\right)^2+\left(z-z_i\right)^2=d_i^2, i=1,2,3,\ldots N$$

An independent simple equation of N-1 can be formulated by removing the quadratic equation after subtracting the ith and i + 1th equations. In this case, N must be equal or larger than 2 at least.

4.5 Proposed Methods to Improve Accuracy of RSSI-Based Positioning

4.5.1 Area Division and Utilization of Reference Node(s)

First, divide the space into nine areas and measure the variables relevant to each area. Then, determine the area in which the target object stays by using reference node(s). The position of the target object can be estimated by using these node(s) based on the area determined (Xiao-Sheng et al., 2011).

4.5.2 Using Cumulative Probability Distribution

Figure 25 shows using cumulative probability distribution. By using the signal strengths of the target object (Mobile) and the reference node(Ref) calculate the length L and draw a circle with the radius of L. This circle represents the maximum probable points on which the target object may be placed. As in the figure on the right, the point deviating from the circle

Figure 25. Using cumulative probability distribution

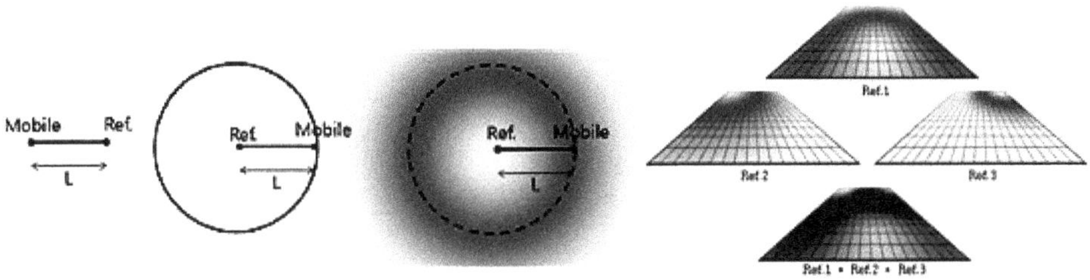

generate a probability distribution corresponding the deviated distance.The target's position will be stochastically estimated by considering the generated probability distribution of each reference node. Since this method uses cumulative probability distribution, the margin of error in calculations due to the environmental changes can be reduced (Yao, 2013).

4.5.3 Utilization of Center of Mass of Convex Polygon

The circle might appear in various forms as shown in the figure on the right when performing a trilateration so that the factor(ei), which calibrates the error, should be added as below:

$$\delta_1^2 = \left(x - x_1\right)^2 + \left(y - y_1\right)^2 + e_1$$

$$\delta_2^2 = \left(x - x_2\right)^2 + \left(y - y_2\right)^2 + e_2$$

$$\delta_3^2 = \left(x - x_3\right)^2 + \left(y - y_3\right)^2 + e_3$$

During the past performance evaluation, the coordinates calculated by trilaterating the three nodes closest to the tag transmitting location in a multi-sensor network environment had been considered as the best ones. Accordingly, we perform a similar trilateration by selecting the three strongest signals after comparing the received RSSI values.

As shown in Figure 26, trilateration has been performed with the three nodes selected based on each level of signal strength. Then, using the estimated positions, draw a convex polygon. The center of mass of this polygon will be

Figure 26. Tetrahedron

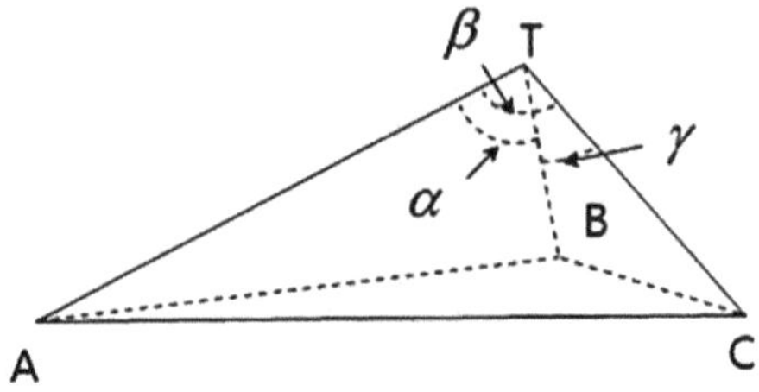

considered as the position of the target object (Tsuzuki, Areni & Yamada, 2012).

4.5.4 Trilateration by Using Tetrahedron

The reference nodes A, B, C and the target T comprise a tetrahedron as shown in Figure 26 and the equations below.

$$V = \frac{AT \cdot BT \cdot CT}{6}\sqrt{1 + \left(2 \cdot \cos\alpha \cdot \cos\beta \cdot \cos\gamma\right) - \left(\cos^2\alpha + \cos^2\beta + \cos^2\gamma\right)}$$

$$\cos\alpha = \frac{AT^2 + BT^2 - AB^2}{2 \cdot AT \cdot BT} \text{ (The second law of cosines)}$$

Here, since we know the lengths of six sides of the tetrahedron, Alpha, Beta and Gamma cosines can be also calculated. The fact that the volume V is a real number means that the value within the root has to be a positive number and this also concludes that all the absolute values of cosines are smaller than one. This case can be represented as the figure on the far right. Thus, the form of this figure can be distinguished with the volume V. It is clear that the absolute value of the volume becomes larger when the length of a certain side exhibit a large margin of error.

The author of this study (Tsuzuki, Areni & Yamada, 2012) calibrates the minimum and maximum values by separating the intersecting forms (not similar to above mentioned form) of three circles as the one whose cosine is smaller than -1 or larger than 1.

Additionally, the same author has proposed a more reliable positioning method called "Iterative Weighted Centroid Algorithm (IWCA)" which is performed as below:

Figure 27.

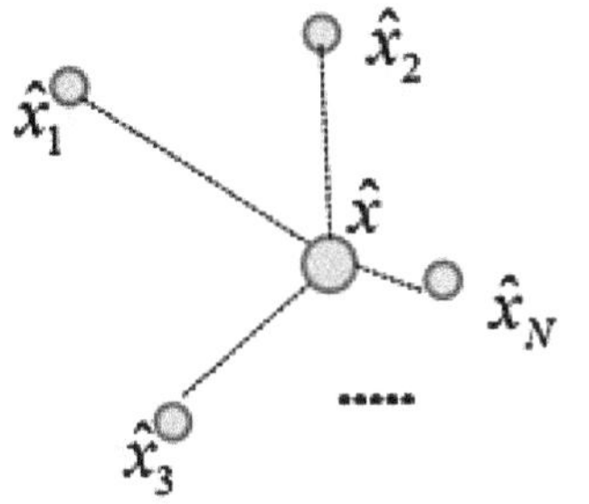

$$\hat{x} = \sum_{i=1}^{N} w_i \times \hat{x}_i \qquad w_i = \frac{(\hat{r}_i)^2}{\sum_{k=1}^{N} (\hat{r}_i)^2}$$

1. Divide N estimated distance values into the N subsets that have N-1 elements.
2. If N-1>3, divide the previously obtained subsets again to generate the new subsets that have N-2 elements. If N-1=3, estimate the position by trilateration.
3. By multiplying the position estimated with trialateration by the weighted value obtained with the estimated distance values, determine the positions of respective 4 subsets having 4 elements. Again, with the resulting values, calculate the positions of subsets having 5 elements. Repeat the process until the number of N has been reached.
4. x is the position estimated finally and Xi refers the estimated position excluding the ith node. wi is of course, the relevant weighted value (PlanetLab, n.d.).

4.5.5 Particle Filter-Based Positioning Algorithm

Kalman filter is the one most widely used for tracking moving objects. If the system assumes a linear shape, the Kalman filter is the best solution. However, if the system is non-linear, there's no optimal filters for it yet. Alternative sub-optimal filters include extended Kalman filter, multi-model filter, particle filters or the theoretically-approached filters. Although the particle filters have a demerit of requiring much more computations or having some difficulties in determining the optimal number of particles, they can provide an approximated error distribution rather precisely (GENI, n.d.).

4.6 Linking External Camera(s) With Android

In this practice, we attempt to configure a system which can monitor the indoor positions by using the cameras described in Figure 28. The camera-image transmission process is explained below.

A camera is linked to the server and the system's imaging process application stores the free-frames captured by the camera in the buffer after compressing them with RAW or JPEG while the camera is still operating. The same application also regularly receives videos from the camera and store them in the buffer. At the same time, the broadcasting server reads the buffer containing compressed images and then transmits the images to the client accessing the server. The Android-based smart phone attempts to establish a connection with the broadcasting server through the LTE or Wi-

Figure 28. The bird's eye view of the target system

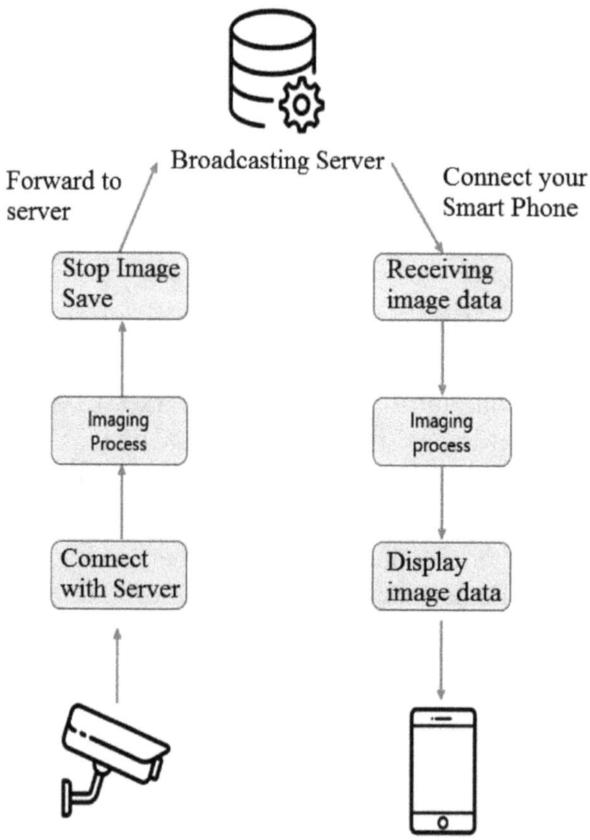

Fi network and if the server accepts client's request, the compressed images will be transmitted to the client. The image processing application embedded in an Android smart phone displays the images received from the server on a phone screen.

REFERENCES

GENI. (n.d.). *Global Environment for Network Innovations (GENI)*. Retrieved from: http://www.geni.net/

Huh, J. H., & Seo, K. (2014). PLC-based smart grid home network system design and implementation using OPNET simulation. *Journal of Multimedia and Information Systems*, *1*(2), 113–117.

PlanetLab. (n.d.). Retrieved from: http://www.planet-lab.org/

Tsuzuki, S., Areni, S. A., & Yamada, Y. (2012). A feasibility study of 1gbps plc system assuming a high-balanced DC power-line channel. *2012 IEEE International Symposium on Power Line Communications and Its Applications*, 386-391. doi:10.1109/ISPLC.2012.6201334

Xiao-Sheng, L., Liang, Z., Yan, Z., & Dianguo, X. (2011). Performance analysis of power line communication network model based on spider web. *2011 IEEE 8th International Conference on Power Electronics ECCE Asia*, 953-959.

Yao, W. M. (2013). *Increasing scalability in network simulation and testbed experiments* (Ph.D thesis). Purdue University, West Lafayette, IN.

Appendix 2

1. INTRODUCTION

The government of the Republic of Korea (ROK) started a large-scale Smart Grid Project in Jeju Island focusing on the 5 technical areas to prepare for the introduction of Smart Grid System. The goal of the Smart Grid Test Bed in the Island was to find the optimal usage of smart grid technology and develop additional necessary technologies by testing the available smart grid system.

The preparatory works were carried in 2010 and actual operation in a demonstration site began in 2011. The tested areas were Smart Power Grid (SPG), Smart Place (SP), Smart Transportation (ST), Smart Renewable (SR), and Smart Electric Service (SES). Jeju Island offers: superior new and renewable energy resources (Total power generation: 29,9993MWh); power supply capability of 800MW; peak power consumption of 624MW; 284,909 consumers; 103 power cable routes ; line length of 8,162 C-km; HDVC link with land power system (150MW x 2); and independent power system network, so that it was the optimal place for testing. Moreover, the fact that diverse type of smart grids could be implemented on the island, and that the island had an advantage of being able to respond quickly to the legal and institutional problems that might arise during or after the establishment as the island was designated as a special self-governing province justified the attraction of Smart Grid Test Bed.

2. SMART POWER GRID (SPG)

The efficiency of the Smart Power Grid was tested. That is, how efficiently and intelligently the grid can be operated by integrating the power devices used in the power transmission, transformation and distribution system with the operating system was tested. Currently, the test bed includes the reactive power management system, satellite network-based voltage-phase management system, intelligent power transmission system that deals with an active telemetrics that uses power line ball sensors, digital substation adopting IED (IEC61850), and the distribution system with an intelligent power quality monitoring terminal and integrated SCADA function.

2.1 System Configuration and Installation Status

The SPG management system has been referred as the Power Grid Operation & Management System (PGOMS) and the power transmission, substation automation, and intelligent distribution systems will be integrated with this PGOMS afterwards along with the application function that will be used for the operation of the grid. Here the integration means to design an integrated form of individually configured HMIs and manage each DB by integrating them. The SP, SR and ST operation systems will be connected to the Total Operation Center (TOC) to provide operation data and PGOMS will share the data with other smart grid systems through this TOC to operate the power grid efficiently. The PGOMS-based smart power grid is then linked with the E-service to facilitate Smart Renewable, Smart Place, Smart Transportation systems.

2.2 Intelligent Power Transmission

The intelligent power transmission includes: a reactive power management system that stabilizes the power grid's voltage by controlling the power output of 3 power stations as well as the voltage controllers and reactive power devices of 10 substations; an active telemetrics system that remotely measures the information such as currents, temperatures and wind speeds by installing ball sensors to control current levels based on the information. The reactive power management system allows stable maintenance of a reference bus voltage by controlling the reactive power output of 3 power stations by actively responding to the fluctuations in the bus voltage set at the power station. The voltage-phase management system compares the reference voltage-phase with the voltage-phases of 10 substations simultaneously and determines whether the system is stable or not.

2.3 Digital Power Substation

The digital power substation system is an automated substation operation system and includes: a system that converts protocols between IED, DNP3.0 and IEC61850, all of which are used for surveillance, controlling, protection and interlocking of power facilities within the substation; the gateways for interface between KEPCO SCADA RTUs; and the HMI for various data monitoring and controlling activities. The targeted testing site was the Seongsan Substation in Jeju Island and the descriptions of test are as follows: for the HMI, system maneuvering and suspension, data surveillance and measurement, control processing, alarm/event have been verified with IEC61860 standard. For the gateway IEC61850 Client communication service test, DNP communication test, and redundancy-route test were conducted and verified. 26 IEDs including 154kV transmission line IED, bus protection IED, main transformer IED and 22.9kV distribution line IED have been installed and their performance was verified by installing them in parallel with existing power facility. As a result of verification, it was confirmed that all of them worked normally.

2.4 Intelligent Distribution

The intelligent distribution system manages the distribution efficiently by remotely monitors and controls the apparatuses like the switches on the distribution lines in real time and performs the functions such as failure management, loss minimization, load balancing, failure management of main transformer bank, power flow calculations, distribution line voltage management and selection of optimal position of automatic switch. Verified systems are the intelligent terminal and lightning currents and arrester monitors.

An intelligent terminal has functions such as voltage/current measurement, failure detection, measurement of fault current level and direction, power-quality monitoring and their measurement accuracy has increased from ±3.0% to ±1.0% compared to existing automated distribution terminals. Here, a bi-directional protection function which can determine faulty sections on the line linked to the distributed power supply by monitoring the power qualities such a sag, swell, interruption and harmonics online and comparing the directions of fault currents, has been added. The current installation status is three, 22.9kV distribution line reclosers and 14 switches. All the system components including the main and terminal apparatuses and applications have been verified.

The lightning detection and arrester condition monitoring apparatus are installed on the overhead distribution lines to measure the scale and waveform of in-flow lightning electricity and monitor the leakage current in real time. The general structure of this monitoring apparatus includes current measuring circuit, power supply, separate communications device, and storage battery charging apparatus, as well as high current and micro current detection sensors. All of these are contained in an enclosure that can be installed outdoors. The installation area was selected by considering the frequency of lightning strikes and their distribution. This apparatus is installed on the 22.9kV distribution lines to monitor lightning currents and leakage currents of arresters and send the results to the main device in real time. 20 of this system were installed and it was verified that all their functions worked fine.

3. SMART PLACE (SP)

The Smart Place technology aims to optimize energy efficiency and power facilities through Advanced Metering Infrastructure (AMI)-based demand response. Currently, various types of demand response models are being tested to the optimized way of using energies in houses and buildings to develop a model that is beneficial to both power suppliers and consumers.

3.1 System Configuration and Installation Status

The data specifying the amount of power produced and used by the appliances within households/building or electric vehicle charging are sent to the smart place operation center where the data and customer information are used to provide some additional services (e.g., failure management system, consumer/ mobile portals and network management system, etc.) With the data link between the center and TOC, the real-time pricing (RTP) is performed. AMI has been installed at 560 households among the total of 570 households along with the integrated meters (power + water) which were installed at 125 households. Additionally installed devices are 11 DCUs and 150 modems.

3.2 Power Transaction

To achieve economic both power purchase and supply, a forecasting model for power demands and trading prices has been put to the test. The past demand patterns (LP data), weather information, and the management system of predictor variables (e.g. status of resources in reduced demands and other forecasts) were used for the transaction.

3.3 Demand Response (DR)

Power consumption efficiency based on real time pricing, participation level in DR market, as well as improvement of stability through direct loads control have been verified. For the power consumption efficiency based on real time pricing, a voluntary DR was subjected to the verification by notifying a retail price decided based on the transaction price in the wholesale market to the customers while the participation level in the DR market was put to the test by selling the reduced-demand power resource at the wholesale market and distributing the profits or incentives to the participants afterwards. The improvement in the stability was verified by confirming that the system's stability was established through securing of reserve powers at the time of an emergency. Here, likewise, the incentives were provided to the consumers.

3.4 Additional Services

The Outage Management System (OMS), Consumer/Mobile portals, Network Management System (NMS) have been verified. In an instance of low-voltage-oriented failure, the power failure/outage system automatically recognizes the blackout areas and attempts to reduce the number of these areas and shorten the recovery time. The consumer/Mobile portals provide current and expected future power consumptions along with their prices, as well as the CO2 reduction information. The NMS monitors the conditions of communications in real time and analyzes the communication problem types.

4. SMART TRANSPORTATION (ST)

In the field of Smart Transportation, more EVs are running on the roads following the rapid development of vehicle batteries and charging mechanisms (e.g., rapid/moderate/home charging systems). An average battery capacity is 16kMh and the capacity and charging time of each charging device is 50kW, 7.7kW, and 3.3kW and 20~30, 4~6 hours, and 6~8 hours, respectively. Also, various additional user-convenient services like the real-time payment service, vehicle operating system, and information provision systems for location of charging stations, pricing, number of chargers, and accidents history are being offered.

4.1 System Configuration and Installation Status

The Smart Transportation System is comprised of the power supply system that supplies power to the charging station, charging stations, the CIM system for customer care and billing, and infrastructure operation/management system for operation of EVs and charging infrastructure.

4.2 Verifications

The charging infrastructure being verified is comprised of a ST operation center and charging station. In the operation center, charging station operation/management and monitoring, charging rates notification, payment/ billing information acquisition, power transaction and demand-resources management, and other additional services are carried out, while the charging station monitors operational status of chargers and obtains the information regarding the payments and billings, user-statistics. There are 32 chargers in Jeju for the testing purpose. The details of tests involve charging efficiency of power chargers (charging rate/hour), durability of chargers, management of communications and data, and emergency control capability (remote control in an event of emergency stop).

5. SMART RENEWABLE (SR)

In this field, stabilization of output of Distributed Energy Resources (DER), enhancement of power quality through improvement of power factor, power output stabilization using Energy Storage System (ESS), maneuvering/ protection/interaction of distributed power systems linked to the power system in an event of outage recovery and operation of independent distributed power have been studied and verified.

5.1 System Configuration and Installation Status

The SR system is comprised of following 5 elements: Power Conditioning System (PCS) that controls power status; Battery Management System (BMS) that controls battery charging and discharging operations; Static Synchronous Compensator (STATCOM) which compensates the power quality; Energy management System (EMS) that controls distributed power supplies based on the operational information; and Distributed Resource Interconnection Management System (DRIMS) that monitors and controls the changes in voltage levels of linked distributed power supplies.

5.2 Verifications

Surveillance through EMS, control function; forecasting function of long and short-term power generation volume through the wind power generation prediction system and provision function of transaction history; controlling and monitoring functions for charging/discharging statuses through ESS; adjustment function of power output levels for battery charging/discharging; and control function of power factor and voltage using STATCOM have been verified.

6. SMART ELECTRIC SERVICE (SES)

The target subjects for verification in this field include Real-Time Pricing (RTP), exchange of operational information between PG, SP, ST, SR and SES using the Common Information Model (CIM) and construction of an integrated platform, and establishment of TOC. The role of TOC is monitoring of operational information of smart grid's elements and their power uses, and the power transaction activities including real-time meter rates.

6.1 System Configuration and Installation Status

4 verification projects against SPG, SP, ST and SR are associated to the TOC and the other verifications required for these projects will go through TOC as well. For instance, when the data regarding the volume of power generated with the new and renewable resources are required by SPG to operate the

power system, SR will send the information to the TOC and they deliver the same to SPG. For such a datalink, an integrated platform which utilizes CIM (IEC61968, 61970) has been constructed. The customer service portal that provides information regarding power usages, real-time rates, and test bed operation status is established along with the integrated security control system.

6.2 Verification

The 5-minute-cycle real time market tests are in progress. It is the shortest-term market pricing in the world. The forecasting program used for verification performs short-term forecastings (6 hours later, 10-minute renewal with an interval of 5 minutes) and middle-term forecastings (48 hours later, 1-hour renewal with an interval of 1 hour) and aims to provide the forecasting under the error rate of 20% in a 24-hour forecasting period. This program will be used as a supporting program for the market participants' early (1 day) biddings, reserve power production planning by the TOC, and real-time emergency supplying of powers. Also, a variety of demand-side distributed power supplies (e.g., EVs, heat-pumps, smart appliances, photovoltaic, wind power, fuel cells, batteries, and DR resources) will be connected to a power network so that the TOC will be able to establish a foundation for the stable power operation in an ever complexifying power systems through cutting-edge technologies while improving the energy efficiency in national-level and reducing the carbon emissions by promoting competitiveness in the power markets by attracting active market participation of demand-side distributed power supplies. For this, the testing process is in progress for the task of finding the best way to establish connections between SPG, SP, ST, SR, and SES through TOC. The information provision function for market information, CO2 emission information, and blackout information is also being under verification process.

Appendix 3

1. APP DEVELOPER

Currently, I often encounter the incidents involving a contract for app development. There are cases that seek my advice if there are any special legal issues before entering into an agreement but many more cases are the ones that asking me to review the dispute between contractor (hereafter 'client') and app developer (hereafter 'developer') and make judgment on whose argument is more valid. Therefore, while I deal with such cases, I have arranged, in my own way, a list of precautions to be considered when finalizing the contract.

1.1 You Must Draw Up the Contract

It is very surprising to know that there are significant number of cases where neither the client nor the developer prepares a contract sheet when they are discussing the development of an app.

There are also many cases where the client who is requesting an app development to a certain developer asks about how much would it cost and until when will it be completed and just pay for it if the conditions are met.

Clients are most likely to make such a request to the developers introduced by their acquaintances so that they are often reluctant to draw up a contract as they think it would be disrespectful to do so, or they do not wish to give an impression that they are preparing for the troubles.

Since verbal contracts are valid under the current law, no particular problems will come up as long as the app has been developed to the client's satisfaction and the payment was paid in due time. However, if a dispute arises between the two without any written contract, both parties could suffer from financial or time losses.

Thus, what I would like to advise first is that you must prepare a written contract even if you have entered a contract with a close developer or with minimal amount of money. This is absolutely necessary to avoid any unnecessary disputes.

1.2 Enter Your Requirements for the App to be Developed

First, as a client, what is the most important thing when making a contract? Of course the answer is getting a superb app that meets his/her expectations. If so, the client should pay his/her attention when preparing a contract:

First, if there are any specific requirements or specifications for the commissioned app, they must include in the contract. The most serious conflict between client and developer is associated with the degree of completion. It will be possible that the developer would think that he has performed all his obligation with the level of work he/she achieved, whereas the client makes a judgment that the developer's work does not satisfy his/her expectations. For this reason, if there are elements like menus functions that must be included in the contract, they must be specified in the contract in details to minimize possible disputes.

Second, you must determine the scope of development and specify it in the contract.

For example, make clear that the app is either for iOS or for Android, or both. In some cases, you can limit the scope by mentioning a certain device(s) or a platform(s).

Third, you have to enter an exact date or period of completion for the developer to keep. It is advised to include indemnity clause(s) should the development has failed or was incomplete as it is not easy to prove the damages caused by the delays in development. For instance, you will be able to put in a claim for the damages without proving them if you specify a specific amount (e.g., "Monthly 000 won or 00% of the development cost when the development has been delayed").

Finally, the terms related with maintenance, repairs or upgrades must be included. This is a common and most frequent problem when OS has been upgraded or a certain function(s) which was not anticipated at the time of request needs to be added. Especially, for the gaming apps, there are many cases where items, events and quests are needed to be added. Unless you are planning to entrust maintenance, repair and upgrading works to the other agent or company, you need to specify such terms on the contract with a specific mandatory period(s) and the cost(s) involved.

1.3 You Must Specify the Payment Terms for Development Cost

Getting a payment is the most important matter of concern for the developers. For this, you should pay your attention when preparing the contract:

First, the deadline of payment. That is, specify until when the payment(s) must be made. Usually, a specific sum of money is paid when the agreement has been made and the rest of the amount will be paid when developed app has pass the review and formally registered on App Store or Play Store.

For your information, currently there are instances of receiving the additional payments proportional to the sales amount of developed app. If this is the case, the followings should be considered as well: You need to specify whether the 'sales amount' includes the sales achieved through advertisement or by the paid items, in addition to the sales of app itself, and the same is calculated after deducting commissions that should be paid to Apple or Google or otherwise.

Second, the method of payment. The most reliable method of payment is to specify that the payment will be considered to be completed when it has been remitted to the developer's account designated on the contract.

2. THINGS THAT THE MANAGING DIRECTORS SHOULD AVOID DOING AT THE SW COMPANIES

2.1 The Interview With the CEO of Ewoosoft, Republic of Korea, and What It Means

The author would like to introduce Ewoosoft operating in the ROK briefly. This company is an affiliated company of Vatech Networks specializing in dental imaging. Ewoosoft is responsible for the development of softwares for the dental clinics, especially X-ray imaging-based diagnostic softwares.

As these softwares are being used by most of the dentist, those who are subscribing ZDnet would have consulted with their doctors based on the results obtained by Ewoosoft's diagnostic software after completing X-ray imaging.

The reason for this explanation is to add some sort of reality to the software development culture which will be introduced by the author. Surely it is possible for you to feel some distance in this introduction as the corporate cultures can be quite different. A certain culture can be regarded as strange and unreal to those who has not experienced it. However, in our company, Ewoosoft's corporate culture is considered as a natural and necessary major element required to compete with global software companies.

First, ignoring of developer's estimation of development period. This is happening at many companies so that such a conduct cannot be unilaterally regarded as a managing director's fault.

There are many similar cases. For example, it is common that the managers often compel developers to finish their development much earlier than the desirable deadline giving a variety of reasons. They do listen to developer's argument and tell them to advance the schedule no matter what based on groundless expectation that the schedule could be delayed. Some of the managers think that if they do not push hard, developers will not work hard and even try to avoid necessary overtimes.

Although some projects may succeed or the company could profit from shortened development period, eventually everyone will lose out if such a practice settles in, not to mention regression of company's development culture. If company owners attempt to shorten the schedule by force, the developers might lie about their expected period of development needed, extending it longer.

On the other hand, sometime the developers present the development period without any convincing grounds so that the managers ignores the presented schedule and just try to shorten it. No one can be a winner in this situation. Managers will fail the project as they will not be able to make a business forecast while the developers suffer from endless overtimes.

To avoid such a situation, the developers should present the schedule based on a detailed analysis of required specification and other requirements from the clients. Then, the managers should trust developers by acknowledging the expected schedule so that the developer will try to set up an exact schedule. If a manger needs to shorten the period, he/she should use a reasonable method to do so. Asking the developers to work overtime at night can be one of the methods but customary overtimes usually ends with more losses. The reasonable means of reducing the development time include reduction of the number of functions and focusing on a key function, establishment of phased development schedule, use of expert consultants and partial utilization of already commercialized modules.

It takes much time for such a trusted relationship to be established, regardless of development methodology but everyone should know that it is the best way to finish the project.

Second, changing the agreed requirements following the manager's preferences. There are many Korean companies where managing directors have a powerful authority that they can overturn anything they want. They can change the shape of an impending product or change colors to their liking. It is common in the field of software designing.

The role of a manager is different in each project but it is quite certain that he/she is not an autocrat, just a stakeholder. In most projects, the role of a manager is to establish an overall vision or strategy. Bill Gates deeply involved himself in his early projects but he was just assuming the role of Chief Architect.

In the project, the managing director should deliver his own vision(s) and strategic requirements, especially in early stages as it is very difficult and not adequate to change them later. If the changes should be made, they must be made based on a reasonable ground.

The manager making an arbitrary decisions getting involved in the latter part of development process is a taboo. It could completely destroy the product architecture while discouraging developers and opening a crack between them. In Ewoosoft, the managing director must undergo the Change Control Process if he/she wants change requirements. Nothing will be allowed if the Change Control Board refuses the requested change(s).

This way they will do their best to perform their roles properly and do not attempt to change anything in later stages.

Third, giving instructions or having a discussion with a developer(s) any time they wish. In author's company, no one bothers developers by having a unscheduled meeting or discussion, including CEO. Distracting developers in the middle of their work often leads to unwanted losses. According to Peopleware, one incident of such a conduct will results in 30 minutes of delay. You can imagine what will happen when this has been repeated 3 or 4 times a day.

In order to have a meeting with the developer, one has to make an appointment a few hours prior or fix the time a day before. If it is urgent, we carefully check whether he/she is concentrating on his work or not.

For this reason, messengers ar forbidden in our company, even the Kakao Talk function must be set to silent mode to avoid any distractions. Of course, calls are seldom made to the developers. Such a rule can be very helpful in reducing overtimes as they can focus on the work in their business hours.

Fourth, Asking for a report frequently. The companies with a good information sharing culture document almost of all progress information and share them online so that the managing director can monitor the progress in real time. If it is not a special case, he/she does not ask developers to make a new report. Even though reports may save some of managers' time, employees have to consume away hundreds of hours to make them. The the progress is to be checked by the reports only, they can be stretched in both volume and contents. It is better to see as much information as possible directly and ask for a report when it is absolutely necessary. To achieve this, a proper sharing culture must be settled in.

So far, four things that the managing directors, CEOs, and managers should not do have been introduced. Then, do they not have much things to do? They should be concentrating on establishing the company visions, goals and strategies, in addition to making company rules and appropriate cultures while recruiting talented workers and continuously training the employees for to upgrade their skills. The author is a CEO, architect, and an expert of software internationalization so that he involves himself directly with the issues pertaining Software Engineering, architecture and internationalization operation. He participates in discussions and presents his opinions but only as a stakeholder as mentioned earlier but never tries to put the heat on any things that have been agreed upon. If he wants, he will have to follow the procedure.

Those companies that follow above mentioned 4 Not-to-Do's will be able to compete with global software companies. Individual skills of developers alone will not bring about competitiveness. The development culture is what makes a software company strong. Such a culture cannot be created right away through giving free lunches, allowing free attire, introducing some novel engineering methodologies. The key to change hearts and habits of company members, and as this is a rather difficult task, the head of an organization should set an example by changing himself/herself.

Appendix 4

JAVA SOURCE CODE

```
//////////////
CIAgentState.java
//////////////
package ciagent;
import java.io.*;
public class CIAgentState implements Serializable{
         public static final int UNINITIATED = 0;
         public static final int INITIATED = 1;
         public static final int ACTIVE = 2;
         public static final int SUSPENDED = 3;
         public static final int UNKNOWN = 4;
         private int state
```

/**

* Creates a <code>CIAgentState</code> object.

*/

```
         public CIAgentState() {
                  state = UNINITIATED



         /**
          * Sets the state.
          *
          * @param state the state of an agent
          */
         public synchronized void setState(int state) {
                  this.state = state;
```

/**
* Retrieves the state.
*
* @return an integer that represents the state
*/

```
        public int getState() {
                return state
```

/**
* Coverts the state to a printable format.
*
* @return a formatted String that represents the state
*/

```
        public String toString() {
                switch (state) {
                        case UNINITIATED: {
                                return "Uninitiated"

                        case INITIATED: {
                                return "Initiated"

                        case ACTIVE: {
                return "Active"

                        case SUSPENDED: {
                return "Suspended"

                        case UNKNOWN: {
                return "Unknown"

         return "Unknown"
         }
AgentPlatform.java
```

```
//////////////
package ciagent;
import java.util.*;
public interface AgentPlatform {
```

/**

* Returns a list of registered agents.

*

* @return the Vector object that contains the agents

*/

```
          public Vector getAgents();
```

/**

* Returns an agent with the specified name.

*

* @return the CIAgent object or null if not found

*/

```
          public CIAgent getAgent(String name);
CIAgent.java
////////////////////////
package ciagent;
import java.util.*;
import java.awt.*;
//import javax.swing.*;
import java.beans.*;
import java.io.*;
import android.util.Log;
import android.widget.TextView;
import android.os.Handler
import android.os.Looper
import android.os.HandlerThread
import android.os.Message
public abstract class CIAgent implements CIAgentEventListener,
Serializable {
          public static final int DEFAULT_SLEEPTIME = 15000; //
15 seconds
          public static final int DEFAULT_ASYNCTIME = 1000; //
1 second
          protected String name // the agent's name
          private CIAgentState state = new CIAgentState();
          private CIAgentTimer timer = new CIAgentTimer(this);
          transient private Vector listeners = new Vector(); //
list of listeners
```

```
          transient private CIAgentEventQueue eventQueue = new
CIAgentEventQueue();
          transient private PropertyChangeSupport changes = new
PropertyChangeSupport(this);
          private boolean traceOn = false
          protected int traceLevel = 0;
          protected AgentPlatform agentPlatform = null // agent
platform reference
          protected Vector children = new Vector() ; //
contained agents (if any)
          protected CIAgent parent = null ; // container agent
(if any)



          TextView textView


  /**
  * Creates a <code>CIAgent</code> instance.
  */

           public CIAgent() {
           this("CIAgent");
           }



  /**
  * Creates a CIAgent with specified name.
  *
  * @param name the String name given to this agent
  */

           public CIAgent(String name) {
           this.name = name;
           timer.setAsyncTime(DEFAULT_ASYNCTIME); // set
asynchTime
           timer.setSleepTime(DEFAULT_SLEEPTIME); // set
sleepTime
           state.setState(CIAgentState.UNINITIATED);
           }
```

```
/**
* Sets the name of this agent.
*
* @param newName the String that contains the new name of this agent
*/

        public void setName(String newName) {
        String oldName = name
        name = newName;
        changes.firePropertyChange("name", oldName, name);
        }

/**
* Retrieves the name of this agent.
*
* @return the String that contains the name of this agent
*/

        public String getName() {
        return name
        }

        // working with text view
        public void SetTextView (TextView textView) {
                this.textView = textView;
        }

        public void SetText (final String strText) {
                if (strText == null) return
                if (textView == null) return

                textView.post(new Runnable() { @Override
public void run() { textView.append(strText); } });
        }

/**
```

* Sets the current state of this agent (UNINITIATED, INITIATED, ACTIVE,
* SUSPENDED, UNKNOWN).
*
* @param state the new state of this agent
*
* @param newState the int object
*
*/

```
protected void setState(int newState) {
int oldState = state.getState();
changes.firePropertyChange("state", oldState,
newState);
this.state.setState(newState);
}
```

/**
* Retrieves the current state of this agent (UNINITIATED, INITIATED, ACTIVE,
* SUSPENDED, UNKNOWN).
*
* @return the state of this agent
*/

```
public CIAgentState getState() {
return state
}
```

/**
* Sets the agent sleep time (in milliseconds) for autonomous processing. Note
* that this has no effect on a running agent.
*
* @param sleepTime the sleep time (in milliseconds) for this agent
*/

```
public void setSleepTime(int sleepTime) {
        Log.d ("Debug", "CIAgent setSleepTime");
timer.setSleepTime(sleepTime);
}
```

/**
* Retrieves the agent sleep time (in milliseconds).
*
* @return the sleep time (in milliseconds) for this agent
*/

```
public int getSleepTime() {
return timer.getSleepTime();
}
```

/**
* Sets the time (in milliseconds) for asynchronous event processing. Note
that
* this has no effect on a running agent.
*
* @param asyncTime the time (in milliseconds) for asynchronous event
processing
*/

```
public void setAsyncTime(int asyncTime) {
timer.setAsyncTime(asyncTime);
}
```

/**
* Retrieves the time (in milliseconds) for asynchronous event processing.
*
* @return the time (in milliseconds) for asynchronous event processing
*/

```
public int getAsyncTime() {
return timer.getAsyncTime();
}
```

/**
* Sets the trace level.
*
* @param traceLevel the integer that represents the trace level
*/

```
public void setTraceLevel(int traceLevel) {
this.traceLevel = traceLevel;
}
```

/**
* Retrieves the trace level.
*
* @return the trace level of this agent
*/

```
public int getTraceLevel() {
return traceLevel
}
```

/**
* Sets the agent platform for this agent.
*
* @param agentPlatform the AgentPlatform object for this agent
*/

```
public void setAgentPlatform(AgentPlatform
agentPlatform) {
this.agentPlatform = agentPlatform;
}
```

/**
* Retrieves the agent platform.
*
* @return the agent platform of this agent
*/

```
public AgentPlatform getAgentPlatform() {
return agentPlatform
}
```

/**
* Retrieves the other agents running on the same agent platform
*
* @return a vector of agents running on this platform, null if no platform
*/

```
public Vector getAgents() {
if (agentPlatform == null) {
return null
} else {
return agentPlatform.getAgents();
}
}
```

/**
* Retrieves an agent running on the same agent platform
*
* @return a CIAgnet running on this platform, null if no match
*/

```
public CIAgent getAgent(String name) {
if (agentPlatform == null) {
return null
} else {
return agentPlatform.getAgent(name);
}
}
```

/**
* Retrieves a formatted string for display of this agent's current task.
*
* @return the String that represents the current task
*/

```
public abstract String getTaskDescription();
```

/**
* Retrieves a list of CIAgent objects contained by this agent.
*
* @return a Vector of CIAgent objects contained by this agent
*/

```
public Vector getChildren() {
return (Vector)children.clone() ;
}
```

/**
* Sets the parent CIAgent for this agent.

*
* @param parent the CIAgent object that is the parent of this agent
*/

```
public void setParent(CIAgent parent) {
this.parent = parent ;
}
```

/**
* Retrieves the parent CIAgent object.
*
* @return a CIAgent object or null if no parent is defined
*/

```
public CIAgent getParent() {
return parent ;
}
```

/**
* Uses introspection on this bean to get the customizer class (if any).
*
* @return the customizer class for this agent bean
*/

```
            public Class getCustomizerClass() {
            Class customizerClass = null
//            try {
//            BeanInfo beanInfo = Introspector.getBeanInfo
(this.getClass());
//            BeanDescriptor beanDescriptor = beanInfo.
getBeanDescriptor();
//
//            customizerClass = beanDescriptor.
getCustomizerClass();
//            } catch (IntrospectionException exc) {
//            System.out.println("Can't find customizer bean
property " + exc);
//            }
            return customizerClass;
            }
```

/**
* Uses introspection on this bean to get the display name (defaults to the

* class name).
*
* @return the display name for this agent bean
*/

```
            public String getDisplayName() {
            String name = null
//            try {
//            BeanInfo beanInfo = Introspector.getBeanInfo
(this.getClass());
//            BeanDescriptor beanDescriptor = beanInfo.
getBeanDescriptor();
//



//            name = (String) beanDescriptor.getValue
("DisplayName");
//            } catch (IntrospectionException exc) {
//            System.out.println("Can't find display name bean
property " + exc);
//            }
//
//            if (name == null) {
//            name = this.getClass().getName(); // default to
class name
//            }
            return name;
            }
            /**
            * Resets this agent so that it is in a known state.
            */
            public void reset() {}
```

/**
* Initializes this agent for processing.
*/

```
            public abstract void initialize();
```

/**
* Starts the agent timer and asynchronous event processing and sets the agent
* state to ACTIVE.

*/

```
public synchronized void startAgentProcessing() {
timer.startTimer();
setState(CIAgentState.ACTIVE);
}
```

/**
* Stops the agent timer and asynchronous event processing and sets the agent
* state to UNKNOWN.
*/

```
public synchronized void stopAgentProcessing() {
// stop processing async events and timer events
// change agent state to unknown
timer.quitTimer();
setState(CIAgentState.UNKNOWN);
}
```

/**
* Temporarily stops the agent timer so that the autonomous behavior is
* suspended and sets the agent state to SUSPENDED.
*/

```
public void suspendAgentProcessing() {
// turn asynch event processing off
// turn timer event processing off
timer.stopTimer();
setState(CIAgentState.SUSPENDED);
}
```

/**
* Resumes agent processing of the timer and asynchronous events and sets the
* agent state to ACTIVE.
*/

```
public void resumeAgentProcessing() {
// turn asynch event processing on
// turn timer event processing on
timer.startTimer();
```

```
        setState(CIAgentState.ACTIVE);
        }
```

/**
* Provides the synchronous processing done by this agent.
*/

```
        public abstract void process();
```

/**
* Provides the asynchronous, autonomous behavior of this agent that
* occurs periodically, depending on the sleep time for this agent.
*/

```
        public abstract void processTimerPop();
```

/**
* Processes all events on the asynchronous event queue periodically,
* depending on the asynchronous event time for this agent.
*/

```
        public void processAsynchronousEvents() {
        CIAgentEvent event = null
        while ((event = eventQueue.getNextEvent()) != null)
{
        // System.out.println("CIAgent: " + name + "
dispatching an Async event");
                Log.d ("Debug", "CIAgent
processAsynchronouseEvent");
        processCIAgentEvent(event);
        }
        }
```

/**
* Performs synchronous event processing for this agent.
*
* @param event the CIAgentEvent object
*/

```
        public void processCIAgentEvent (CIAgentEvent event)
{
        }
```

/**
* Posts an event to this agent's event queue.
*
* @param event the CIAgentEvent to be posted
*/

```
        public void postCIAgentEvent (CIAgentEvent event) {
        eventQueue.addEvent(event);
        }
```

/**
* Adds a listener for CIAgent events.
*
* @param listener the CIAgentEventListener to be added
*/

```
        public synchronized void addCIAgentEventListener
(CIAgentEventListener listener) {




//                    Log.d ("Debug", "CIAgent
addCIAgentEventListener");
        listeners.addElement(listener)
        }
```

/**
* Removes a listener for CIAgent events.
*
* @param listener the CIAgentEventListener to be removed
*/

```
        public synchronized void removeCIAgentEventListener
(CIAgentEventListener listener) {
        listeners.removeElement(listener);
        }
```

/**
* Delivers the CIAgent event to all registered listeners.
*

* @param e the CIAgentEvent to be sent to all listeners
*/

```
		protected void notifyCIAgentEventListeners
(CIAgentEvent e) {
		Vector l;
//		  Log.d ("Debug", "CIAgent
notifyCIAgentEventListeners");
		synchronized (this) {
		l = (Vector) listeners.clone();
		}
		for (int i = 0; i < l.size(); i++) { // deliver the
event
//				Log.d ("Debug", "CIAgent
notifyCIAgentEventListeners clone");
		((CIAgentEventListener) l.elementAt(i)).
processCIAgentEvent(e);
		}
		l.clear();
		}
```

/**
* Adds a listener for PropertyChange events.
*
* @param listener the PropertyChangeListener to be added
*/

```
		public synchronized void addPropertyChangeListener
(PropertyChangeListener listener) {
		changes.addPropertyChangeListener(listener);
		}
```

/**
* Removes a listener for PropeortyChange events.
*
* @param listener the PropertyChangeListener to be removed
*/

```
		public synchronized void
removePropertyChangeListener (PropertyChangeListener listener)
{
		changes.removePropertyChangeListener(listener);
		}
```

/**
* Sends a trace event to all registered listeners.
*
* @param msg the String that is the message portion of the trace event
*/

```
public void trace(String msg) {
// create a data event
CIAgentEvent event = new CIAgentEvent(this, "trace",
msg);
// and send it to any registered listeners
notifyCIAgentEventListeners(event);

//
}
```

/**
* Adds a CIAgent to this agent and sets its parent member to point to
* this agent.
*
* @param child the CIAgent object to be added to this object as a child
*/

```
public void addAgent(CIAgent child) {
children.addElement(child) ;
child.setParent(this) ;
}
```

/**
* Removes a CIAgent from this agent.
*
* @param child the CIAgent object to be removed from this agent
*
*/

```
public void removeAgent(CIAgent child) {
children.removeElement(child) ;
}
```

/**
* De-serialize the object from the specified input stream by
* re-initializing the object's transient variables, de-serializing

* the object with defaultReadObject(), and then hooking up the
* de-serialized stuff to the re-initialized stuff.
*
*
* @param ois the ObjectInputStream object from which this object
* is to be read
*
* @exception ClassNotFoundException if any class file is not found
* @exception IOException on any IO error
*/

```
          private void readObject (ObjectInputStream
theObjectInputStream) throws ClassNotFoundException,
IOException {
          // Restore the transient variables.
          changes = new PropertyChangeSupport(this);
          listeners = new Vector();
          eventQueue = new CIAgentEventQueue() ;
          // Restore the rest of the object.
          theObjectInputStream.defaultReadObject();
          }
CIAgentEvent.java
////////////////////////
package ciagent;
import android.util.Log;
public class CIAgentEvent extends java.util.EventObject {
         private Object argObject = null
          private String action = null
          private static final long serialVersionUID = 1L;
```

/**
* Creates an extremely simple <code>CIAgentEvent</code> object.
*
* @param source the Object that is a source of this event
*/

```
          public CIAgentEvent(Object source) {
          super(source);
          }
```

/**
* Creates a more complex <code>CIAgentEvent</code> object that includes an

```
* object related to the event.
*
* @param source the Object that is a source of this event
* @param argObject the Object that is used to supply additional data about
* this event
*/

          public CIAgentEvent(Object source, Object argObject)
{
          this(source);
          this.argObject = argObject;
          }

/**
* Creates a more complex <code>CIAgentEvent</code> object that
includes an
* object related to the event and an action.
*
* @param source the Object that is a source of this event
* @param action the String that represents the action for this event
* @param argObject the Object that is used to supply additional data about
* this event
*/

          public CIAgentEvent(Object source, String action,
Object argObject) {
          this(source);
          this.action = action;
          this.argObject = argObject;
          }

/**
* Retrieves the object related to this event.
*
* @return the argument Object (if any)
*/

          public Object getArgObject() {
          return argObject
          }

/**
```

* Retrieves the action related to this event.
*
* @return the action (if any)
*/

```
        public String getAction() {
        return action
        }
```

/**
* Converts the event to a string formatted for display or printing.
*
* @return the formatted <code>String</code> object
*/

```
        public String toString() {
                Log.d("Debug", "CIAgentEvent toString()");
        StringBuffer buf = new StringBuffer();
        buf.append("CIAgent ");
        buf.append("source: " + source);
        buf.append("action: " + action);
        buf.append("argObject: " + argObject);
        return buf.toString();
        }
CIAgentEventListener.java
///////////////////
package ciagent;
public interface CIAgentEventListener extends java.util.
EventListener
```

/**
* Processes the event.
*
* @param e the CIAgentEvent to be processed
*/

```
        public void processCIAgentEvent(CIAgentEvent e);
```

/**
* Adds the event to the asynchronous event queue.
*
* @param e the event to be added to the event queue

*/

```
            public void postCIAgentEvent(CIAgentEvent e);
CIAgentEventQueue.java
///////////////////
package ciagent;
import java.util.*;
import java.io.*;
public class CIAgentEventQueue {
          private Vector eventQueue
```

/**
* Creates an event queue.
*/

```
            public CIAgentEventQueue() {
            eventQueue = new Vector();
            }
```

/**
* Adds an event to the end of the queue.
*
* @param event the CIAgentEvent to be added to the queue
*/

```
            public synchronized void addEvent(CIAgentEvent
event) {
            eventQueue.addElement(event)
            }
```

/**
* Retrieves the first element from the queue (if any) after removing it from
* the queue.
*
* @return the first event on the queue
*/

```
            public synchronized CIAgentEvent getNextEvent() {
            if (eventQueue.size() == 0) {
            return null
            } else {
            CIAgentEvent event = (CIAgentEvent) eventQueue.
elementAt(0);
```

```
eventQueue.removeElementAt(0);
return event;
}
}
```

/**

* Retrieves the first element from the queue (if any) without removing it.

*

* @return the first event on the queue

*/

```
public synchronized CIAgentEvent peekEvent() {
if (eventQueue.size() == 0) {
return null
} else {
return (CIAgentEvent) eventQueue.elementAt(0);
}
}
CIAgentTimer.java
///////////////////
package ciagent;
import java.util.*;
import java.io.*;
public class CIAgentTimer implements Runnable, Serializable{
        private CIAgent agent // owner agent
        private int sleepTime = 1000; // mSec - default to 1
second
        private boolean timerEnabled = true
        private int asyncTime = 500; // mSec - default to 0.5
seconds
        transient private Thread runnit = new Thread(this);
        private boolean quit = false // used to exit thread
run() method
        private boolean debug = false // turn on to see timer
vars
```

/**

* Creates a timer, specifying the agent that owns the timer.

*

* @param agent the CIAgent that owns this timer object

*

*/

```
        public CIAgentTimer(CIAgent agent) {
                this.agent = agent;
```

```
/**
 * Sets the time (in milliseconds) that determines how often an agent performs
 * the autonomous behavior that is defined in its <code>processTimerPop</
code>
 * method.
 *
 * @param sleepTime the amount of sleep time in milliseconds
 */

        public void setSleepTime(int sleepTime) {
                this.sleepTime = sleepTime;


/**
 * Retrieves the sleep time (in milliseconds).
 *
 * @return the amount of sleep time in milliseconds
 */

        public int getSleepTime() {
         return sleepTime

/**
 * Sets the time (in milliseconds) that determines how often the agent
processes
 * asynchronous events on its event queue.
 *
 * @param asyncTime the interval time in milliseconds
 */


        public void setAsyncTime(int asyncTime) {
         this.asyncTime = asyncTime;

/**
 * Retrieves the asynchronous interval time (in milliseconds).
 *
 * @return the asychronous interval time in milliseconds
 */
```

```
public int getAsyncTime() {
return asyncTime
}
```

/**

* Starts the timer thread.

*/

```
public void startTimer() {
timerEnabled = true
if (!runnit.isAlive()) {
runnit.start(); // start on first time only
}
}
```

/**

* Indicates that timer events events should not be fired.

*/

```
public void stopTimer() {
timerEnabled = false
}
```

/**

* Indicates that the timer thread should be ended.

*/

```
public void quitTimer() {
quit = true
}
```

/**

* Processes the asynchronous events and autonomous timer events periodically,

* with the interval based on the sleep time and/or the asynchronous event time.

*/

```
public void run() {
// NOTE: sleepTime and asyncTime must be set before
// startTimerEvents() is called
long startTime = 0;
long curTime = 0;
```

```
            if (debug) {
            startTime = new Date().getTime(); // get start time
in mSec
            curTime = startTime; // current time in mSec
            }
            if (sleepTime < asyncTime) {
            asyncTime = sleepTime
            }
            int numEventChecks = sleepTime / asyncTime
            if (debug) {
            System.out.println("sleepTime= " + sleepTime
+ " asyncTime= " + asyncTime + "numEventChecks= " +
numEventChecks);
            }
            while (quit == false) {
            try {
            for (int i = 0; i < numEventChecks; i++) {
            Thread.sleep(asyncTime);
            if (debug) {
            curTime = new Date().getTime(); // current time in
mSec
            System.out.println("async events timer at " +
(curTime - startTime));
            }
            if (quit) {
            break // exit loop
            }
            agent.processAsynchronousEvents();
            }
            if (timerEnabled && (quit == false)) {
            if (debug) {
            curTime = new Date().getTime(); // current time in
mSec
            System.out.println("timer event at " + (curTime -
startTime));
            }
            agent.processTimerPop(); // call the timer
processing method
            }
            } catch (InterruptedException e) {}
            }
            }
```

```
/**
* De-serialize the object from the specified input stream by
* re-initializing the object's transient variables, de-serializing
* the object with defaultReadObject(), and then hooking up the
```

```
* de-serialized stuff to the re-initialized stuff.
*
*
* @param theObjectInputStream from which this object is to be read
* <p>
* @exception ClassNotFoundException
* if any class file is not found
* @exception IOException
* on any IO error
*/

          private void readObject (ObjectInputStream
theObjectInputStream) throws ClassNotFoundException,
IOException {
          // Restore the transient variables.
          runnit = new Thread(this);
          // Restore the rest of the object.
          theObjectInputStream.defaultReadObject();
          }
BasicNegotiation.java
///////////////////
package marketplace;
import java.util.*;
public class BasicNegotiation {
         protected Offer offer // current offer for this item
          protected long strikePrice // price where we make a
deal
          protected long lastOffer // last price we offered
          protected Offer prevOffer // previous offer for this
item
          protected int iteration // # of back and forth
offers

/**
* Creates a <code>BasicNegotiation</code> object
* used to add items to inventory or wishList.
*
* @param item the String object
* @param strikePrice the long object
*
*/
```

```
        BasicNegotiation(String item, long strikePrice) {
        offer = new Offer(item);
        this.strikePrice = strikePrice;
        lastOffer = 0;
        iteration = 0;
        }

/**
* Creates a <code>BasicNegotiation</code> object
* used for a new negotiation.
*
* @param offer the Offer object that is part of the negotiation
* @param strikePrice the long value that is the strike price in the
* negotiation
*
*/

        BasicNegotiation(Offer offer, long strikePrice) {
        this.offer = offer;
        lastOffer = 0;
        iteration = 0;
        strikePrice = strikePrice
        }

/**
* Sets the current offer to the new offer.
*
* @param newOffer the Offer object that contains the new offer
*
*/

        public void newOffer(Offer newOffer) {
        prevOffer = offer
        offer = newOffer;
        }

/**
* Retrieves the item from the offer that is part of this negotiation.
*
* @return the String object that represents the item
*
*/
```

```
            String getItem() {
            return offer.item
            }
BestBuyerAgent.java
////////////////
package marketplace;
import java.util.*;
import ciagent.*;
import rule.*;
public class BestBuyerAgent extends BuyerAgent implements
Effector{
           protected BooleanRuleBase rb = new
BooleanRuleBase("BestBuyer");
            protected RuleVariable offerDelta
            protected RuleVariable spread
            protected RuleVariable firstOffer
            protected Offer offer // the current offer
```

/**

* Creates a <code>BestBuyerAgent</code> object.

*

*/

```
            public BestBuyerAgent() {
            this("BestBuyer");
            }
```

/**

* Creates a <code>BestBuyerAgent</code> object with the given name.

*

* @param name the String object that contains the name

*

*/

```
            public BestBuyerAgent(String name) {
            super(name);
            }
```

/**

* Initializes the rule base for this buyer agent.

*

*/

```
            public void initialize() {
            initBestBuyerRuleBase();
            super.initialize(); // call base class
initialization
            }
```

/**

* Method effector used by action rules in rule base to make or accept offers

*

* @param obj the Object object that implements the effector

* @param eName the String object that contains the effector name

* @param args the String object that contains the effector arguments

*

* @return the long value that represents the status of the effector execution

*

*/

```
            public long effector(Object obj, String eName,
String args) {
            if (eName.equals("make-offer")) {
            // rule base decided to counter-offer
            long delta = (new Long(offerDelta.getValue())).
longValue();
            current.lastOffer = offer.price - delta; //
            BuySellMessage answer = new BuySellMessage("make-
offer", offer.item + " " + offer.id + " " + current.lastOffer,
offer.item, msg.sender, offer.item, name);
            CIAgentEvent e = new CIAgentEvent(this,
"processMessage", answer);
            notifyCIAgentEventListeners(e); // respond through
Facilitator
            return 0;
            }
            if (eName.equals("accept-offer")) {
            // rule base decided to accept the offer
            // just resend it to the seller
            BuySellMessage answer = new BuySellMessage ("make-
offer", msg.content, msg.replyWith, msg.sender, offer.item,
name);
            CIAgentEvent e = new CIAgentEvent(this,
"processMessage", answer);
            notifyCIAgentEventListeners(e); // respond through
Facilitator
            return 0;
            }
```

```
        return 1; // unknown effector method
        }
```

/**

* Uses the rule base and forward chaining to decide whether to accept

* an offer or make a counter offer.

*

* @param offer the Offer object that contains the current offer

* @param msg the BuySellMessage object

*

*/

```
        void negotiate(Offer offer, BuySellMessage msg) {
        // figure out if 1st time or second time or other
        // compute spread
        // let rule base figure out our response
        // if deltaoffer is 0 then we should accept
        rb.reset(); // allow all rules to fire, clear vars
        this.offer = offer; // make visible to rule base
effectors
        if (offer.price >= current.strikePrice) {
        long delta = offer.price - current.strikePrice
        if (delta < 25) {
        spread.setValue("<25");
        } else if (delta < 50) {
        spread.setValue("25-50");
        } else if (delta >= 50) {
        spread.setValue(">50");
        }
        } else {
        spread.setValue("0"); // meets our price, accept
        }
        if (current.prevOffer.id.equals("")) {
        firstOffer.setValue("yes");
        } else {
        firstOffer.setValue("no");
        }
        offerDelta.setValue(null);
        rb.forwardChain(); // inference
        if (offerDelta.getValue() == null) {
        trace(name + " rule base couldn't decide what to
do.\n");
        }
        }
```

/**

* Initializes the BestBuyer rule base.
*
*/

```
            public void initBestBuyerRuleBase() {
            offerDelta = new RuleVariable(rb, "offerDelta");
            offerDelta.setLabels("0 25 50 100");
            firstOffer = new RuleVariable(rb, "firstOffer");
            firstOffer.setLabels("yes no");
            spread = new RuleVariable(rb, "spread");
            spread.setLabels("<25 25-50 >50");
            // Note: at this point all variables values are NULL
            Condition cEquals = new Condition("=");
            Condition cNotEquals = new Condition("!=");
            // define rules
            Rule first = new Rule(rb, "first", new
Clause(firstOffer, cEquals, "yes"),
            new Clause(offerDelta, cEquals, "75")); // counter
            Rule second1 = new Rule(rb, "second1", new Clause[]{
new Clause(firstOffer, cEquals, "no"),
            new Clause(spread, cEquals, ">50") },
            new Clause(offerDelta, cEquals, "50")); // counter
            Rule second2 = new Rule(rb, "second2", new Clause[]{
new Clause(firstOffer, cEquals, "no"),
            new Clause(spread, cEquals, "25-50") },
            new Clause(offerDelta, cEquals, "25")); // counter
            Rule second3 = new Rule(rb, "second3", new Clause[]{
new Clause(firstOffer, cEquals, "no"),
            new Clause(spread, cEquals, "<25") },
            new Clause(offerDelta, cEquals, "0")); // accept
            // action rule
            Rule accept = new Rule(rb, "accept", new
Clause(offerDelta, cEquals, "0"),
            new EffectorClause("accept-offer", "0"));
            // action rule
            Rule counter = new Rule(rb, "counter", new
Clause(offerDelta, cNotEquals, "0"),
            new EffectorClause("make-offer", null)); // use
offerDelta
            // define this object as effector implementor
            rb.addEffector(this, "make-offer");
            rb.addEffector(this, "accept-offer");
            }
BestSellerAgent.java
////////////////
package marketplace;
import java.util.*;
import ciagent.*;
```

```
import rule.*;
public class BestSellerAgent extends SellerAgent implements
Effector{
        protected BooleanRuleBase rb = new
BooleanRuleBase(“BestSeller”);
          protected RuleVariable offerDelta
          protected RuleVariable spread
          protected RuleVariable firstOffer
          protected Offer offer // the current offer

  /**
  * Creates a <code>BestSellerAgent</code> object.
  *
  */

          public BestSellerAgent() {
          this(“BestSeller”);
          }

  /**
  * Creates a <code>BestSellerAgent</code> object with the given name.
  *
  * @param name the String object that contains the name
  *
  */

          public BestSellerAgent(String name) {
          super(name);
          }

  /**
  * Initializes the BestSeller rule base.
  *
  */

          public void initialize() {
          initBestSellerRuleBase();
          super.initialize();
          }

  /**
  * Used by action rules in rule base to make or accept offers.
  *
```

```
* @param obj the Object object implementing the effector
* @param eName the String object that contains the name of the effector
* @param args the String object that contains the arguments for the effector
*
* @return the long object
*
*/
```

```
public long effector(Object obj, String eName,
String args) {
if (eName.equals("make-offer")) {
// rule base decided to counter-offer
long delta = (new Long(offerDelta.getValue())).
longValue();
if (delta == -100) {
current.lastOffer = current.strikePrice
} else {
current.lastOffer = offer.price + delta; //
}
BuySellMessage answer = new BuySellMessage("make-
offer", offer.item + " " + offer.id + " " + current.lastOffer,
offer.item, msg.sender, offer.item, name);
CIAgentEvent e = new CIAgentEvent(this,
"processMessage", answer);
notifyCIAgentEventListeners(e); // respond through
Facilitator
return 0;
}
if (eName.equals("reject-offer")) {
// rule base decided to reject the offer
rejectOffer(offer);
return 0;
}
if (eName.equals("accept-offer")) {
// rule base decided to accept the offer
BuySellMessage answer = new BuySellMessage("accept-
offer", msg.content, msg.replyWith, msg.sender, offer.item,
name);
CIAgentEvent e = new CIAgentEvent(this,
"processMessage", answer);
notifyCIAgentEventListeners(e); // respond through
Facilitator
return 0;
}
return 1; // unknown effector method
}
```

/**
* Uses forward chaining and the rule base to determine whether to
* accept or reject an offer.
*
* @param offer the Offer object that contains the current offer
* @param msg the BuySellMessage object
*
*/

```
          void negotiate(Offer offer, BuySellMessage msg) {
          // figure out if 1st time or second time or other
          // compute spread
          // let rule base figure out our response
          // if deltaoffer is 0 then we should accept
          rb.reset(); // allow all rules to fire, clear vars
          this.offer = offer; // make visible to rule base
effectors
          if (offer.price < current.strikePrice) {
          spread.setValue("<0"); // below asking price
          } else { // above asking price
          long delta = offer.price - current.strikePrice
          if (delta < 25) {
          spread.setValue("0-25");
          } else if (delta < 50) {
          spread.setValue("25-50");
          } else if (delta >= 50) {
          spread.setValue(">50");
          }
     }
          // if buyer echos the lastOffer, then he is agreeing
          if (offer.price == current.lastOffer) {
          spread.setValue(">50"); // accept via rule action
          }
          if (current.iteration == 0) {
          firstOffer.setValue("yes");
          } else {
          firstOffer.setValue("no");
          }
          current.iteration++; // increment iteration count
          if (current.iteration > 10) { // could be randomized
          rejectOffer(offer);
          return // break off negotiations
          }
          rb.forwardChain(); // inference
          if (offerDelta.getValue() == null) {
          trace(name + " rule base couldn't decide what to do
```

```
-- so reject it.\n");
          rejectOffer(offer);
          }
          }
```

/**
* Initializes the BestSeler rule base.
*
*/

```
          public void initBestSellerRuleBase() {
          offerDelta = new RuleVariable(rb, "offerDelta");
          offerDelta.setLabels("-100 0 30 60 100");
          firstOffer = new RuleVariable(rb, "firstOffer");
          firstOffer.setLabels("yes no");
          spread = new RuleVariable(rb, "spread");
          spread.setLabels("<0 0-25 25-50 >50");
          // Note: at this point all variables values are NULL
          Condition cEquals = new Condition("=");
          Condition cNotEquals = new Condition("!=");
          // define rules
          Rule first = new Rule(rb, "first", new
Clause(firstOffer, cEquals, "yes"),
          new Clause(offerDelta, cEquals, "50")); // counter
          Rule second1 = new Rule(rb, "second1", new Clause[]{
new Clause(firstOffer, cEquals, "no"),
          new Clause(spread, cEquals, "25-50") },
          new Clause(offerDelta, cEquals, "40")); // counter
          Rule second2 = new Rule(rb, "second2", new Clause[]{
new Clause(firstOffer, cEquals, "no"),
          new Clause(spread, cEquals, "0-25") },
          new Clause(offerDelta, cEquals, "0")); // accept
          Rule second3 = new Rule(rb, "second3", new Clause[]{
new Clause(firstOffer, cEquals, "no"),
          new Clause(spread, cEquals, ">50") },
          new Clause(offerDelta, cEquals, "0")); // accept
          Rule second4 = new Rule(rb, "second4", new Clause[]{
new Clause(firstOffer, cEquals, "no"),
          new Clause(spread, cEquals, "<0") },
          new Clause(offerDelta, cEquals, "-100")); // reject
          // action rule
          Rule accept = new Rule(rb, "accept", new
Clause(offerDelta, cEquals, "0"), new EffectorClause("accept-
offer", "0"));
          // action rule
          Rule reject = new Rule(rb, "reject", new Clause[]{
new Clause(firstOffer, cEquals, "no"),
```

```
            new Clause(offerDelta, cEquals, "-100") },
            new EffectorClause("reject-offer", "0"));
            // action rule
            Rule counter = new Rule(rb, "counter", new
Clause(offerDelta, cNotEquals, "0"),
            new EffectorClause("make-offer", null)); // use
offerDelta
            // define this object as effector implementor
            rb.addEffector(this, "make-offer");
            rb.addEffector(this, "reject-offer");
            rb.addEffector(this, "accept-offer");
            }
BetterBuyerAgent.java
////////////////
package marketplace;
//import java.util.*;
import ciagent.*;
public class BetterBuyerAgent extends BuyerAgent{
            /**
            * Creates a <code>BetterBuyerAgent</code> object.
            *
            */
            public BetterBuyerAgent() {
            this("BetterBuyer");
            }
```

/**
* Creates a <code>BetterBuyerAgent</code> object with the given name.
*
* @param name the String object that contains the name of the agent
*
*/

```
            public BetterBuyerAgent(String name) {
            super(name);
            }
```

/**
* Negotiates by accepting an offer if it is less than the current strike
* price or making a counter offer that is $50 less than the most recent
* offer.
*
* @param offer the Offer object that contains the current offer
* @param msg the BuySellMessage object

*
*/

```
void negotiate(Offer offer, BuySellMessage msg) {
if (offer.price < current.strikePrice) {
// accept the offer -- by repeating it to the seller
BuySellMessage answer = new BuySellMessage("make-
offer", msg.content, msg.replyWith, msg.sender, offer.item,
name);
CIAgentEvent e = new CIAgentEvent(this,
"processMessage", answer);
notifyCIAgentEventListeners(e); // respond through
Facilitator
} else {
// make a counter offer
current.lastOffer = offer.price - 50; //
BuySellMessage answer = new BuySellMessage("make-
offer", offer.item + " " + offer.id + " " + current.lastOffer,
offer.item, msg.sender, offer.item, name);
CIAgentEvent e = new CIAgentEvent(this,
"processMessage", answer);
notifyCIAgentEventListeners(e); // respond through
Facilitator
}
}
BetterSellerAgent.java
/////////
package marketplace;
import ciagent.*;
public class BetterSellerAgent extends BuyerAgent{
```

/**
* Creates a <code>BetterSellerAgent</code> object.
*
*/

```
public BetterSellerAgent() {
this("BetterSeller");
}
```

/**
* Creates a <code>BetterSellerAgent</code> object with the given name.
*
* @param name the String object that contains the name of the agent
*

*/

```
public BetterSellerAgent(String name) {
super(name);
}
```

/**

* Accepts, rejects or makes a counter offer depending on the offer price.

*

* @param offer the Offer object that contains the current offer

* @param msg the BuySellMessage object

*

*/

```
void negotiate(Offer offer, BuySellMessage msg) {
if (offer.price >= current.strikePrice) {
// accept
BuySellMessage answer = new BuySellMessage("accept-
offer", msg.content, msg.replyWith, msg.sender, offer.item,
name);
CIAgentEvent e = new CIAgentEvent(this,
"processMessage", answer);
notifyCIAgentEventListeners(e); // respond through
Facilitator
return
}
if (offer.price < current.strikePrice) {
long spread = current.strikePrice - offer.price
if (spread > 50) { // reject
BuySellMessage answer = new BuySellMessage("reject-
offer", msg.content, msg.replyWith, msg.sender, offer.item,
name);
CIAgentEvent e = new CIAgentEvent(this,
"processMessage", answer);
notifyCIAgentEventListeners(e); // respond through
Facilitator
} else { // counter offer at strike price
BuySellMessage answer = new BuySellMessage("make-
offer", offer.item + " " + offer.id + " " + current.
strikePrice, msg.replyWith, msg.sender, offer.item, name);
CIAgentEvent e = new CIAgentEvent(this,
"processMessage", answer);
notifyCIAgentEventListeners(e); // respond through
Facilitator
}
return
```

```
        }
        }
BuyerAgent.java
/////////////////
package marketplace;
import java.util.*;
import ciagent.*;
public class BuyerAgent extends CIAgent {
        protected BuySellMessage msg // current message being
processed
        protected BasicNegotiation current
        protected Vector wishList = new Vector();
        protected BasicNegotiation pending = null // item
waiting for seller
        protected Hashtable inventory = new Hashtable(); //
items we have purchased
        protected long totalSpent = 0; // total money spent
on items
        protected Hashtable negotiations = new Hashtable();
// transaction history

/**
* Creates a <code>BuyerAgent</code> object.
*
*/

        public BuyerAgent() {
        this("Buyer");
        }

/**
* Creates a <code>BuyerAgent</code> object with the given name.
*
* @param Name the String object the name of the agent
*
*/

        public BuyerAgent(String name) {
        super(name);
        }

/**
* Registers with the facilitator and initializes its wish list.
*
```

*/

```
        public void initialize() {
        // start a thread running and send an update message
every interval secs
        trace(name + ": initialize() \n");
        msg = new BuySellMessage("register:", name, null,
null, null, null);
        CIAgentEvent e = new CIAgentEvent(this,
"processMessage", msg);
        FacilitatorAgent.register(e); // add this agent to
the Facilitator
        // add items to buy on wish list
        wishList.addElement(new BasicNegotiation("guitar",
100))
        wishList.addElement(new BasicNegotiation("drums",
200))
        wishList.addElement(new BasicNegotiation("guitar",
100))
        setSleepTime(5 * 1000); // process every 5 seconds
        setState(CIAgentState.INITIATED);
        }
```

/**
* Retrieves a description of what this agent is doing.
*
* @return the String object that contains the description
*
*/

```
        public String getTaskDescription() {
        return "Buying stuff..."
        }
```

/**
* Does nothing.
*
*/

```
        public void process() {}
```

/**
* Kicks off netotiations if there are items on the wish list and no
* negotiation is already in progress.

*
*/

```
            public void processTimerPop() {
            if ((wishList.size() > 0) && (pending == null)) {
            current = (BasicNegotiation) wishList.
firstElement();
            pending = current // we have a pending negotiation
            wishList.removeElementAt(0);
            trace(name + " is looking to buy " + current.offer.
item + "\n");

            SetText (name + " is looking to buy " + current.
offer.item + "\n");

            msg = new BuySellMessage("recommend-one", current.
offer.item, null, "Seller", current.offer.item, name);
            CIAgentEvent e = new CIAgentEvent(this,
"processMessage", msg);
            notifyCIAgentEventListeners(e); // signal interested
observer
            }
            trace(name + " has purchased " + inventory.size() +
" items for " + totalSpent + "\n");

            SetText (name + " has purchased " + inventory.size()
+ " items for " + totalSpent + "\n");
            }
```

/**
* Processes a CIAgentEvent.
*
* @param e the CIAgentEvent object that contains the message to be processed
*
*/

```
            public void processCIAgentEvent (CIAgentEvent e) {
            if (traceLevel > 0) {
//            trace(name + ": CIAgentEvent received by " + name
+ " from " + e.getSource() + " with args " + e.getArgObject());
//                        Object obj = e.getArgObject();
//                        if (obj instanceof String) {
//                                    SetText ((String) obj);
//
            }
```

```
            Object arg = e.getArgObject();
            Object action = e.getAction();
            if ((action != null) && (action.
equals("processMessage"))) {
            msg = (BuySellMessage) arg;
            if (traceLevel > 0) {
            msg.display(); // show message contents
            }
            processMessage(msg);
            }
            }
```

/**

* Processes a BuySellMessage received in a CIAgentEvent.

*

* @param msg the BuySellMessage object to be processed

*

*/

```
            public void processMessage(BuySellMessage msg) {
            // facilitator has found a seller for item
            if (msg.sender.equals("Facilitator")) {
            if (msg.performative.equals("tell")) {
            // OK, ask seller what he wants for the item
            BuySellMessage answer = new BuySellMessage("ask",
pending.offer.item, msg.replyWith, msg.content, pending.offer.
item, name);
            CIAgentEvent e = new CIAgentEvent(this,
"processMessage", answer);
            notifyCIAgentEventListeners(e); // respond through
Facilitator
            return
            }
            }
            // seller is denying he has any items to sell
            // this is a response to an "ask"
            if (msg.performative.equals("deny")) {
            trace(name + ": Seller denied our 'ask' about " +
msg.content + "\n");
            wishList.addElement(pending) // put back on wish
list
            pending = null // try again
            return
            }
            Offer offer = new Offer(msg); //
            trace(name + ": Offer from " + offer.sender + ":
content is " + offer.item + " " + offer.id + " " + offer.price
```

```
+ "\n");

            SetText (name + ": Offer from " + offer.sender + ":
content is " + offer.item + " " + offer.id + " " + offer.price
+ "\n");

            if (negotiations.containsKey(offer.id)) {
            current = (BasicNegotiation) negotiations.get(offer.
id);
            } else {
            current = pending // get pending negotiation
            pending = null // no items pending
            negotiations.put(offer.id, current) // place on
active list
            }
            current.newOffer(offer); // update the negotiation
object
            // seller has agreed and sales transaction is
complete
            // transfer item to buyer, seller accepts our last
offer
            if (msg.performative.equals("accept-offer")) {
            trace(name + ": OK -- sale of item " + offer.item +
" with id " + offer.id + " at price of " + offer.price + " is
complete \n");

            SetText (name + ": OK -- sale of item " + offer.item
+ " with id " + offer.id + " at price of " + offer.price + " is
complete \n");

            BuySellMessage answer = new BuySellMessage("tell",
msg.content, msg.replyWith, msg.sender, offer.item, name);
            trace(name + ": " + offer.item + " purchased! \n");




            SetText (name + ": " + offer.item + " purchased!
\n");

            inventory.put(offer.id, current)
            negotiations.remove(offer.id);
            totalSpent += offer.price
            CIAgentEvent e = new CIAgentEvent (this,
"processMessage", answer);
            notifyCIAgentEventListeners(e); // respond through
Facilitator
            }
```

```
            // seller is making a counter offer
            // we can either, make a counter offer, accept, or
reject
            if (msg.performative.equals("make-offer")) {
            negotiate(offer, msg);
            }
            // seller is rejecting our last offer - no counter
offer
            if (msg.performative.equals("reject-offer")) {
            trace(name + ": " + msg.sender + " rejected our last
offer ");

            SetText (name + ": " + msg.sender + " rejected our
last offer ");

            negotiations.remove (offer.id); // remove from
active list
            wishList.addElement (current) // put back on wish
list
            }
            }
```

/**
* Accepts the offer if it is less than the strike price or offers a
* counter offer.
*
* @param offer the Offer object that contains the current offer
* @param msg the BuySellMessage object
*
*/

```
            void negotiate(Offer offer, BuySellMessage msg) {
            if (offer.price < current.strikePrice) {
            // accept the offer -- by repeating it to the seller
            BuySellMessage answer = new BuySellMessage("make-
offer", msg.content, msg.replyWith, msg.sender, offer.item,
name);
            CIAgentEvent e = new CIAgentEvent(this,
"processMessage", answer);
            notifyCIAgentEventListeners(e); // respond through
Facilitator
            } else {
            // make a counter offer
            current.lastOffer = offer.price - 25; //
            BuySellMessage answer = new BuySellMessage("make-
offer", offer.item + " " + offer.id + " " + current.lastOffer,
```

```
offer.item, msg.sender, offer.item, name);
          CIAgentEvent e = new CIAgentEvent(this,
"processMessage", answer);
          notifyCIAgentEventListeners(e); // respond through
Facilitator
          }
          }
BuySellMessage.java
/////////////////
package marketplace;
import java.io.*;
import java.util.*;
public class BuySellMessage {
         protected String performative
          protected String content
          protected String inReplyTo
          protected String language
          protected String ontology
          protected String receiver
          protected String replyWith
          protected String sender
```

/**

* Creates a <code>BuySellMessage</code> object with the given parameters.

*

* @param Performative the String object for the performative

* @param Content the String object for the content

* @param InReplyTo the String object used when replying

* @param Language the String object for the language used in the messages

* @param Ontology the String object for the ontology used in the messages

* @param Receiver the String object that represents the receiver

* @param ReplyWith the String object used when replying

* @param Sender the String object that represents the sender

*

*/

```
          BuySellMessage(String Performative, String Content,
String InReplyTo, String Language, String Ontology, String
Receiver, String ReplyWith, String Sender) {
          performative = Performative;
          content = Content;
          inReplyTo = InReplyTo;
          language = Language;
          ontology = Ontology;
```

```
            receiver = Receiver;
            replyWith = ReplyWith;
            sender = Sender;
            }
```

/**

* Creates a <code>BuySellMessage</code> object with the given parameters.

*

* @param Performative the String object for the performative

* @param Content the String object for the content

* @param InReplyTo the String object used when replying

* @param Receiver the String object that represents the receiver

* @param ReplyWith the String object used when replying

* @param Sender the String object that represents the sender

*

*/

```
            BuySellMessage (String Performative, String Content,
String InReplyTo, String Receiver, String ReplyWith, String
Sender) {
            performative = Performative;
            content = Content;
            inReplyTo = InReplyTo;
            receiver = Receiver;
            replyWith = ReplyWith;
            sender = Sender;
            }
```

/**

* Displays the information contained in the message.

*

*/

```
            public void display() {
            System.out.println("performative: " + performative +
"\n" + "content: " + content + "\n" + "inReplyTo: " + inReplyTo
+ "\n" + "language: " + language + "\n" + "ontology: " +
ontology + "\n" + "receiver: " + receiver + "\n" + "replyWith:
" + replyWith + "\n" + "sender: " + sender + "\n");
            }
FacilitatorAgent.java
////////////
package marketplace;
```

```
import java.util.*;
import android.util.Log;
//import android.view.View.OnClickListener;
//import javax.swing.*;
import ciagent.*;
public class FacilitatorAgent extends CIAgent {
          private static FacilitatorAgent instance = null //
Singleton
          protected Random random = new Random(); // used to
select agents
          protected Hashtable allAgents = new Hashtable(); //
agent name is key, agent object is value
          protected Hashtable communities = new Hashtable();
// domain name is key, vector of agent object is value
          protected BuySellMessage msg // current message
being processed

  /**
  * Creates a <code>FacilitatorAgent</code> object.
  *
  * Note: can't be used as a Java Bean without public constructor
  *
  */

          protected FacilitatorAgent() {
          this("Facilitator");
          }

  /**
  * Creates a <code>FacilitatorAgent</code> object.
  *
  * @param name the String object that contains the name of the facilitator
  *
  */

          protected FacilitatorAgent(String name) {
          super(name);
          }

  /**
  * Initializes the agent by setting the sleep time to 10 seconds.
  *
  */
```

```
public void initialize() {
setSleepTime(10 * 1000); // sleep for 10 seconds
setState(CIAgentState.INITIATED);
}
```

/**
* Clears the agent and communities hashtables.
*
*/

```
public void reset() {
allAgents = new Hashtable(); // clear all agents
communities = new Hashtable(); // clear all
communities
}
```

/**
* Retrieves the task description (null).
*
* @return the String object (null)
*
*/

```
public String getTaskDescription() {
return null
}
```

/**
* Does nothing.
*/

```
public void process() {}
```

/**
* Processes a timer pop by putting out a trace message.
*
*/

```
public void processTimerPop() {
if (traceLevel > 0) trace("Facilitator: active \n");
}
```

/**

* Processes a CIAgentEvent.

*

* @param e the CIAgentEvent object to be processed

*

*/

```
          public void processCIAgentEvent (CIAgentEvent e) {
//                    Log.d ("Debug", "FacilitatorAgent
processCIAgentEvent");
          if (traceLevel > 0) {
                    trace("Facilitator: CIAgentEvent received
by " + name + " from " + e.getSource() + " with args " +
e.getArgObject() + "\n");
          }

          Log.d ("Debug", "FacilitatorAgent
processCIAgentEvent traceLevel > 0");

          Object arg = e.getArgObject();
          Object action = e.getAction();
          if ((action != null) && (action.
equals("processMessage"))) {
          msg = (BuySellMessage) arg;
          if (traceLevel > 0) {
          msg.display(); // show message contents
          }
          route(msg);
          }
          }
```

/**

* In the Singleton design pattern, used to get single instance.

*

* @return the FacilitatorAgent object

*

*/

```
          static public FacilitatorAgent getInstance() {
          if (instance == null) {
          instance = new FacilitatorAgent("Facilitator");
          }
          return instance
          }
```

/**

* Registers the agent contains in the given event.

*

* @param e the CIAgentEvent object that contains the agent to be registered

*

*/

```
            public static synchronized void
register(CIAgentEvent e) {
            if (instance == null) {
            instance = new FacilitatorAgent();
            }
            instance.allAgents.put(((CIAgent) e.getSource()).
getName(), e.getSource())

            ((CIAgent) e.getSource()).addCIAgentEventListener(in
stance);
            instance.addCIAgentEventListener((CIAgent)
e.getSource());
            }
```

/**

* Routes a message to the proper agent.

*

* @param msg the BuySellMessage object that contains the message to be

* routed

*

*/

```
            public synchronized void route(BuySellMessage msg) {
                     Log.d ("Debug", "FacilitatorAgent route");

            CIAgent sender = (CIAgent) allAgents.get(msg.
sender);
            // agent wants to say they can handle questions
concerning content
            if (msg.performative.equals("advertise")) {
            trace("Facilitator: adding " + msg.sender + " to " +
msg.content + " community \n");

            SetText ("Facilitator: adding " + msg.sender + " to
" + msg.content + " community \n");
```

```
            if (communities.containsKey (msg.content)) {
            Vector agents = (Vector) communities.get(msg.
content);
            agents.addElement(sender)
            } else {
            Vector agents = new Vector();
            communities.put(msg.content, agents)
            agents.addElement(sender)
            }
            return
            }
            // agent wants to remove themselves from the
community
            if (msg.performative.equals("unadvertise")) {
            trace("Facilitator: removing " + msg.sender + " from
" + msg.content + " community \n");

            SetText ("Facilitator: removing " + msg.sender + "
from " + msg.content + " community \n");

            if (communities.containsKey(msg.content)) {
            Vector agents = (Vector) communities.get(msg.
content);
            agents.removeElement(sender);
            if (agents.size() == 0) {
            communities.remove(msg.content);
            }
            }
            return
            }
            // agent wants facilitator to recommend an agent to
process content
            // randomly pick one from the list of advertising
seller agents
            if (msg.performative.equals("recommend-one")) {
            String item = msg.content
            if (communities.containsKey(msg.content)) {
            Vector agents = (Vector) communities.get(msg.
content);
            int num = agents.size();
            int index;
            if (num > 1) {
            double rand = random.nextDouble();
            index = (int) (rand * num);
            } else {
```

```
            index = 0;
            }
            CIAgent agent = (CIAgent) agents.elementAt(index);
// first agent?
            msg.performative = "tell"
            msg.content = agent.getName();
            msg.receiver = msg.sender
            msg.sender = name // show facilitator is the sender
of this message
            // need to construct an anwser message and send it
to the sender agent
            trace("Facilitator: Recommended " + agent.getName()
+ " to " + msg.receiver + " for " + item + "\n");




            SetText ("Facilitator: Recommended " + agent.
getName() + " to " + msg.receiver + " for " + item + "\n");

            sender.postCIAgentEvent(new CIAgentEvent(this,
"processMessage", msg));
            } else {
            trace("Facilitator: there are no agents advertising
" + msg.content + "\n");
            SetText ("Facilitator: there are no agents
advertising " + msg.content + "\n");
            }
            return
            }
            trace("Facilitator: routing " + msg.performative + "
message from " + msg.sender + " to " + msg.receiver + "\n");
            SetText("Facilitator: routing " + msg.performative +
" message from " + msg.sender + " to " + msg.receiver + "\n");

            // route the message to the receiver agent
            CIAgent receiver = (CIAgent) allAgents.get(msg.
receiver);
            if (receiver != null) {
            receiver.postCIAgentEvent(new CIAgentEvent(this,
"processMessage", msg));
            } else {
            // should tell sender that receiver could not be
found?
            trace("Facilitator: receiver " + msg.receiver + " is
unknown! \n");
            SetText ("Facilitator: receiver " + msg.receiver + "
is unknown! \n");
```

```
        }
        }
Offer.java
////////////////////
package marketplace;
import java.util.*;
public class Offer {
          protected String sender // who made the offer
           protected String item // what item is offered
           protected String id // the unique object id
           protected long price // the offered sales price
```

/**

* Creates a <code>Offer</code> object that contains the given item.

*

* @param item the String object that contains the item for which an offer

* is being made

*

*/

```
           Offer(String item) {
           sender = ""
           this.item = item;
           id = ""
           price = 0;
           }
```

/**

* Creates a <code>Offer</code> object with the given parameters.

*

* @param sender the String object that contains who is making the offer

* @param item the String object that contains the item

* @param id the String object that contains the item identifier

* @param price the long value for the offer price

*

*/

```
           Offer(String sender, String item, String id, long
price) {
           this.sender = sender;
           this.item = item;
           this.id = id;
           this.price = price;
           }
```

/**

* Creates a <code>Offer</code> object using information in the BuySellMessage.

*

* @param msg the BuySellMessage object that contains information about

* the offer being made

*

*/

```
            Offer(BuySellMessage msg) {
            // split content into item, id, price
            StringTokenizer s = new StringTokenizer(msg.content,
" "); //space delimited
            int num = s.countTokens();
            item = s.nextToken();
            id = s.nextToken();
            price = new Long(s.nextToken()).longValue();
            sender = msg.sender
            }
SellerAgent.java
//////////////
package marketplace;
import java.util.*;
import ciagent.*;
import android.util.Log;
public class SellerAgent extends CIAgent{
            private long seed = 0;
            protected BuySellMessage msg // current message
being processed
            protected BasicNegotiation current
            protected long income = 0; // total money earned
            protected Vector inventory = new Vector(); // items
we have for sale
            protected Hashtable negotiations = new Hashtable();
// transaction history
```

/**

* Creates a <code>SellerAgent</code> object.

*

*/

```
            public SellerAgent() {
            this("Seller");
            }
```

```
/**
* Creates a <code>SellerAgent</code> object with the given name.
*
* @param name the String object that contains the name of the agent
*
*/

        public SellerAgent(String name) {
        super(name);
        }

/**
* Retrieves the task description that indicates what this agent is doing.
*
* @return the String object that contains the task description
*
*/

        public String getTaskDescription() {
        return "Selling stuff..."
        }

/**
* Registers with the facilitator, initalizes its inventory, and advertises
* the items it has for sale.
*
*/

        public void initialize() {

        // start a thread running and send an update message
every interval secs
        trace(name + ": initialize() \n");
        msg = new BuySellMessage("register:", name, null,
null, null, null);
        CIAgentEvent e = new CIAgentEvent(this,
"processMessage", msg);
        FacilitatorAgent.register(e); // add this agent to
the Facilitator

        inventory.addElement(new BasicNegotiation("guitar",
100)) // item and strike price
        inventory.addElement(new BasicNegotiation("drums",
```

```
225))
            inventory.addElement(new BasicNegotiation("guitar",
100)) // item and strike price
            // advertise all items for sale
            Enumeration en = inventory.elements();

            while (en.hasMoreElements()) {
            current = (BasicNegotiation) en.nextElement();
            msg = new BuySellMessage("advertise", current.offer.
item, null, null, current.offer.item, name);
            e = new CIAgentEvent(this, "processMessage", msg);
            notifyCIAgentEventListeners(e); // signal interested
observer
            }
            setSleepTime(15 * 1000); // 15 seconds
            setState(CIAgentState.INITIATED);
            }
            /**
            * Does nothing.
            *
            */
            public void process() {}
```

/**

* Processes a timer pop by displaying a trace message.

*

*/

```
            public void processTimerPop() {
            trace(name + " has " + inventory.size() + " items in
inventory. \n");
            trace(name + " has earned " + income + ".\n");
            }
```

/**

* Processes a CIAgentEvent.

*

* @param e the CIAgentEvent object to be processed

*

*/

```
            public void processCIAgentEvent (CIAgentEvent e) {
                        Log.d ("Debug", "SellerAgent
processCIAgentEvent");
            if (traceLevel > 0) {
```

```
//                       Log.d ("Debug", "SellerAgent1 " +
e.getSource());
//                       Log.d ("Debug", "SellerAgent2 " +
e.getArgObject());
//
//                       Object obj = e.getArgObject();
//                       if (obj instanceof String) {
//                                  SetText ((String) obj);
//



//                       SetText (e.getArgObject().toString());
//             trace(name + ": CIAgentEvent received by " + name
+ " from " + e.getSource() + " with args " + e.getArgObject());
//                       trace(name + ": CIAgentEvent received by
" + name + " from " + " with args ");

           }

           Object arg = e.getArgObject();
           Object action = e.getAction();
           if ((action != null) && (action.
equals("processMessage"))) {
           msg = (BuySellMessage) arg;
           if (traceLevel > 0) {
           msg.display(); // show message contents
           }
           processMessage(msg);
           }
           }
```

/**

* Generates a unique id for an item in inventory.

*

* @return the String object that contains the unique id

*

*/

```
           String genId() {
           seed++;
           return name + seed
           }
```

/**

* Processes the BuySellMessage received in a CIAgentEvent object.

```
*
* @param msg the BuySellMessage object to be processed
*
*/

            public void processMessage (BuySellMessage msg) {
            if (msg.performative.equals("ask")) {
            String item = msg.content
            // see if we have any items left
            if (itemInInventory(item)) {
            // start a new negotiation
            String id = genId();
            current = removeItemFromInventory(item);
            current.offer = new Offer(msg.sender, item, id, 0);
            current.lastOffer = current.strikePrice + 100;
            negotiations.put(id, current)
            BuySellMessage answer = new BuySellMessage("make-
offer", item + " " + id + " " + current.lastOffer, msg.
replyWith, msg.sender, item, name);
            CIAgentEvent e = new CIAgentEvent(this,
"processMessage", answer);
            notifyCIAgentEventListeners(e); // respond through
Facilitator
            } else {
            // deny we have any to sell
            BuySellMessage answer = new BuySellMessage("deny",
item, msg.replyWith, msg.sender, item, name);
            CIAgentEvent e = new CIAgentEvent(this,
"processMessage", answer);
            trace(name + ": deny we have any " + item + " to
sell to " + msg.sender + "\n");
            notifyCIAgentEventListeners(e); // respond through
Facilitator
            }
            return
            }
            Offer offer = new Offer(msg);
            trace(name + ": Offer from " + offer.sender + ":
content is " + offer.item + " " + offer.id + " " + offer.price
+ "\n");

            SetText (name + ": Offer from " + offer.sender + ":
content is " + offer.item + " " + offer.id + " " + offer.price
+ "\n");

            current = (BasicNegotiation) negotiations.get(offer.
id);
```

```
            if (current == null) {
            // we must have sold this item --- reject offer
            BuySellMessage answer = new BuySellMessage("reject-
offer", msg.content, msg.replyWith, msg.sender, offer.item,
name);
            CIAgentEvent e = new CIAgentEvent(this,
"processMessage", answer);
            notifyCIAgentEventListeners(e); // respond through
Facilitator
            return
            }
            current.newOffer(offer);
            // buyer has made an offer, we can accept, make a
counter offer, or reject
            if (msg.performative.equals("make-offer")) {
            negotiate(offer, msg);
            }
            // buyer has received and acknowledged our accept-
offer
            // remove item from active list, add $ to income
            // unadvertise this item with the Facilitator
            if (msg.performative.equals("tell")) {
            negotiations.remove(offer.id); // remove from
negotiations list
            income += offer.price
            BuySellMessage msg2 = new
BuySellMessage("unadvertise", current.offer.item, null, null,
current.offer.item, name);
            CIAgentEvent e = new CIAgentEvent(this,
"processMessage", msg2);
            notifyCIAgentEventListeners(e); // signal interested
observer
            }
            }
```

/**
* Negotiates by accepting or rejecting offers.
*
* @param offer the Offer object for the current offer
* @param msg the BuySellMessage object
*
*/

```
            void negotiate(Offer offer, BuySellMessage msg) {
            if (offer.price > current.strikePrice) {
            // accept
```

```
            BuySellMessage answer = new BuySellMessage("accept-
offer", msg.content, msg.replyWith, msg.sender, offer.item,
name);
            CIAgentEvent e = new CIAgentEvent(this,
"processMessage", answer);
            notifyCIAgentEventListeners(e); // respond through
Facilitator
            return
            }
            if (offer.price < current.strikePrice) {
            // reject
            rejectOffer(offer);
            return
            }
            BuySellMessage answer = new BuySellMessage("make-
offer", offer.item + " " + offer.id + " " + current.lastOffer,
msg.replyWith, msg.sender, offer.item, name);
            CIAgentEvent e = new CIAgentEvent(this,
"processMessage", answer);
            notifyCIAgentEventListeners(e); // respond through
Facilitator
            }
```

/**
* Breaks off a negotiation (for whatever reason), returns the item to
* inventory, and removes the negotiation from the active list.
*
* @param offer the Offer object that contains the offer being rejected
*
*/

```
            void rejectOffer(Offer offer) {
            BuySellMessage answer = new BuySellMessage("reject-
offer", msg.content, msg.replyWith, msg.sender, offer.item,
name);
            CIAgentEvent e = new CIAgentEvent(this,
"processMessage", answer);
            notifyCIAgentEventListeners(e); // respond through
Facilitator
            negotiations.remove(offer.id); // no further
negotiations
            current.offer.id = null // offer id has expired
            inventory.addElement(current) // place item back in
inventory
            }
```

/**

* Checks if an item is in inventory

*

* @param item the String object that contains the item to be checked

*

* @return returns true if the item is in stock

*

*/

```
          boolean itemInInventory(String item) {
          Enumeration en = inventory.elements();
          boolean haveItem = false
          while (en.hasMoreElements()) {
          BasicNegotiation stockItem = (BasicNegotiation)
en.nextElement();
          if (stockItem.getItem().equals(item)) {
          haveItem = true
          break
          } else {
          continue
          }
          }
          return haveItem;
          }
```

/**

* Removes an item from inventory.

*

* @param item the String object that contains the itme to be removed

*

* @return the BasicNegotiation object containing the item from inventory

* or null if we have no items

*

*/

```
          BasicNegotiation removeItemFromInventory (String
item) {
          Enumeration en = inventory.elements();
          BasicNegotiation stockItem = null
          while (en.hasMoreElements()) {
          stockItem = (BasicNegotiation) en.nextElement();
          if (stockItem.getItem().equals(item)) {
          inventory.removeElement(stockItem);
          break
```

```
} else {
stockItem = null // this isn't it
continue
}
}
return stockItem;
}
BooleanRuleBase.java
////////////////////
package rule;
import java.util.*;
import java.io.*;
import java.awt.*;
//import javax.swing.*;
import android.widget.TextView;
import android.util.Log;
public class BooleanRuleBase {
        String name
        Hashtable variableList = new Hashtable(); // all
variables in the rulebase
        Clause clauseVarList[];
        Vector ruleList = new Vector(); // list of all rules
        Vector conclusionVarList // queue of variables
        Rule rulePtr // working pointer to current rule
        Clause clausePtr // working pointer to current
clause
        Stack goalClauseStack = new Stack(); // for goals
(cons clauses) and subgoals
        Hashtable effectors
        Hashtable sensors
        Vector factList
        TextView textArea1
```

```
/**
* Sets the text area for the display of information.
*
* @param txtArea the JTextArea to be used for the display of information
*/
```

```
        public void setDisplay(TextView txtArea) {
        textArea1 = txtArea;
        }
```

```
/**
* Creates a <code>BooleanRuleBase</code> with the given name.
*
```

* @param name the String that contains the name of the rule base
*/

```
public BooleanRuleBase(String name) {
this.name = name;
}
```

/**
* Adds the given text to the trace information.
*
* @param text the String that contains the text to be displayed as part of the
* trace information
*/

```
public void trace(String text) {
        Log.d("Testing", "BooleanRuleBase trase");
if (textArea1 != null) {

        textArea1.append(text);
}
}
```

/**
* Displays all variables and their values in the given text area.
*
* @param textArea the JTextArea where the variables and their values are
* displayed
*/

```
public void displayVariables (TextView textArea) {
Enumeration en = variableList.elements();

Log.d("Testing", "BooleanRuleBase
displayVariables");

while (en.hasMoreElements()) {
RuleVariable temp = (RuleVariable) en.nextElement();
textArea.append("\n" + temp.name + " value = " +
temp.value);
}
}
```

/**

* Displays all the rules and facts in text format in the given text area.
*
* @param textArea the JTextArea where the rules and facts are displayed
*/

```
          public void displayRules (TextView textArea) {
                      Log.d("Testing", "BooleanRuleBase
displayRules");
          textArea.append("\n" + name + " Rule Base: " +
"\n");
          Enumeration en = ruleList.elements();
          while (en.hasMoreElements()) {
          Rule temp = (Rule) en.nextElement();
          temp.display(textArea);
          }
          // now display the facts
          if (factList != null) {
          en = factList.elements();
          while (en.hasMoreElements()) {
          Fact temp = (Fact) en.nextElement();
          temp.display(textArea);
          }
          }
          }
```

/**
* Dislays all rules in the conflict set.
*
* @param ruleSet the Vector that contains the conflict set of rules
*/

```
          public void displayConflictSet(Vector ruleSet) {
          trace("\n" + " -- Rules in conflict set:\n");
          Enumeration en = ruleSet.elements();
          while (en.hasMoreElements()) {
          Rule temp = (Rule) en.nextElement();
          trace(temp.name + "(" + temp.numAntecedents() + "),
");
          }
          }
```

/**
* Resets the rule base for another round of inferencing.
*/

```
public void reset() {
trace("\n --- Setting all " + name + " variables to
null");
Enumeration en = variableList.elements();
while (en.hasMoreElements()) {
RuleVariable temp = (RuleVariable) en.nextElement();
temp.setValue(null);
}
if (factList != null) {
en = factList.elements();
while (en.hasMoreElements()) {
Fact temp = (Fact) en.nextElement();
temp.fired = false
}
}
Enumeration enum2 = ruleList.elements();
while (enum2.hasMoreElements()) {
Rule temp = (Rule) enum2.nextElement();
temp.reset(); // set fired flag to false
}
}
```

/**
* Determines if the given goal is true by finding the set of rules that
* refer to the goal in their consequent clause and then testing the
* antecedent clauses to determine if the rule should fire. If an antecedent
* clause cannot be tested because a variable is unknown, that variable becomes
* the new goal variable and the process recurses until the given goal is found
* to be true or false.
*
* @param goalVarName the String that contains the name fo the goal variable
*/

```
public void backwardChain(String goalVarName) {
RuleVariable goalVar = (RuleVariable) variableList.
get(goalVarName);
Enumeration goalClauses = goalVar.clauseRefs.
elements();
while (goalClauses.hasMoreElements()) {
Clause goalClause = (Clause) goalClauses.
nextElement();
if (goalClause.consequent == false) {
continue
```

```
            }
            goalClauseStack.push(goalClause)
            Rule goalRule = goalClause.getRule();
            Boolean ruleTruth = goalRule.backChain(); // find
rule truth value
            if (ruleTruth == null) {
            trace("\nRule " + goalRule.name + " is null, can't
determine truth value.");
            } else if (ruleTruth.booleanValue() == true) {
            // rule is OK, assign consequent value to variable
            goalVar.setValue(goalClause.rhs);
            goalVar.setRuleName(goalRule.name);
            goalClauseStack.pop(); // clear item from subgoal
stack
            trace("\nRule " + goalRule.name + " is true, setting
" + goalVar.name + ": = " + goalVar.value);
            if (goalClauseStack.empty() == true) {
            trace("\n +++ Found Solution for goal: " + goalVar.
name);
            break // for now, only find first solution, then
stop
            }
            } else {
            goalClauseStack.pop(); // clear item from subgoal
stack
            trace("\nRule " + goalRule.name + " is false, can't
set " + goalVar.name);
            }
            } // endwhile
            if (goalVar.value == null) {
            trace("\n +++ Could Not Find Solution for goal: " +
goalVar.name);
            }
            }

    /**
    * Finds the set of rules that can fire when forward chaining.
    *
    * @param test the boolean value that determines if the antecedents should be
    * tested (<code>true</code>) or just checked (<code>false</code>)
    * @return the Vector of rules that can fire
    */

            public Vector match(boolean test) {
            Vector matchList = new Vector();
            Enumeration en = ruleList.elements();
```

```
            while (en.hasMoreElements()) {
            Rule testRule = (Rule) en.nextElement();
            if (test) {
            testRule.check(); // test the rule antecedents
            }
            if (testRule.truth == null) {
            continue
            }
            // fire the rule only once for now
            if ((testRule.truth.booleanValue() == true) &&
(testRule.fired == false)) {
            matchList.addElement(testRule)
            }
            }
            displayConflictSet(matchList);
            return matchList;
            }
```

/**
* Selects a rule to fire when forward chaining, based on specificity.
*
* @param ruleSet the Vector of rules that can fire
*
* @return the selected rule
*/

```
            public Rule selectRule(Vector ruleSet) {
            Enumeration en = ruleSet.elements();
            long numClauses;
            Rule nextRule;
            Rule bestRule = (Rule) en.nextElement();
            long max = bestRule.numAntecedents();
            while (en.hasMoreElements()) {
            nextRule = (Rule) en.nextElement();
            if ((numClauses = nextRule.numAntecedents()) > max)
{
            max = numClauses;
            bestRule = nextRule;
            }
            }
            return bestRule;
            }
```

/**
* Fires rules, generating new data, based on the initial set of data. The

* process repeats with the new data until no more rules and be fired.
*/

```
            public void forwardChain() {
            Vector conflictRuleSet = new Vector();
            // first test all rules, based on initial data
            conflictRuleSet = match(true); // see which rules
can fire
            while (conflictRuleSet.size() > 0) {
            Rule selected = selectRule(conflictRuleSet); //
select the "best" rule
            selected.fire(); // fire the rule, do the consequent
action, update all clauses and rules
            conflictRuleSet = match(false); // see which rules
can fire
            }
            }
```

/**
* Adds an effector and effector name to the rule base.
*
* @param obj the Object that contains the effector
* @param effectorName the String that contains the name of the effector
*/

```
            public void addEffector(Object obj, String
effectorName) {
            if (effectors == null) {
            effectors = new Hashtable();
            }
            effectors.put(effectorName, obj)
            }
```

/**
* Retrieves the effector associated with the given name.
*
* @param effectorName the String that contains the name of the effector to be
* be retrieved
*
* @return the effector object associated with the given name
*/

```
        public Object getEffectorObject(String effectorName)
{
        return effectors.get(effectorName);
        }
```

/**
* Adds a sensor and sensor name to the rule base.
*
* @param obj the Object that contains the sensor
* @param sensorName the String that contains the name of the sensor
*/

```
        public void addSensor(Object obj, String sensorName)
{
        if (sensors == null) {
        sensors = new Hashtable();
        }
        sensors.put(sensorName, obj)
        }
```

/**
* Retrieves the sensor associated with the given name.
*
* @param sensorName the String that contains the name of the sensor to be
* be retrieved
*
* @return the sensor object associated with the given name
*/

```
        public Object getSensorObject (String sensorName) {
        return sensors.get(sensorName);
        }
```

/**
* Initializes the facts associated with this rule base.
*/

```
        public void initializeFacts() {
        if (factList != null) {
        Enumeration en = factList.elements();
        while (en.hasMoreElements()) {
        Fact fact = (Fact) en.nextElement();
        fact.asserts(this);
```

```
}
}
}
```

/**
* Adds a fact to this rule base.
*
* @param fact the Fact to be added to the rule base
*/

```
public void addFact(Fact fact) {
if (factList == null) {
factList = new Vector();
}
factList.addElement(fact)
}
```

/**
* Adds a variable to this rule base
*
* @param variable the Variable to be added to the rule base
*/

```
public void addVariable (RuleVariable variable) {
variableList.put(variable.getName(), variable)
}
```

/**
* Retrieves the list of variables.
*
* @return a Hashtable object that contains the list of variables
* defined in this ruleset
*/

```
public Hashtable getVariables() {
return (Hashtable) variableList.clone();
}
```

/**
* Retreives all the variables referenced in the consequent of any rule in
* the rule base.
*

* @return the Vector that contains the goal variables
*/

```
public Vector getGoalVariables() {
Vector goalVars = new Vector();
Enumeration en = variableList.elements();
while (en.hasMoreElements()) {
RuleVariable ruleVar = (RuleVariable)
en.nextElement();
Vector goalClauses = ruleVar.clauseRefs
if ((goalClauses != null) && (goalClauses.size() !=
0)) {
goalVars.addElement(ruleVar)
}
}
return goalVars;
}
```

/**
* Retrives the variable associated with the given name.
*
* @param name the String that contains the name of the variable
*
* @return the variable associated with the given name
*/

```
public RuleVariable getVariable(String name) {
if (variableList.containsKey(name)) {
return (RuleVariable) variableList.get(name);
}
return null
}
```

/**
* Sets the value of the variable associated with the given name.
*
* @param name the String that contains the name of the variable
* @param value the String object that contains
*
* @return the variable associated with the given name
*/

```
            public void setVariableValue(String name, String
value) {
            RuleVariable variable = getVariable(name);
            if (variable != null) {
            variable.setValue(value);
            } else {
            System.out.println("BooleanRuleBase: Can't set
value, variable " + name + " is not defined!");
            }
            }
Clause.java
/////////////
package rule;
import java.util.*;
import java.io.*;
public class Clause {
          Vector ruleRefs
            RuleVariable lhs
            String rhs
            Condition cond
            boolean consequent // true if clause is a consequent
clause
            Boolean truth // states = null(unknown), true or
false
```

/**
* Creates a clause with the given rule variable on the left-hand side, the
* condition, and string value on the right-hand side.
*
* @param lhs the RuleVariable that makes up the left-hand side of this clause
* @param cond the Condition which tests equality, greater than, or less than
* @param rhs the String that makes up the right-hand side of this clause
*/

```
            public Clause (RuleVariable lhs, Condition cond,
String rhs) {
            this.lhs = lhs;
            this.cond = cond;
            this.rhs = rhs;
            lhs.addClauseRef(this);
            ruleRefs = new Vector();
            truth = null
            consequent = false // default is antecedent
            }
```

```
/**
* Creates a clause.
*/

        public Clause() {}

/**
* Retrieves the <code>Clause</code> in a format that can be displayed.
*
* @return the <code> String representation of this clause
*/

        public String toString() {
        return lhs.name + cond.toString() + rhs + " "
        }

/**
* Registers the given rule with this clause.
*
* @param ref the Rule to be registered with this clause
*/

        public void addRuleRef(Rule ref) {
        ruleRefs.addElement(ref)
        }

/**
* Performs a test of the clause. This method handles String and numeric
values.
* It does not perform a lot of error checking!
*
* @return a Boolean <code>true</code> if the logical condition is true,
<code>false
* </code> if the logical condition is false, or <code>null</code> if the
* clause is a consequent clause or the left-hand side of the clause is
* unbound.
*/

        public Boolean check() {
        if (consequent == true) {
        return truth = null
```

```
}
if (lhs.value == null) {
return truth = null // can't check truth, if
variable value is undefined
} else {
Double lhsNumericValue = null
Double rhsNumericValue = null
boolean bothNumeric = true
try {
lhsNumericValue = Double.valueOf(lhs.value);
rhsNumericValue = Double.valueOf(rhs);
} catch (Exception e) {
// if exception is thrown, assume one or both values
are NOT numeric
// treat them as String values
bothNumeric = false
}
switch (cond.index) {
case 1:
if (bothNumeric) {
truth = new Boolean(lhsNumericValue.compareTo
(rhsNumericValue) == 0)
} else {
truth = new Boolean(lhs.value.equalsIgnoreCase(rhs))
}
break
case 2:
if (bothNumeric) {
truth = new Boolean (lhsNumericValue.compareTo
(rhsNumericValue) > 0)
} else {
truth = new Boolean(lhs.value.compareTo(rhs) > 0)
}
break
case 3:
if (bothNumeric) {
truth = new Boolean(lhsNumericValue.compareTo
(rhsNumericValue) < 0)
} else {
truth = new Boolean(lhs.value.compareTo(rhs) < 0)
}
break
case 4:
if (bothNumeric) {
truth = new Boolean(lhsNumericValue.compareTo
(rhsNumericValue) != 0)
} else {
truth = new Boolean(lhs.value.compareTo(rhs) != 0)
}
```

```
break
}
return truth
}
}
```

/**

* Determines if the clause is a consequent clause.

*/

```
public void setConsequent() {
consequent = true
}
```

/**

* Retrieves the rule that owns this clause.

*

* @return the owning rule if this clause is a consquent. Otherwise returns

* <code>null</code>.

*/

```
public Rule getRule() {
if (consequent == true) {
return (Rule) ruleRefs.firstElement();
} else {
return null
}
}
Condition.java
///////////
package rule;
import java.util.*;
import java.io.*;
public class Condition {
    int index
    String symbol
```

/**

* Creates a <code>Condition</code> with the given conditional test.

*

* @param symbol the String that represents the conditional test (=, <, >, or !=)

*/

```
public Condition(String symbol) {
this.symbol = symbol;
if (symbol.equals("=")) {
index = 1;
} else if (symbol.equals(">")) {
index = 2;
} else if (symbol.equals("<")) {
index = 3;
} else if (symbol.equals("!=")) {
index = 4;
} else {
index = -1;
}
}
```

```
/**
* Converts the <code>Condition</code> to a format that can be displayed.
*
* @return the <code>Condition</code> in string format
*/
```

```
public String toString() {
return symbol
}
Effector.java
//////////////
package rule;
import java.util.*;
import java.io.*;
public abstract interface Effector {
```

```
/**
* Defines the effector method that must be implemented by
* every class that can be called as an effector.
*
* @param obj the Object that implements an effector
* @param eName the String object that contains the name of the effector
* @param args the String object that contains the parameters for the effector
*
* @return the long value (usually a status code)
*
*/
```

```
          public long effector(Object obj, String eName,
String args);
EffectorClause.java
//////////////////////
package rule;
import java.util.*;
import java.io.*;
public class EffectorClause extends Clause{
         Effector object // object to call
          String effectorName // method to call
          String arguments // parameters to pass
```

/**

* Creates an <code>EffectorClause</code> object with the given name
* and argument string.
*
* @param eName the String object that contains the name of the effector
* @param args the String object that contains arguments for the effector
*
*/

```
          public EffectorClause(String eName, String args) {
          ruleRefs = new Vector();
          truth = new Boolean(true) // always true
          consequent = true // must be consequent
          effectorName = eName;
          arguments = args;
          }
```

/**

* Returns a display string for the effector.
*
* @return the String object that contains a displayable string
*
*/

```
          public String display() {
          return "effector(" + effectorName + "," + arguments
+ ") "
          }
          // call the effector method on the target object
```

/**

```
* Peforms the effector clause.
*
* @param rb the BooleanRuleBase object that this clause is
* part of
*
* @return the Boolean object that contains the value <code>true</code>
*
*/

          public Boolean perform(BooleanRuleBase rb) {
          object = (Effector) (rb.getEffectorObject(effectorN
ame));
          object.effector(this, effectorName, arguments);
          return truth // always true
          }
Fact.java
///////////////
package rule;
import java.util.*;
import java.io.*;
import java.awt.*;
//import javax.swing.*;
import android.widget.TextView;
import android.util.Log;
public class Fact {
         BooleanRuleBase rb1
          String name
          Clause fact //only 1 clause allowed
          Boolean truth // states = (null=unknown, true, or
false)
          boolean fired = false

/**
* Creates a <code>Fact</code> object as part of the given rule base
* with the specified name and the fact clause.
*
* @param Rb the BooleanRuleBase object that this fact belongs to
* @param Name the String object that contains the name of the fact
* @param f the Clause object that contains the fact
*
*/

          Fact(BooleanRuleBase Rb, String Name, Clause f) {
          rb1 = Rb;
```

```
name = Name;
fact = f;
rb1.addFact(this); // add self to fact list
truth = null
}
// assert the fact
```

/**
* Asserts the fact in the given rule base.
*
* @param rb the BooleanRuleBase object in which this fact is being
* asserted
*
*/

```
// ---check---
public void asserts(BooleanRuleBase rb) {
if (fired == true) {
return // only assert once
}
rb1.trace("\nAsserting fact " + name);
truth = new Boolean(true)
fired = true
if (fact.lhs == null) {
// it's an effector
((EffectorClause) fact).perform(rb1); // call the
effector method
} else {
// set the variable value and update clauses
fact.lhs.setValue(fact.rhs);
// now retest any rules whose clauses just changed
}
}
```

/**
* Adds display information about this fact to the given text
* area.
*
* @param textArea the JTextArea object to which the fact information
* is added
*
*/

```
            void display(TextView textArea) {
                        Log.d("Debug", "Fact display");
            textArea.append(name + ": ");
            textArea.append(fact.toString() + "\n");
            }
Rule.java
////////////
package rule;
import java.util.*;
import java.io.*;
import java.awt.*;
//import javax.swing.*;
import android.widget.TextView;
import android.util.Log;
public class Rule {
          BooleanRuleBase rb
           String name
           Clause antecedents[]; // allow up to 4 antecedents
for now
           Clause consequent //only 1 consequent clause allowed
           Boolean truth // states = (null=unknown, true, or
false)
           boolean fired = false

   /**
   * Creates a <code>Rule</code> for the given rule base, with the given
name,
   * antecedent and consequent.
   *
   * @param rb the BooleanRuleBase associated with this rule
   * @param name the String that contains the name of the rule
   * @param lhs the antecedent clause of this rule
   * @param rhs the consequent clause of this rule
   */

           public Rule(BooleanRuleBase rb, String name, Clause
lhs, Clause rhs) {
           this.rb = rb;
           this.name = name;
           antecedents = new Clause[1];
           antecedents[0] = lhs;
           lhs.addRuleRef(this);
           consequent = rhs;
           rhs.addRuleRef(this);
           rhs.setConsequent();
           rb.ruleList.addElement(this) // add self to rule
```

```
list
          truth = null
          }
```

/**

* Creates a <code>Rule</code> for the given rule base, with the given name,

* one or more antecedents and consequent.

*

* @param rb the BooleanRuleBase associated with this rule

* @param name the String that contains the name of the rule

* @param antecedents the antecedent clauses of this rule

* @param lhsClauses the Clause[] object

* @param rhs the consequent clause of this rule

*/

```
          public Rule(BooleanRuleBase rb, String name,
Clause[] lhsClauses, Clause rhs) {
          this.rb = rb;
          this.name = name;
          antecedents = new Clause[lhsClauses.length];
          for (int i = 0; i < lhsClauses.length i++) {
          antecedents[i] = lhsClauses[i];
          antecedents[i].addRuleRef(this);
          }
          consequent = rhs;
          rhs.addRuleRef(this);
          rhs.setConsequent();
          rb.ruleList.addElement(this) // add self to rule
list
          truth = null
          }
```

/**

* Retrieves the number of antecedent clauses in this rule.

*

* @return the number of antecedents

*/

```
          long numAntecedents() {
          return antecedents.length
          }
```

```
/**
* Retests all clauses and rules that reference a variable in a forward
* chaining system.
*
* @param clauseRefs the Vector that contains the clauses that reference a
* certain variable
*/
```

```
public static void checkRules(Vector clauseRefs) {
Enumeration en = clauseRefs.elements();
while (en.hasMoreElements()) {
Clause temp = (Clause) en.nextElement();
Enumeration enum2 = temp.ruleRefs.elements();
while (enum2.hasMoreElements()) {
((Rule) enum2.nextElement()).check(); // retest the
rule
}
}
}
```

```
/**
* Tests the antecedent clauses in this rule in a forward chaining system.
*
* @return <code>true</code> if all the antecedent clauses are true,
<code>false
* </code> if any of the antecedent clauses are false, or <code>null
* </code> if the truth value of any of the antecedent clauses cannot be
* determined
*/
```

```
Boolean check() {
rb.trace("\nTesting rule " + name);
for (int i = 0; i < antecedents.length i++) {
if (antecedents[i].truth == null) {
return truth = null
}
if (antecedents[i].truth.booleanValue() == true) {
continue
} else {
return truth = new Boolean(false) //don't fire this
rule
}
} // endfor
return truth = new Boolean(true) // could fire this
```

```
rule
            }
```

/**

* Fires a rule in a forward chaining system by performing the consequent clause,

* updating all clauses where any changed variables are referenced, and updating

* all the rules that contain those clauses.

*/

```
            void fire() {
            rb.trace("\nFiring rule " + name);
            truth = new Boolean(true)
            fired = true
            if (consequent.lhs == null) {
            // it's an effector
            ((EffectorClause) consequent).perform(rb); // call
the effector method
            } else {
            // set the variable value and update clauses
            consequent.lhs.setValue(consequent.rhs);
            // now retest any rules whose clauses just changed
            checkRules(consequent.lhs.clauseRefs);
            }
            }
```

/**

* Determines if a rule is true by recursively trying to prove its

* antecedent clauses are true or by determining that any of its antecedent

* clauses are false.

*

* @return <code>true</code> if the antecedent clauses of this rule can be

* proven true, or <code>false</code> if any of the antecedents

* are false.

*/

```
            Boolean backChain() {
            rb.trace("\nEvaluating rule " + name);
            for (int i = 0; i < antecedents.length i++) { //
test each clause
            if (antecedents[i].truth == null) {
            rb.backwardChain(antecedents[i].lhs.name);
            }
```

```
            if (antecedents[i].truth == null) { // we couldn't
prove true or false
            antecedents[i].lhs.askUser(); // so ask user for
help
            truth = antecedents[i].check(); // redundant?
            } // endif
            if (antecedents[i].truth.booleanValue() == true) {
            continue // test the next antecedent (if any)
            } else {
            return truth = new Boolean(false) // exit, if any
are false
            }
            } // endfor
            return truth = new Boolean(true) // all antecedents
are true
            }
```

/**

* Displays this rule in text format in the given text area.

*

* @param textArea the JTextArea where the rule is displayed

*/

```
            void display(TextView textArea) {
                    Log.d("Debug", "Rule display");
            textArea.append(name + ": IF ");
            for (int i = 0; i < antecedents.length i++) {
            Clause nextClause = antecedents[i];
            textArea.append(nextClause.toString());
            if ((i + 1) < antecedents.length) {
            textArea.append("\n AND ");
            }
            }
            textArea.append("\n THEN ");
            textArea.append(consequent.toString() + "\n");
            }
```

/**

* Resets the rule, allowing it to be fired again.

*/

```
            void reset() {
            fired = false
            } // allow rule to fire again
RuleVariable.java
/////////////
```

```
package rule;
import java.util.*;
import java.awt.*;
//import javax.swing.*;
public class RuleVariable extends Variable{
         protected BooleanRuleBase rb
          protected Vector clauseRefs // clauses which refer
to this var
          protected String promptText // used to prompt user
for value
          protected String ruleName // if value is inferred,
null = user provided
```

/**
* Creates a new rule variable object with specified name.
*
* @param rb the BooleanRuleBase object this rule variable belongs to
* @param name the String object that contains the name of the variable
*/

```
          public RuleVariable(BooleanRuleBase rb, String name)
{
          super(name);
          this.rb = rb;
          rb.addVariable(this);
          clauseRefs = new Vector();
          }
```

/**
* Set the value of the variable and retests all clauses that refer to
* this variable.
*
* @param value the String object that contains the new value
*/

```
          public void setValue(String value) {
          this.value = value;
          updateClauses();
          }
```

/**
* Prompts the user to provide a value for a variable during backward
* inferencing (set to null if user cancels out of dialog).

```
*
* @return the String object that contains the user response
*/

            public String askUser() {
            String answer = null
            // position dialog over parent dialog
//            JFrame frame = new JFrame();
//            RuleVarDialog dlg = new RuleVarDialog (frame, "Ask
User for Value", true);
//
//            dlg.setLocation(200, 200);
//            dlg.setData(this);
//            dlg.show();
//            answer = dlg.getData(); // retrieve user value
//            setValue(answer); // set value, update clauses
            return value
            }

/**
* Adds a reference to a rule clause.
*
* @param ref the Clause object to be added
*/

            public void addClauseRef(Clause ref) {
            clauseRefs.addElement(ref)
            }

/**
* Checks all rule clauses that refer to this variable.
*/

            public void updateClauses() {
            Enumeration en = clauseRefs.elements();
            while (en.hasMoreElements()) {
            ((Clause) en.nextElement()).check(); // retest the
truth condition
            }
            }

/**
* Sets the rule name.
*
```

```
* @param ruleName the String object that contains the name
*
*/

        public void setRuleName(String ruleName) {
        this.ruleName = ruleName;
        }

/**
* Sets the prompt text for this variable.
*
* @param prompText the String object that contains the prompt text
*
*/

        public void setPromptText(String prompText) {
        this.promptText = prompText;
        }

/**
* Retrieves the prompt text for this variable.
*
* @return the String object that contains the prompt text
*
*/

        public String getPromptText() {
        return promptText
        }

/**
* Not used.
*
* @param inValue the String object
*
*/

        public void computeStatistics(String inValue) {}
        ;

/**
```

```
* Not used.
*
* @param inValue the String object
* @param outArray the float[] object
* @param inx the int object
*
* @return the int object
*
*/

          public int normalize(String inValue, float[]
outArray, int inx) {
          return inx;
          }
Variable.java
//////////
package rule;
import java.util.*;
import java.io.*;
public abstract class Variable {
         protected String name
          protected String value
          // used by categorical only
          protected Vector labels
          protected int column

/**
* Creates a variable object.
*/

          public Variable() {}
          ;

/**
* Creates a variable object with specified name.
*
* @param name the String object that contains the variable name
*/

          public Variable(String name) {
          this.name = name;
          value = null
          }
```

```
/**
* Retrieves the variable name.
*
* @return the String object that contains the variable name
*
*/

        public String getName() {
        return name
        }

/**
* Sets the value of the variable.
*
* @param value the String object that contains the value for the variable
*
*/

        public void setValue(String value) {
        this.value = value;
        }

/**
* Retrieves the value of the variable.
*
* @return the String object that contains the value
*
*/

        public String getValue() {
        return value
        }

/**
* Sets the labels for the variable from the given string.
*
* @param newLabels the String object that contains the labels, separated
* by spaces
*
*/
```

```
public void setLabels(String newLabels) {
labels = new Vector();
StringTokenizer tok = new StringTokenizer(newLabels,
" ");
while (tok.hasMoreTokens()) {
labels.addElement(new String(tok.nextToken()))
}
}
```

/**

* Retrieves the label at the specified index.

*

* @param index the integer of the index

*

* @return the String object that contains the label

*

*/

```
public String getLabel(int index) {
return (String) labels.elementAt(index);
}
```

/**

* Retrieves all the labels for this variable.

*

* @return the Vector object that contains the labels

*

*/

```
public Vector getLabels() {
return (Vector) labels.clone();
}
```

/**

* Retrieves a string containing all the labels separated by spaces.

*

* @return the String object that contains the labels

*/

```
public String getLabelsAsString() {
String labelList = new String();
Enumeration en = labels.elements();
```

```
while (en.hasMoreElements()) {
labelList += en.nextElement() + " "
}
return labelList;
}
```

/**
* Retrieves the index of the given label.
*
* @param label the String object that contains the label whose index is returned
*
* @return the integer value of index or -1 if not found
*/

```
public int getIndex(String label) {
int index = -1;
if (labels == null) {
return index;
}
for (int i = 0; i < labels.size(); i++) {
if (label.equals((String) labels.elementAt(i))) {
index = i;
break
}
}
return index;
}
```

/**
* Checks if the variable is a categorical variable.
*
* @return the boolean true if the variable is categorical (has labels)
* or false if it is not categorical
*
*/

```
public boolean categorical() {
if (labels != null) {
return true
} else {
return false
}
}
```

```
/**
* Retrieves the name of the variable.
*
* @return the String object that contains the name
*
*/

            public String toString() {
            return name
            }

/**
* Sets the column (in a DataSet).
*
* @param column the integer that represents the column
*
*/

            public void setColumn(int column) {
            this.column = column;
            }

/**
* Computes statistics for the variable.
*
* @param inValue the String object used as input to the computation
*
*/

            public abstract void computeStatistics(String
inValue);

/**
* Normalizes the variable.
*
* @param inValue the String object containing the input value
* @param outArray the float[] object that contains the normalized value
* @param inx the integer that represents the index
*
* @return the integer return value
```

```
  *
  */

          public abstract int normalize(String inValue,
float[] outArray, int inx);

  /**
  * Retrieves the normalized size of the variable.
  *
  * @return the integer value (the normalized size)
  *
  */

          public int normalizedSize() {
          return 1; // return default value of 1
          }

  /**
  * Retrieves the decoded value at the given index
  *
  * @param act the float[] object that contains the values to be decoded
  * @param index the integer index value
  *
  * @return the String object that contains the decoded value
  *
  */

          public String getDecodedValue(float[] act, int
index) {
          return String.valueOf(act[index]);
          }
/////////
mainUI.java
/////////
package com.example.apppicachu;
import android.app.Activity;
import android.content.Context;
import android.content.Intent;
import android.net.Uri;
import android.net.wifi.WifiInfo;
import android.net.wifi.WifiManager;
import android.os.Bundle;
import android.os.Handler;
```

```
import android.os.HandlerThread;
import android.util.Log;
import android.view.Menu;
import android.view.MenuItem;
import android.view.View;
import android.view.View.OnClickListener;
import android.widget.Button;
import android.widget.EditText;
import android.widget.TextView;
import android.widget.Toast;
import java.awt.*;
//import java.awt.event.*;
//import javax.swing.*;
import ciagent.*;
import marketplace.*;
import android.text.method.ScrollingMovementMethod;
public class Picachu extends Activity implements
CIAgentEventListener{
          String mac_new;
          String mac_sto;




          FacilitatorAgent facilitator;
          BuyerAgent basicBuyerAgent;
          BetterBuyerAgent intermedBuyerAgent;
          BestBuyerAgent advancedBuyerAgent;
          SellerAgent basicSellerAgent;
          BetterSellerAgent intermedSellerAgent;
          BestSellerAgent advancedSellerAgent;
          public static final int SUMMARY = 0;  // trace level
          public static final int DETAILS = 1;  // trace level
          private int traceLevel = SUMMARY;

          private TextView txtView1;
          private TextView txtView2;




          @Override
          protected void onCreate(Bundle savedInstanceState) {
                    super.onCreate(savedInstanceState);
                    setContentView(R.layout.activity_picachu);
                    //By Hyojin
//                    WifiManager wifimanager_new =
```

```
(WifiManager)getSystemService(Context.WIFI_SERVICE);
//                      WifiInfo wifiinfo_new = wifimanager_new.
getConnectionInfo();
//                      mac_new=wifiinfo_new.getMacAddress();
                    final Handler handler = new Handler();
                    final String TAG = "WIFI state";
//                      new Thread(new Runnable(){
//                                @Override
//                                public void run(){
//                                          while(true){
//                                                    handler.
post(new Runnable(){
//
public void run(){
//
if(mac_new==mac_sto)
//
{
//
Log.d(TAG,"Mac ==");
//
}
//
{
//
Log.d(TAG, "Mac !=");
//
}
//
//                                                              }
//                                                    });
//                                          }
//                                }
//                       }).start();
                    /*
                    GlobalVal Global = (GlobalVal)
getApplication();
                    if(wiThread == null)
                    {
                              wiThread=new Thread(new
WifiThread());
                              wiThread.start();
                    }
                    else
                    {
                              wiThread.interrupt();
                              wiThread = null;
                    }
                     */
```

```
                    Button btn_url = (Button)findViewById(R.
id.Button_url);
                    txtView1 = (TextView)findViewById (R.id.
textView1);
                    txtView2 = (TextView)findViewById (R.id.
textView2);

//                    txtView1.setMovementMethod(new
ScrollingMovementMethod());
//                    txtView2.setMovementMethod(new
ScrollingMovementMethod());
//                    txtView1.setEnabled(false);
//                    txtView2.setEnabled(false);




                    btn_url.setOnClickListener(new
OnClickListener() {
                              @Override
                              public void onClick(View v) {
//                                    Test();
                                  Start();
                              }
                    });
          }

          public void Start() {
                    int traceLevel = SUMMARY;  // default is
summary
              traceLevel = DETAILS;
              Log.d ("Testing", "Start");

              facilitator = FacilitatorAgent.getInstance();  //
singleton
              facilitator.reset ();  // clear it out
              facilitator.SetTextView (txtView1);
              facilitator.setTraceLevel (traceLevel);
              facilitator.addCIAgentEventListener (this);
              facilitator.initialize();
              facilitator.startAgentProcessing();

              basicSellerAgent = new SellerAgent();
              basicSellerAgent.SetTextView (txtView2);
              basicSellerAgent.setTraceLevel(traceLevel);
              basicSellerAgent.addCIAgentEventListener(this);
// for tracing
```

```
            basicSellerAgent.initialize();
            basicSellerAgent.startAgentProcessing();

            basicBuyerAgent = new BuyerAgent();
            basicBuyerAgent.SetTextView (txtView2);
            basicBuyerAgent.setTraceLevel(traceLevel);
            basicBuyerAgent.addCIAgentEventListener(this);
// for tracing
            basicBuyerAgent.initialize();
            basicBuyerAgent.startAgentProcessing();
//              txtView1.post(new Runnable() { @Override public
void run() { txtView1.append("Start MarketPlace"); } });
//              txtView1.append("Start MarketPlace");
        }

        @Override
        public boolean onCreateOptionsMenu (Menu menu) {
                // Inflate the menu; this adds items to the
action bar if it is present.
                getMenuInflater().inflate(R.menu.picachu,
menu);
                return true;
        }
        @Override
        public boolean onOptionsItemSelected(MenuItem item) {
                // Handle action bar item clicks here. The
action bar will
                // automatically handle clicks on the Home/
Up button, so long
                // as you specify a parent activity in
AndroidManifest.xml.
                int id = item.getItemId();
                if (id == R.id.action_settings) {
                        return true;
                }
                return super.onOptionsItemSelected (item);
        }
```

/**
* Used by agents to display messages in marketplace pane.
*
* @param msg the String object that contains the message to be displayed
*

*/

```
            synchronized void trace(String msg) {
//                      txtView1.append(msg + "\n");
//                      txtView1.setCaretPosition(txtView1.
getText().length());
            }
```

/**

* Used by the Facilitator agent to display msgs in top pane.

*

* @param msg the String object that contains the message to be displayed

*

*/

```
            synchronized void traceFacilitator(String msg) {
//                      txtView2.append(msg + "\n");
//                      txtView2.setCaretPosition(txtView2.
getText().length());

                    Log.d ("Main Menu", msg);
            }
```

/**

* Processes a CIAgentEvent if it contains a trace message.

*

* @param event the CIAgentEvent object to be processed

*

*/

```
            public void processCIAgentEvent(CIAgentEvent event)
{
              Object source = event.getSource();
              Object arg = event.getArgObject();
              Object action = event.getAction();
              if ((action != null) && (action.equals("trace")))
{
                if (((arg != null) && (arg instanceof String)))
{
                  if (source instanceof FacilitatorAgent) {
                    traceFacilitator((String) arg);
                  } else {
                    trace((String) arg);  // display the msg
                  }
```

```
            }
        }
    }

/**
 * Method postCIAgentEvent
 *
 * @param event the CIAgentEvent object
 *
 */

    public void postCIAgentEvent(CIAgentEvent event) {
      processCIAgentEvent(event);  // don't queue, just
process
    }
}
```

Appendix 5

IDS

EventHandler

```
EventIPButton.java
//////////////////////////////////////////////////////////////
//////////////////////////////////////////////////////////////
/////////////
package EventHandler;
import java.awt.event.ActionEvent;
import java.awt.event.ActionListener;
import java.io.IOException;
import javax.swing.JTextField;
import Modules.ConnectURL;
public       class EventIPButton implements ActionListener {
   JTextField ipCombo1 = new JTextField();
   public EventIPButton(JTextField ipCombo) {
      ipCombo1 = ipCombo;
   }
   public void actionPerformed(ActionEvent e) {
      ConnectURL cu = new ConnectURL();
      try {
            cu.getLocationXY(ipCombo1.getText().toString());
//  getSelectedItem()).toString());
            cu.PrintMap();
      } catch (IOException e1) {
            e1.printStackTrace();
      }
   }
}
//////////////////////////////////////////////////////////////
////////////////////////////////////////
EventTracert
//////////////////////////////////////////////////////////////
```

```
///////////////////////////////////////
EventTracert.java
package EventHandler;
import java.awt.Dimension;
import java.awt.Graphics;
import java.awt.Image;
import java.awt.TextArea;
import java.awt.Toolkit;
import java.awt.event.ActionEvent;
import java.awt.event.ActionListener;
import javax.swing.JLabel;
import javax.swing.JPanel;
import javax.swing.JTextArea;
import javax.swing.JTextField;
import javax.swing.SwingConstants;
import Modules.TracertRouterNode;
public class EventTracert implements ActionListener {
   private final int SCROLLBARS_NONE = 3;
   JPanel p = new JPanel();
   JTextField ip = new JTextField();
   boolean flg = false;
   public EventTracert(JPanel routerGraph, JTextField ipT) {
      p = routerGraph;
      ip = ipT;
   }
   public void actionPerformed(ActionEvent arg0) {
      TracertRouterNode t = new TracertRouterNode();
      JPanel pS = new JPanel();
      p.add("South", pS);
      TextArea ta = new TextArea("", 30, 0, SCROLLBARS_NONE);
      FThread th = new FThread(ta, t);
      RThread th1 = new RThread(t);
      pS.setPreferredSize(new Dimension(1095, 100));
      YThread th2 = new YThread(t, pS);
      th.start();
      th1.start();
      th2.start();
      // pS.add(new JLabel(" In put Fig ", JLabel.RIGHT));
      ta.setEditable(false);
      p.add("Center", ta);
      p.repaint();
   }
   class RThread extends Thread {
      TracertRouterNode t;
      public RThread(TracertRouterNode t) {
            this.t = t;
      }
      public void run() {
            t.setTracertIP(ip.getText());
```

```
        }
    }
    class YThread extends Thread {
        TracertRouterNode t;
        JPanel p;
        Toolkit tk = Toolkit.getDefaultToolkit();
        Image router;
        Image host;
        Image server;
        int sizeX;
        int sizeY;
        int setX;
        int i = 1;
        int size;
        int pre_setX;
        int Loca = 0;
        public YThread(TracertRouterNode t, JPanel pS) {
                this.t = t;
                this.p = pS;
        }
        public void run() {
                router = tk.getImage(".\/Image/caution.jpg");
                i = 0;
                host = tk.getImage(".\/Image/caution.jpg");
                server = tk.getImage(".\/Image/caution.jpg");
                setX = 30;
                while (true) {
                        Graphics g = p.getGraphics();
                        if (flg) {
                                size = t.GetSizeOfTracertArray();
                                sizeX = p.getWidth();
                                sizeY = p.getHeight();
                                for (int i = 0; i < size; i++) {
                                        setX = (sizeX / 100) + Loca;
                                        System.out.println(setX);
                                        System.out.println(sizeX);
                                        System.out.println(Loca);
                                        if (i == 0) {
                                                g.drawImage(host, setX,
sizeY / 4, p);
                                        } else if (i == size-1) {
                                                g.drawImage(server, setX,
sizeY / 4, p);
                                        } else
                                                g.drawImage(router, setX,
sizeY / 4, p);
//                                      g.drawImage(router[i], setX,
sizeY / 4, p);
                                        if (i > 0) {
```

```
                                        g.drawLine(pre_setX + 30,
(sizeY / 4) + 10, setX,
                                                    (sizeY / 4)
+ 10);
                                  }
                                  pre_setX = setX;
                                  Loca = (900 * (i + 1)) / size;
                            }
                            break;
                      }
                }
          }
    }
    class FThread extends Thread {
          TextArea ta;
          TracertRouterNode t;
          String result = "";
          public FThread(TextArea ta, TracertRouterNode t) {
                this.ta = ta;
                this.t = t;
          }
          public void run() {
                while (true) {
                      if(t.GetSizeOfQueue() > 1){
                            result += t.RemoveElementQ() + "\n";
                            System.out.println(result);
                            ta.setText(result);
                            // System.out.println(setX);
                            // System.out.println(i);
                            try {
                                  sleep(1000);
                            } catch (InterruptedException e) {
                                  // TODO Auto-generated catch
block
                                  e.printStackTrace();
                            }
                            // p.repaint();
                            flg = t.GetFlag();
                            // Loca+=50;
                      } else if (t.GetSizeOfQueue() == 0 && flg ==
true){
                            System.out.println("End");
                            return;
                      }
                }
          }
    }
}
///////////////////////////////////////////////////////////////
```

```
////////////////////////////////////
Modules
/////////////////////////////////////////////////////////////////
////////////////////////////////////
CheckStrangeIP.java
package Modules;
import java.io.BufferedReader;
import java.io.FileNotFoundException;
import java.io.FileReader;
import java.io.IOException;
public class CheckStrangeIP {
   static String[][] DangerC1 = new String[220][15];
   static String[][] DangerC2 = new String[220][15];
   static String[][] DangerH1 = new String[656][15];
   static String[][] DangerH2 = new String[656][15];
   static String[] filename = { "china", "hongkong" };
   int loop;
   public void ReadStrange() {
      for(int a=0; a<220; a++)
            for(int b = 0; b <15 ; b++){
                  DangerC1[a][b] = "";
                  DangerC2[a][b] = "";
            }
      for(int a = 0; a<220; a++)
            for(int b = 0; b< 15 ; b++){
                  DangerH1[a][b] = "";
                  DangerH2[a][b] = "";
            }
      for (loop = 0; loop < filename.length; loop++) {
            try {
                  BufferedReader reader = new
BufferedReader(new FileReader(
                              filename[loop] + ".txt"));
                  String tmp = "";
                  int i, j, k;
                  i = j = k = 0;
                  if (loop == 0) {
                        while ((tmp = reader.readLine()) !=
null) {
                              while (tmp.charAt(i) != '-')
                                    DangerC1[j][k++] += tmp.
charAt(i++);
                              i += 2;
                              k = 0;
                              while (i < tmp.length())
                                    DangerC2[j][k++] += tmp.
charAt(i++);
                              i = 0;
                              j++;
```

```
                              k = 0;
                        }
                  } else {
                        while ((tmp = reader.readLine()) !=
null) {
                              while (tmp.charAt(i) != '-')
                                    DangerH1[j][k++] += tmp.
charAt(i++);
                              i += 2;
                              k = 0;
                              while (i < tmp.length())
                                    DangerH2[j][k++] += tmp.
charAt(i++);
                              i = 0;
                              j++;
                              k = 0;
                        }
                  }
            } catch (FileNotFoundException e) {
                  e.printStackTrace();
            } catch (IOException e) {
                  e.printStackTrace();
            }
      }
   }
}
////////////////////////////////////////////////////////////////
//////////////////////////////////////
CheckStrangeIP
////////////////////////////////////////////////////////////////
//////////////////////////////////////
CheckStrangeIP.java
package Modules;
import java.io.BufferedReader;
import java.io.FileNotFoundException;
import java.io.FileReader;
import java.io.IOException;
public class CheckStrangeIP {
   static String[][] DangerC1 = new String[220][15];
   static String[][] DangerC2 = new String[220][15];
   static String[][] DangerH1 = new String[656][15];
   static String[][] DangerH2 = new String[656][15];
   static String[] filename = { "china", "hongkong" };
   int loop;
   public void ReadStrange() {
      for(int a=0; a<220; a++)
            for(int b = 0; b <15 ; b++){
                  DangerC1[a][b] = "";
                  DangerC2[a][b] = "";
```

```
            }
        for(int a = 0; a<220; a++)
            for(int b = 0; b< 15 ; b++){
                DangerH1[a][b] = "";
                DangerH2[a][b] = "";
            }
        for (loop = 0; loop < filename.length; loop++) {
            try {
                BufferedReader reader = new
BufferedReader(new FileReader(
                        filename[loop] + ".txt"));
                String tmp = "";
                int i, j, k;
                i = j = k = 0;
                if (loop == 0) {
                    while ((tmp = reader.readLine()) !=
null) {
                        while (tmp.charAt(i) != '-')
                            DangerC1[j][k++] += tmp.
charAt(i++);
                        i += 2;
                        k = 0;
                        while (i < tmp.length())
                            DangerC2[j][k++] += tmp.
charAt(i++);
                        i = 0;
                        j++;
                        k = 0;
                    }
                } else {
                    while ((tmp = reader.readLine()) !=
null) {
                        while (tmp.charAt(i) != '-')
                            DangerH1[j][k++] += tmp.
charAt(i++);
                        i += 2;
                        k = 0;
                        while (i < tmp.length())
                            DangerH2[j][k++] += tmp.
charAt(i++);
                        i = 0;
                        j++;
                        k = 0;
                    }
                }
            } catch (FileNotFoundException e) {
                e.printStackTrace();
            } catch (IOException e) {
                e.printStackTrace();
```

```
            }
        }
    }
}
//////////////////////////////////////////////////////////////////
//////////////////////////////////////
Flooding
//////////////////////////////////////////////////////////////////
//////////////////////////////////////
Flooding.java
package Modules;
import java.util.LinkedList;
import java.util.Queue;
class Flooding{
   public static double synC, icmpC, udpC, sendC, recvC,
strangeC, synAckC;
   public static String[] strangeIpInfo = new String[7];
   Queue<Double> synQueue = new LinkedList<Double>();
   Queue<Double> udpQueue = new LinkedList<Double>();
   double synFromQ, sum_SYN, average_SYN, diff_SYN,
squareOfDiff_SYN, sumOfSquare_SYN, standardDeviation_SYN;
   double udpFromQ, sum_UDP, average_UDP, diff_UDP,
squareOfDiff_UDP, sumOfSquare_UDP, standardDeviation_UDP;
   double sum_SynAck;

   Flooding(){
      synC = icmpC = udpC = sendC = recvC = strangeC = synAckC
= 0;
      synFromQ = sum_SYN = average_SYN = diff_SYN =
squareOfDiff_SYN = sumOfSquare_SYN = standardDeviation_SYN = 0;
      udpFromQ = sum_UDP = average_UDP = diff_UDP =
squareOfDiff_UDP = sumOfSquare_UDP = standardDeviation_UDP = 0;

   }
   class thread extends Thread{
      public void run() {
            try {
                  Thread.sleep(1000);
            } catch (InterruptedException e) {
                  e.printStackTrace();
            }
            SetQueue();
            if(synQueue.size() > 5)  GetStandardDeviation_
SYN();
            if(udpQueue.size() > 5)  GetStandardDeviation_
UDP();
            CheckStopPoint();
            IsFloodingAttack();
      }
```

```
    };
    private void SetQueue(){
        synQueue.offer(synC);
        sum_SYN += synC;
        udpQueue.offer(udpC);
        sum_UDP += udpC;
        sum_SynAck += synAckC;
    }

    private void GetStandardDeviation_SYN(){
        synFromQ = synQueue.remove();
        sum_SYN -= synFromQ;
        average_SYN = sum_SYN/5;
        diff_SYN = synFromQ - average_SYN;
        squareOfDiff_SYN = diff_SYN*diff_SYN;
        sumOfSquare_SYN =+ squareOfDiff_SYN;
        standardDeviation_SYN = Math.sqrt(sumOfSquare_SYN/5);
    }

    private void GetStandardDeviation_UDP(){
        udpFromQ = udpQueue.remove();
        sum_UDP -= udpFromQ;
        average_UDP = sum_UDP/5;
        diff_UDP = udpFromQ - average_UDP;
        squareOfDiff_UDP = diff_UDP*diff_UDP;
        sumOfSquare_UDP =+ squareOfDiff_UDP;
        standardDeviation_UDP = Math.sqrt(sumOfSquare_UDP/5);
    }

    private void CheckStopPoint() {
        if(standardDeviation_SYN > 15 || standardDeviation_UDP >
15){
                PacketParsing.stpChck = false;
        }else if(standardDeviation_SYN < 15 || standardDeviation_
UDP < 15){
                PacketParsing.stpChck = true;
        }
    }

    private void IsFloodingAttack(){
        if(standardDeviation_SYN > 50 || standardDeviation_UDP >
50){
                if(synFromQ > 1000 || udpFromQ > 1000){
                        tray.ShowMessageListener1();
                }else if((synFromQ > 700 && synFromQ < 1000) ||
(udpFromQ > 700 && udpFromQ < 1000)){
                        tray.ShowMessageListener2();
                }else if(synFromQ > 200 || udpFromQ > 200){
                        tray.ShowMessageListener3();
```

```
            }
        }
    }

    public void reset(){
        thread th = new thread();
        th.start();
        synC = icmpC = udpC = sendC = recvC = strangeC = synAckC
= 0;
    }
}
////////////////////////////////////////////////////////////////
/////////////////////////////////////
Node
////////////////////////////////////////////////////////////////
/////////////////////////////////////
Node.java
package Modules;
public class Node {
    private String srcIP, dstIP;
    private int Count, SCount, UCount, ICount;
    Node(){
        Count = 0;
        SCount = 0;
        UCount = 0;
        ICount = 0;
    }
    public void incICount(){
        ICount++;
    }
    public void incSCount(){
        SCount++;
    }
    public void incUCount(){
        UCount++;
    }
    public void incCount(){
        Count++;
    }
    public void setICount(int st){
        ICount = st;
    }
    public void setSCount(int st){
        SCount = st;
    }
    public void setUCount(int st){
        UCount = st;
    }
    public void setCount(int st){
```

```
        Count = st;
    }
    public void setsrcIP(String st){
        srcIP = st;
    }
    public void setdstIP(String st){
        dstIP = st;
    }
    public String getsrcIP(){
        return srcIP;
    }
    public String getdstIP(){
        return dstIP;
    }
    public int getCount(){
        return Count;
    }
    public int getICount(){
        return ICount;
    }
    public int getSCount(){
        return SCount;
    }
    public int getUCount(){
        return UCount;
    }
}
////////////////////////////////////////////////////////////////
/////////////////////////////////////
PacketCapture
////////////////////////////////////////////////////////////////
/////////////////////////////////////
PacketCapture.java
package Modules;
import java.io.IOException;
import java.net.InetAddress;
import java.net.NetworkInterface;
import jpcap.*;
public class PacketCapture {
    //public static void main( String[ ] args ) throws
NumberFormatException, IOException {
    PacketCapture() throws NumberFormatException, IOException {
        jpcap.NetworkInterface[] devices = JpcapCaptor.
getDeviceList();
        PacketParsing p = new PacketParsing();
        InetAddress address = InetAddress.getLocalHost();
        NetworkInterface ni = NetworkInterface.
getByInetAddress(address);
        byte[] mac = ni.getHardwareAddress();
```

```
        int count= 0;
        int i;

        for(i =0; i < devices.length ; i++){
              for(int j =0 ; j <mac.length; j ++){
                    if(mac[j] != devices[i].mac_address[j]){
                          count = 0 ;
                          break;
                    }
                    count++;
              }
              if(count == 6)
                    break;

        }
        try {
              JpcapCaptor captor=JpcapCaptor.
openDevice(devices[i], 65535, false, 100);
              captor.loopPacket( -1, p);
        } catch (IOException e) {
              e.printStackTrace();
        }
    }
}
////////////////////////////////////////////////////////////////
//////////////////////////////////////
PacketParsing
////////////////////////////////////////////////////////////////
//////////////////////////////////////
PacketParsing.java
package Modules;
import java.io.BufferedReader;
import java.io.BufferedWriter;
import java.io.FileNotFoundException;
import java.io.FileReader;
import java.io.FileWriter;
import java.io.IOException;
import java.io.InputStreamReader;
import java.net.InetAddress;
import java.net.UnknownHostException;
import java.sql.Date;
import java.text.SimpleDateFormat;
import java.util.ArrayList;
import java.util.GregorianCalendar;
import javax.swing.JOptionPane;
import jpcap.PacketReceiver;
import jpcap.packet.*;
public class PacketParsing implements PacketReceiver {
   public String srcIp = null;
```

```
    public String dstIp = null;
    public short protocol = 0;
    public byte version = 0;
    public static ArrayList<Node> NList = new ArrayList<Node>();
    long time = System.currentTimeMillis();
    SimpleDateFormat dayTime = new SimpleDateFormat("yyyy-MM-dd
hh:mm:ss");
    String date = dayTime.format(new Date(time));
    static double Scount = 0;
    static boolean flag = false;
    static boolean stpChck = true;
    PacketParsing() throws NumberFormatException, IOException {
        // BufferedWriter wt = new BufferedWriter(new
FileWriter("out2.txt",
        // true));
        // wt.close();
        // BufferedReader reader = new BufferedReader(new
        // FileReader("out2.txt"));
        // String tmp = "", tmp2 = "";
        // Reads the file and stores/save it in the NList.
        // while ((tmp = reader.readLine()) != null) {
        // int i = 0;
        //
        // Node n = new Node();
        //
        // while (tmp.charAt(i++) != '\t')
        // tmp2 += tmp.charAt(i - 1);
        // n.setsrcIP(tmp2);
        // tmp2 = "";
        //
        // while (tmp.charAt(i++) != '\t')
        // tmp2 += tmp.charAt(i - 1);
        // n.setdstIP(tmp2);
        // tmp2 = "";
        //
        // while (tmp.length() > i)
        // tmp2 += tmp.charAt(i++);
        // n.setCount(Integer.parseInt(tmp2));
        // tmp2 = "";
        //
        // NList.add(n);
        // i = NList.size();
        // }
        // reader.close();
    }
    @Override
    public void receivePacket(Packet arg0) {
        Packet p = arg0;
        String myAddress = null;
```

```
      try {
            myAddress = InetAddress.getLocalHost().
getHostAddress();
      } catch (UnknownHostException e) {
            // TODO Auto-generated catch block
            e.printStackTrace();
      }
      // Casting the IP Packet type to a String.
      srcIp = ((IPPacket) p).src_ip.getHostAddress();
      dstIp = ((IPPacket) p).dst_ip.getHostAddress();
      protocol = ((IPPacket) p).protocol;
      version = ((IPPacket) p).version;
      if (srcIp.equals(myAddress)) {
            Flooding.sendC++;
      } else if (dstIp.equals(myAddress)) {
            if (protocol == 17) {
                  Flooding.udpC++;
            } else if (protocol == 6) {
                  TCPPacket tcpP = (TCPPacket) arg0;
                  System.out.println("Destination: " + srcIp +
" Purpose: " + dstIp);
                  System.out.println(tcpP.syn + " " + tcpP.
ack);

                  if (tcpP.syn) {
                        Flooding.synC++;

                        if (tcpP.ack & tcpP.syn == true) {

                              Flooding.synAckC++;
                        }
                  }
            //    System.out.println("syn : " + Flooding.synC
+ " synAck : " +Flooding.synAckC);
            } else if (protocol == 1) {
                  Flooding.icmpC++;
            }
            Flooding.recvC++;
      } else {
            // If the user's IP is not associated, the
function terminates.
            return;
      }
      // System.out.println(srcIp);
      int i = 0;
      for (i = 0; i < NList.size(); i++) {
            if (NList.get(i).getsrcIP().equals(srcIp)
                        && NList.get(i).getdstIP().
equals(dstIp)) {
```

```
                        // The two lines below are removed when
removing a Node.
                        NList.get(i).incCount();
                        if (protocol == 17) {
                              NList.get(i).incUCount();
                        } else if (protocol == 6) {
                              TCPPacket a = (TCPPacket) arg0;
                              if (a.syn)
                                    NList.get(i).incSCount();
                        }
                        break;
                  }
            }
            if (i == NList.size() || NList.isEmpty()) {
                  Node n = new Node();
                  n.setsrcIP(srcIp);
                  n.setdstIP(dstIp);
                  n.incCount();
                  NList.add(n);
            }
            // Read the Strange IP from the following two files and
compare them with the current packet IP address src IP.
            if (stpChck == true) {
                  String tmp = "";
                  String filename = "";
                  int s, j, k, c, a, b, d, r;
                  long s1, s2, src, s3, s4;
                  s1 = s2 = src = s3 = s4 = 0;
                  s = j = k = c = a = b = d = r = 0;
                  boolean isHongkong = false;
                  boolean isChina = false;
                  // The following while statement reads a file one
line at a time and compares it with srcIp.
                  String tp = "";
                  // System.out.println("Start check");
                  for (s = 0; s < 656; s++) {
                        while (c != 3) {
                              if (!isHongkong) {
                                    while (true) {
                                          if (CheckStrangeIP.
DangerC1[s][k].equals("."))
                                                break;
                                          tp += CheckStrangeIP.
DangerC1[s][k];
                                          k++;
                                    }
                                    s1 = Integer.parseInt(tp);
                                    tp = "";
                                    while (true) {
```

```
                                        if (CheckStrangeIP.
DangerC2[s][a].equals(".")) 
                                                break;
                                        tp += CheckStrangeIP.
DangerC2[s][a];
                                        a++;
                                }
                                s2 = Integer.parseInt(tp);
                                tp = "";
                        }
                        if (!isChina) {
                                while (true) {
                                        if (CheckStrangeIP.
DangerH1[s][b].equals("."))
                                                break;
                                        tp += CheckStrangeIP.
DangerH1[s][b++];
                                }
                                s3 = Integer.parseInt(tp);
                                tp = "";
                                while (true) {
                                        if (CheckStrangeIP.
DangerH2[s][d].equals("."))
                                                break;
                                        tp += CheckStrangeIP.
DangerH2[s][d++];
                                }
                                s4 = Integer.parseInt(tp);
                                tp = "";
                        }
                        while (true) {
                                if (srcIp.charAt(r) == '.')
                                        break;
                                tp += srcIp.charAt(r++);
                        }
                        src = Integer.parseInt(tp);
                        tp = "";
                        if ((s1 <= src && s2 >= src)) {
                                isChina = true;
                                isHongkong = false;
                                k++;
                                a++;
                                r++;
                        }
                        // else if ((s3 <= src && s4 >= src))
{
                        // flg2 = true;
                        // flg1 = false;
                        // b++;
```

```
                    // d++;
                    // r++;
                    // }
                    else {
                          isChina = false;
                          isHongkong = false;
                          k = a = b = d = r = 0;
                          break;
                    }
                    c++;
              }
              c = 0;
              // Flg has a value of true if srcIp is
strange IP.
              // In this case, the UI is displayed and the
contents are stored in the "err" file.
              if (((isChina == true) && (isHongkong ==
false))
                          || ((isChina == false) &&
(isHongkong == true))) {
                    Flooding.strangeC++;
                    String warn = "";
                    int tmp1;
                    String count;
                    for (tmp1 = 0; tmp1 < NList.size();
tmp1++) {
                          if (NList.get(tmp1).getsrcIP().
equals(srcIp)) {
                                tmp1 = NList.get(tmp1).
getCount();
                                break;
                          }
                    }
                    if (isChina)
                          filename = "China";
                    else if (isHongkong)
                          filename = "Hongkong";
                    // It is the part that stores the
text to be displayed in the UI.
                    count = Integer.toString(tmp1);
                    Flooding.strangeIpInfo[0] = date;
                    Flooding.strangeIpInfo[1] = srcIp;
                    Flooding.strangeIpInfo[3] = filename;
                    Flooding.strangeIpInfo[5] = count;
                    // If the flag is true, add run to
the GUI.
                    flag = true;
                    // Write to the file.
                    BufferedWriter outer;
```

```
                        try {
                              outer = new BufferedWriter(new
FileWriter("err.txt",
                                          true));
                              outer.write(warn);
                              outer.newLine();
                              outer.close();
                        } catch (IOException e) {
                              // TODO Auto-generated catch
block
                              e.printStackTrace();
                        }
                        break;
                  }
                  isChina = false;
                  isHongkong = false;
            }
      }
      try {
            // Open the file out2 and overwrite the contents
of the NList.
            BufferedWriter log = new BufferedWriter(new
FileWriter(
                        "PacketLog.txt"));
            for (i = 0; i < NList.size(); i++) {
                  log.write(NList.get(i).getsrcIP() + "\t"
                              + NList.get(i).getdstIP() +
"\t"
                              + NList.get(i).getCount());
                  log.newLine();
            }
            log.close();
      } catch (IOException e) {
            System.exit(1);
      }
*/ }
}
/////////////////////////////////////////////////////////////
////////////////////////////////////////
PrintPacketInfoTable
/////////////////////////////////////////////////////////////
////////////////////////////////////////
PrintPacketInfoTable.java
package Modules;
import java.awt.BorderLayout;
import java.awt.Button;
import java.awt.Dimension;
import java.awt.Panel;
import java.awt.event.ActionEvent;
```

```
import java.awt.event.ActionListener;
import java.io.IOException;
import java.sql.Date;
import java.text.SimpleDateFormat;
import javax.swing.ImageIcon;
import javax.swing.JLabel;
import javax.swing.JScrollPane;
import javax.swing.JTabbedPane;
import javax.swing.JTable;
import javax.swing.JTextField;
import javax.swing.table.DefaultTableModel;
   class PrintPacketInfoTable extends Thread {
      Panel p = new Panel();
      String[][] packInf = new String[100][7];
      int c = 0;
      public void run() {
            UI.tablePane.addTab("Bad Packet Information", new
ImageIcon("./Image/analysis.jpg"), p);

            p.setLayout(new BorderLayout());
            p.setPreferredSize(new Dimension(1300, 120));
            Panel ipP = new Panel();
            p.add("South", ipP);
            ipP.setLayout(new BorderLayout());
            JTextField ipCombo = new JTextField();
            //                JComboBox ipCombo = new
JComboBox();
            ipP.add("Center", ipCombo);
            Button bt = new Button("ENTER");
            ipP.add("East", bt);
            bt.addActionListener(new EventHandler.
EventIPButton(ipCombo));
            ipP.setPreferredSize(new Dimension(200, 20));
            ipCombo.isDisplayable();
            ipP.add("West",
                        new JLabel(
                                    "*GRAPHIC BASED IDS*"
                                                + "
"
                                                + "
"
                                                + "
"
                                                + "
Input IP address to search "));
            Object column[] = { "Time of occurrence", "Type",
"IP address", "Level of danger", "Area", "Quantity" };
            final DefaultTableModel model;
            model = new DefaultTableModel();
```

```
            for (int i = 0; i < 6; i++)
                  model.addColumn(column[i]);
            PacketParsing.Scount = 1;
            JTable table = new JTable(model);
            p.add("North", table);
            JScrollPane scroll = new JScrollPane(table);
            p.add(scroll);
            while (true) {
                  int i;
                  int rowNum;
                  String IP;
                  String tmp1;
                  boolean row = false;
                  boolean isFlooding = false;
                  SimpleDateFormat dayTime = new
SimpleDateFormat("yyyy-MM-dd hh:mm:ss");
                  long time = System.currentTimeMillis();
                  if(PacketParsing.stpChck == false){
                        rowNum = model.getRowCount();
                        for (i = 0; i <= rowNum; i++) {
                              if (rowNum != 0) {
                                    IP = packInf[i][2];
                                    for(int iter = 0; iter <
PacketParsing.NList.size(); iter++){
                                          if (PacketParsing.
NList.get(iter).getsrcIP().equals(IP)) {

                                                packInf[c]
[5] = Integer.toString((PacketParsing.NList.get(iter).
getSCount())+(PacketParsing.NList.get(iter).getUCount()));

                                                isFlooding =
true;
                                                model.
setValueAt(packInf[c][5], i, 5);
                                                break;
                                          }
                                    }
                              } else
                                    break;
                        }
                        if(isFlooding == false){
                              for(int iter = 0; iter <
PacketParsing.NList.size(); iter++){
                                    if(PacketParsing.NList.
get(iter).getSCount() > 200 || PacketParsing.NList.get(iter).
getUCount() > 200){
                                          packInf[c][0] =
dayTime.format(new Date(time));
```

```
                                        packInf[c][1] = "A
lot of packets";
                                        packInf[c][2] =
PacketParsing.NList.get(iter).getsrcIP(); // IP
                                        packInf[c][3] =
"Warning";
                                        packInf[c][4] =
"Republic of Korea";
                                        packInf[c][5] =
Integer.toString((PacketParsing.NList.get(iter).
getSCount())+(PacketParsing.NList.get(iter).getUCount()));
                                        model.
addRow(packInf[c]);
                                        c++;
                                    }
                              }
                              isFlooding = true;
                        }
                  }
                  if (PacketParsing.flag) {
                        rowNum = model.getRowCount();
                        tmp1 = Flooding.strangeIpInfo[1];
                        for (i = 0; i <= rowNum; i++) {
                              if (rowNum != 0) {
                                    IP = packInf[i][2];
                                    if (tmp1.equals(IP)) {
                                          packInf[c][5] =
Flooding.strangeIpInfo[5];
                                          row = true;
                                          model.
setValueAt(packInf[c][5], i, 5);
                                          break;
                                    }
                              } else
                                    break;
                        }
                        if (row == false) {
                              packInf[c][0] = Flooding.
strangeIpInfo[0];
                              packInf[c][1] = "DDoS packet";
                              packInf[c][2] = Flooding.
strangeIpInfo[1]; // IP
                              packInf[c][3] = "Warning";
                              packInf[c][4] = Flooding.
strangeIpInfo[3];
                              packInf[c][5] = Flooding.
strangeIpInfo[5];
                              //
ipCombo.addItem(packInf[c][2]);
```

```
                                ipCombo.isDisplayable();
                                model.addRow(packInf[c]);
                                c++;
                            }
                            PacketParsing.flag = false;
                    }
                }
            }

    }
////////////////////////////////////////////////////////////////
//////////////////////////////////////
Router
////////////////////////////////////////////////////////////////
//////////////////////////////////////
Router.java
package Modules;
public class Router {
    String myIp, subAddr;
    String next;
    short count;
    Router(String myIp, String subAddr, String next){
        this.myIp = myIp;
        this.subAddr = subAddr;
        this.next = next;
    }
    public void addNext(String next){
        this.next = this.next + 'A' + next;
    }
    public void addHost(){
        this.count++;
    }
    public void setSub(String sub){
        this.subAddr = sub;
    }
}
////////////////////////////////////////////////////////////////
//////////////////////////////////////
TracertRouterNode
////////////////////////////////////////////////////////////////
//////////////////////////////////////
TracertRouterNode.java
package Modules;
import java.io.BufferedReader;
import java.io.IOException;
import java.io.InputStream;
import java.io.InputStreamReader;
import java.util.ArrayList;
import java.util.LinkedList;
```

```
import java.util.Queue;
import java.net.*;
public class TracertRouterNode {
   // private String srcIP;
   // private String result;
   // Tracert Return value via Tracert (Type: IP)
   private ArrayList<String> TracertIP = new
ArrayList<String>();
   private Queue<String> resultQ = new LinkedList<String>();
   private boolean flg = false;

   public TracertRouterNode(){
      //          result = "";
   }
   public int GetSizeOfQueue(){
      return this.resultQ.size();
   }
   public String RemoveElementQ(){
      return this.resultQ.remove();
   }

   public boolean GetFlag(){
      return flg;
   }

   public int GetSizeOfTracertArray(){
      return TracertIP.size();
   }
   // Tracert is the Subnet operation value for the result
value.
   // private ArrayList<String> tSubnet = new
ArrayList<String>();
   // public void test() throws UnknownHostException {
   //       int i;
   //       for (i = 0; i < PacketParsing.NList.size(); i++) {
   //             if (PacketParsing.NList.get(i).getdstIP()
   //                         .equals(InetAddress.
getLocalHost().getHostAddress())
   //                         && PacketParsing.NList.get(i).
getCount() >= 10) {
   //                   // System.out.println("daaaaaaa");
   //                   System.out.println(PacketParsing.
NList.get(i).getsrcIP());
   //                   setTracertIP(PacketParsing.NList.
get(i).getsrcIP());
   //             }
   //       }
   //
   //       // System.out.println("dddddddddd");
```

```
    //        for (i = 0; i < TracertIP.size(); i++)
    //                System.out.println("Circuit (Round) : " + i
+ " : " + TracertIP.get(i));
    //
    //        calSubnet();
    // }
    // public void setSrcIP(String st) {
    //        srcIP = st;
    // }
    //
    // public String getSrcIP() {
    //        return srcIP;
    // }
    /*
    public String GetResult(){
        return result;
    }
     */
    public void setTracertIP(String ip) {
        Process p;
        String traceCmd = "tracert -d -h 4";
        String command = traceCmd + " " + ip;
        try {
                p = Runtime.getRuntime().exec(command);
                readResult(p.getInputStream());
                p.destroy();
        } catch (IOException e) {
                // TODO Auto-generated catch block
                e.printStackTrace();
        }
        //                  BufferedReader br = new
BufferedReader(new InputStreamReader(p.getInputStream()));
        //                  String line = null;
        //                  while ((line = br.readLine()) !=
null) {
        //                          {
        //                                  result += line +"\n";
        //                          }
        //                  }
        //                  System.out.println(result);
    }
    private void readResult(InputStream inputStream) {
        String tmp = "";
        String line = null;
        int count = 0;
        int i;
        int c1 = 0;
        int t = 0;
        BufferedReader br = new BufferedReader(new
```

```
InputStreamReader(
                        inputStream));
        int gg = 0;
        try {
                while ((line = br.readLine()) != null) {
                        resultQ.add(line);
                        //                              result += line
+"\n";
                        count++;
                        // this.doResultLine(line+"\n");
                        //                              System.out.
println(line);
                        if (count >= 4) {
                                // isFist_ip = Integer.valueOf(line.
charAt(2));
                                int c, d;
                                for (i = 0, c = 0, d = 0; i < line.
length(); i++) {
                                        if (line.charAt(i) == 's') {
                                                c++;
                                        } else if (c == 3 && line.
charAt(i) == ' ') {
                                                d++;
                                        } else if (c == 3 && d == 2) {
                                                tmp += line.charAt(i);
                                                // System.out.
println(line.charAt(i));
                                        } else if (line.charAt(i) == '
요') {
                                                System.out.println("no
signal");
                                                c1++;
                                                tmp = "No Signal";
                                                if (c1 == 2)
                                                        break;
                                        } else if (line.charAt(i) == '
추') {
                                                if (line.charAt(i + 2) ==
'을')
                                                        break;
                                        }
                                }
                                if (!tmp.equals("")) {
                                        // if(isFist_ip > 1)
                                        TracertIP.add(tmp);
                                }
                        }
                        tmp = "";
                }
```

```
            count = 0;
            this.TracertIP.add(tmp);
            //                  System.out.println("Array 
Size"+TracertIP.size());
            //
            //                  for(i = 0; i < TracertIP.
size(); i++ )
            //                  {
            //                        System.out.
println(TracertIP.get(i));
            //                  }
      } catch (IOException e) {
            e.printStackTrace();
      } finally {
            try {
                  br.close();
            } catch (IOException e) {
                  e.printStackTrace();
            }
      }
      // if (this.isInterrupted())
      // System.out.println("*** Canceld ***");
      // calSubnet();
   }
   // public void calSubnet() {
   //       String str_ip = "";
   //       int first_ip = 0;
   //
   //       for (int i = 0; i < TracertIP.size(); i++) {
   //             if (!TracertIP.get(i).equals("No Signal")) {
   //                   String[] strTmp = TracertIP.get(i).
split("\\.");
   //                   // StrTmp [1] is the next IP value.
   //                   first_ip = Integer.valueOf(strTmp[0]);
   //
   //                   // A, B, C Class division
   //                   if (first_ip > 0 && (first_ip <= 
126)) {
   //                         // 255.0.0.0 와 XOR algorithmx
   //                         str_ip = String.valueOf(first_
ip & 255) + "." + strTmp[1]
   //                                     + "." + strTmp[2] 
+ "." + strTmp[3];
   //                   } else if (first_ip > 127 && (first_
ip <= 191)) {
   //                         // 255.255.0.0 와 XOR algorithm
   //                         str_ip = String.valueOf(first_
ip & 255)
   //                                     + "."
```

```
    //                                              + String.
valueOf((Integer.valueOf(strTmp[1]) & 255))
    //                                              + "." + strTmp[2]
+ "." + strTmp[3];
    //                      } else if (first_ip > 191 && (first_
ip <= 223)) {
    //                            // 255.255.255.0 와 XOR
algorithm
    //                            str_ip = String.valueOf(first_
ip & 255)
    //                                              + "."
    //                                              + String.
valueOf((Integer.valueOf(strTmp[1]) & 255))
    //                                              + "."
    //                                              + String.
valueOf((Integer.valueOf(strTmp[2]) & 255))
    //                                              + "." + strTmp[3];
    //
    //                      }
    //                      tSubnet.add(str_ip);
    //              } else
    //                      tSubnet.add("*");
    //              System.out.println(str_ip);
    //
    //      }
    // }
}
//////////////////////////////////////////////////////////////////
/////////////////////////////////////
tray
//////////////////////////////////////////////////////////////////
/////////////////////////////////////
tray.java
package Modules;
import java.awt.*;
import java.awt.event.*;
import java.io.IOException;
public class tray {
    static TrayIcon trayIcon;

    public static boolean flag = false;
    static class ShowMessageListener implements ActionListener {
        TrayIcon trayIcon;
        String title;
        String message;
        TrayIcon.MessageType messageType;
        ShowMessageListener(
                    TrayIcon trayIcon,
                    String title,
```

```
                    String message,
                    TrayIcon.MessageType messageType) {
              this.trayIcon = trayIcon;
              this.title = title;
              this.message = message;
              this.messageType = messageType;
       }
       public void actionPerformed(ActionEvent e) {
              trayIcon.displayMessage(title, message,
messageType);
       }
   }
   public static void main(String[] args) throws
NumberFormatException, IOException {
       final UI ui = new UI();
       ui.setVisible(false);
       CheckStrangeIP st = new CheckStrangeIP();
       st.ReadStrange();
       final SystemTray tray = SystemTray.getSystemTray();
       final Image image = Toolkit.getDefaultToolkit().
getImage("tray.gif");
       final PopupMenu popup = new PopupMenu();
       trayIcon = new TrayIcon(image, "Easy network analysis",
popup);

       Runnable runner = new Runnable() {
              public void run() {
                     if (SystemTray.isSupported()) {
                            MenuItem item = new
MenuItem("example");
                            item.addActionListener(new
ActionListener() {
                                   public void
actionPerformed(ActionEvent e) {
                                          ui.setVisible(true);
                                   }
                            });
                            popup.add(item);
                            item = new MenuItem("End");
                            item.addActionListener(new
ActionListener() {
                                   public void
actionPerformed(ActionEvent e) {
                                          tray.remove(trayIcon);
                                          System.exit(1);
                                   }
                            });
                            popup.add(item);
                            try {
```

```
                                    tray.add(trayIcon);
                              } catch (AWTException e) {
                                    System.err.println("Can't add
to tray");
                              }
                        } else {
                              System.err.println("Tray
unavailable");
                        }
                  }
            };
            EventQueue.invokeLater(runner);
            new PacketCapture();
      }
      public static void ShowMessageListener1() {
            trayIcon.displayMessage("Step: Serious " , " DDoS attack
is serious.", TrayIcon.MessageType.WARNING);
      }
      public static void ShowMessageListener2() {
            trayIcon.displayMessage("Step: Danger" , "The DDoS
attack is judged.", TrayIcon.MessageType.WARNING);
      }
      public static void ShowMessageListener3() {
            trayIcon.displayMessage("Step: Warning" , "DDoS attack
is suspected.", TrayIcon.MessageType.WARNING);
      }
}
/////////////////////////////////////////////////////////////////
////////////////////////////////////////
UI
/////////////////////////////////////////////////////////////////
////////////////////////////////////////
UI.java
package Modules;
import java.awt.BorderLayout;
import java.awt.Button;
import java.awt.Color;
import java.awt.Dimension;
import java.awt.Graphics;
import java.awt.GridLayout;
import java.awt.Image;
import java.awt.Label;
import java.awt.Panel;
import java.awt.Toolkit;
import java.awt.event.ActionEvent;
import java.awt.event.ActionListener;
import java.beans.EventHandler;
import java.io.File;
import java.io.IOException;
```

```
import java.net.UnknownHostException;
import java.sql.Date;
import java.text.SimpleDateFormat;
import javax.swing.DefaultComboBoxModel;
import javax.swing.ImageIcon;
import javax.swing.JButton;
import javax.swing.JComboBox;
import javax.swing.JFrame;
import javax.swing.JLabel;
import javax.swing.JPanel;
import javax.swing.JScrollPane;
import javax.swing.JTabbedPane;
import javax.swing.JTable;
import javax.swing.JTextArea;
import javax.swing.JTextField;
import javax.swing.event.TableModelEvent;
import javax.swing.event.TableModelListener;
import javax.swing.table.DefaultTableModel;
import org.jfree.chart.ChartFactory;
import org.jfree.chart.ChartPanel;
import org.jfree.chart.JFreeChart;
import org.jfree.chart.axis.ValueAxis;
import org.jfree.chart.plot.XYPlot;
import org.jfree.data.time.Millisecond;
import org.jfree.data.time.TimeSeries;
import org.jfree.data.time.TimeSeriesCollection;
import org.jfree.ui.RectangleEdge;
import EventHandler.EventTracert;
import sun.tools.jar.Main;
public class UI extends JFrame {
   public static boolean flag = false;
   Flooding F = new Flooding();
   JTabbedPane tPane = new JTabbedPane();
   JTabbedPane tPane2 = new JTabbedPane();
   JTabbedPane tPane4 = new JTabbedPane();
   JTabbedPane tPane5 = new JTabbedPane();
   static JTabbedPane tablePane = new JTabbedPane();
   static boolean flag1 = false;
   Label infoSend = new Label("0 /sec", Label.RIGHT);
   Label infoRecv = new Label("0 /sec", Label.RIGHT);
   Label infoSnR = new Label("00 /sec", Label.RIGHT);
   Label infoTotal = new Label("00", Label.RIGHT);
   Label info_tf = new Label("0 ", Label.RIGHT);
   Label info_tcp = new Label("0  %", Label.RIGHT);
   Label info_udp = new Label("0  %", Label.RIGHT);
   Label info_icmp = new Label("0  %", Label.RIGHT);
   Label info_synAck = new Label("0  %", Label.RIGHT);
   SetPacketNumber packetNumber = new SetPacketNumber();
   PrintPacketInfoTable packetTable = new
```

```
PrintPacketInfoTable();
    public TimeSeries series, series1, series2, tcp, udp, icmp,
synAck, stip;
    private int oldIx = -1;
    public UI() {
        setDefaultCloseOperation(JFrame.HIDE_ON_CLOSE);
        flag = tray.flag;
        this.create();
        this.setLocation(20, 0);
        this.setSize(1200, 740);
        this.setTitle("Networking Analysis");
        ImageIcon im = new ImageIcon("./image/main_icon.png");
        this.setIconImage(im.getImage());
        this.setResizable(false);
        packetNumber.start();
        packetTable.start();
        // lth.start();
    }
    public void UI1() {
    }
    private void create() {
        JPanel mainPanel = new JPanel();
        add(mainPanel);
        mainPanel.setLayout(new BorderLayout());
        JPanel graph = new JPanel();
        JPanel routerGraph = new JPanel();
        Panel packetInformation = new Panel();
        mainPanel.add("Center", tPane);
        mainPanel.add("East", packetInformation);
        mainPanel.add("South", tablePane);
        printTrafficFrame(graph);
        printInformationFrame(packetInformation);
        setRouterFrame(routerGraph);
    }
    // Graph part
    public void printTrafficFrame(JPanel p) {
        JPanel mainFrame = new JPanel();
        // Graph Conf
        this.series1 = new TimeSeries("Send", Millisecond.
class);
        this.series2 = new TimeSeries("Receive", Millisecond.
class);
        this.tcp = new TimeSeries("TCP      ", Millisecond.
class);
        this.udp = new TimeSeries("UDP      ", Millisecond.
class);
        this.icmp = new TimeSeries("ICMP     ", Millisecond.
class);
        this.synAck = new TimeSeries("Syn&Ack", Millisecond.
```

```
class);
      this.stip = new TimeSeries("Receive", Millisecond.
class);
      final TimeSeriesCollection dataset1 = new
TimeSeriesCollection(this.series1);
      final TimeSeriesCollection dataset2 = new
TimeSeriesCollection(this.series2);
      final TimeSeriesCollection dataset3 = new
TimeSeriesCollection(this.tcp);
      final TimeSeriesCollection dataset5 = new
TimeSeriesCollection(this.synAck);
      final TimeSeriesCollection dataset4 = new
TimeSeriesCollection(this.stip);
      final JFreeChart chart = createChart(dataset1);
      final JFreeChart chart1 = createChart1(dataset3);
      final JFreeChart chart2 = createChart2(dataset4);
      final ChartPanel chartPanel = new ChartPanel(chart);
      final ChartPanel chartPanel1 = new ChartPanel(chart1);
      final ChartPanel chartPanel2 = new ChartPanel(chart2);
      chartPanel.setPreferredSize(new java.awt.Dimension(950,
165));
      chartPanel1.setPreferredSize(new java.awt.Dimension(950,
165));
      chartPanel2.setPreferredSize(new java.awt.Dimension(950,
165));
      dataset1.addSeries(series2);
      dataset3.addSeries(udp);
      dataset3.addSeries(icmp);
      dataset3.addSeries(synAck);
      tPane.addTab("Traffic Graph", new ImageIcon("./Image/
graph.jpg"), p);
      p.setLayout(new BorderLayout());
      JPanel packetP = new JPanel();
      packetP.setPreferredSize(new Dimension(800, 170));
      JPanel dosP = new JPanel();
      dosP.setPreferredSize(new Dimension(800, 180));
      JPanel signitureP = new JPanel();
      signitureP.setPreferredSize(new Dimension(800, 170));
      // Packet Graph Source Part to add.
      packetP.add(chartPanel);
      dosP.add(chartPanel1);
      signitureP.add(chartPanel2);
      p.add("North", packetP);
      p.add("Center", dosP);
      p.add("South", signitureP);
   }
   public void setRouterFrame(JPanel routerGraph){
      tPane.addTab("Router Graph", new ImageIcon("./Image/
route.jpg"), routerGraph);
```

```
        routerGraph.setLayout(new BorderLayout());
        //-------- 패널 상단 ----------
        JPanel pN = new JPanel();
        routerGraph.add("North", pN);
        pN.setLayout(new BorderLayout());
        JButton bt = new JButton("Packet receive path
traceback.");
        pN.add("East", bt);
        JTextField ip = new JTextField();
        pN.add("Center", ip);
        pN.add("West", new JLabel("  Put IP address to search
", JLabel.RIGHT));
        //              TracertRouterNode t = new
TracertRouterNode();
        //              t.setTracertIP("28.128.23.4");
        //-------- 패널 중간 ----------
        JLabel ta = new JLabel();
        //              JTextArea ta = new JTextArea();
        //              ta.setText(t.GetResult());
        routerGraph.add("Center", ta);
        //-------- 패널 하단 ----------
//      JPanel pS = new JPanel();
//      routerGraph.add("South", pS);
//      pS.add(new JLabel(" It is a place to put pictures.",
JLabel.RIGHT));
        //              mThread th = new mThread(routerGraph);
        //              th.start();
        bt.addActionListener(new EventTracert(routerGraph, ip));
        //              routerGraph.repaint();
        /*
        bt.addActionListener(new ActionListener(JPanel j){
                public void actionPerformed(ActionEvent arg0) {
                        TracertRouterNode t = new
TracertRouterNode();
                        t.setTracertIP(ip.getText());
                        JTextArea ta = new JTextArea();
                        ta.setText(t.GetResult());
                        routerGraph.add("Center", ta);
                        //                              TracertRouterNode
n = new TracertRouterNode();
                        //                              try {
                        //                                      n.test();
                        //                              } catch
(UnknownHostException e1) {
                        //                                      // TODO
Auto-generated catch block
                        //
e1.printStackTrace();
                        //                              }
```

```
            }
        });*/
    }
    // class mThread extends Thread{
    //        JPanel routerGraph;
    //        public mThread(JPanel routerGraph){
    //              this.routerGraph = routerGraph;
    //        }
    //        public void run(){
    //              Image img, img1;
    //              Toolkit tk = Toolkit.getDefaultToolkit();
    //              img = tk.getImage(“./Image/router.jpg”);
    //              img1 = tk.getImage(“./Image/host.jpg”);
    //
    //              JPanel j = new JPanel();
    //
    //              while(true){
    //                    if(routerGraph.isShowing()){
    //                          //
int x = 20, y = 210;
    //                          Graphics g = routerGraph.
getGraphics();
    //                          //
g.drawImage(img1, x-20, y-30, routerGraph);
    //                          //
g.drawLine(x+50, y, x+80, y);
    //                          //
routerGraph.add(“Center”, j);
    //
    //
    //
    //                          int width = routerGraph.
getWidth()/2;
    //                          int height = routerGraph.
getHeight()/2;
    //                          int x=0, y=0, num=2;
    //                          //It is the leftmost circle
drawing.
    //                          g.drawImage(img1, width/5-5,
height-15, routerGraph);
    ////                              g.drawOval(width/5-5,
height-5, 10, 10);
    //                          //Then left circle.
    //                          g.drawImage(img1, width*2/5-5,
height-15, routerGraph);
    ////                              g.drawOval(width*2/5-5,
height-5, 10, 10);
    //                          //Then left circle.
    //                          g.drawImage(img1, width*3/5-5,
```

```
height-15, routerGraph);
   ////                                g.drawOval(width*3/5-5,
height-5, 10, 10);
   //                           x = routerGraph.getWidth()/10;
   //                           y = routerGraph.getHeight()/2;
   //                           //선 그리기
   //                           g.drawLine(width, height, x,
y);
   //
   //                           //It occurs in one circle per
second, and every three seconds in the middle and below.
   //                           for(int i=0; i<80; i++){
   //                                //The height is changed
by changing the height three times.
   //                                height = (int)(Math.
sin(Math.PI*(i/3)/3)*200)+routerGraph.getHeight()/2;
   //                                g.drawOval(width-5,
height-5, 10, 10);
   //
   //                                //Draw a line connecting
the third circle and the fourth circle (one of the top three)
if it is the third circle.
   //                                if(i%3 == 0)
   //
g.drawLine(routerGraph.getWidth()*3/10, routerGraph.
getHeight()/2, width, height);
   //
   //                                x = (int)(Math.cos(Math.
PI*i/(80/2.0))*65)+width;
   //                                y = (int)(Math.sin(Math.
PI*i/(80/2.0))*65)+height;
   //                                g.drawLine(width, height,
x, y);
   //
   //                                g.drawOval(x-5, y-5, 10,
10);
   //
   //                           }
   //                           //Wait for 1 second.
   ////                              try{Thread.sleep(1000);}
   ////                              catch(Exception e){}
   //                           //Increase user 1
   //                           num++;
   ////                              routerGraph.repaint();
   //
   //
   //                      }
   //                 }
   //            }
```

```
   // }
   private JFreeChart createChart(final TimeSeriesCollection
dataset) {
      final JFreeChart result = ChartFactory.
createTimeSeriesChart(
                  "Send & Receive", "Time", "Count", dataset,
true, true, false);
      result.getLegend().setPosition(RectangleEdge.RIGHT);
      final XYPlot plot = result.getXYPlot();
      plot.setBackgroundPaint(Color.white);
      plot.setDomainGridlinePaint(Color.lightGray);
      plot.setRangeGridlinePaint(Color.lightGray);
      // X axis time length 60 seconds domain axis
dynamically...
      ValueAxis axis = plot.getDomainAxis();
      axis.setAutoRange(true);
      axis.setFixedAutoRange(150000.0); // 60 seconds
      // Y axis
      ValueAxis range = plot.getRangeAxis();
      range.setRange(0, 1200);
      return result;
   }
   // Flooding Graph
   private JFreeChart createChart1(final TimeSeriesCollection
dataset) {
      final JFreeChart result1 = ChartFactory.
createTimeSeriesChart(
                  "Flooding", "Time", "Count", dataset, true,
true, false);
      result1.getLegend().setPosition(RectangleEdge.RIGHT);
      final XYPlot plot = result1.getXYPlot();
      plot.setBackgroundPaint(Color.white);
      plot.setDomainGridlinePaint(Color.lightGray);
      plot.setRangeGridlinePaint(Color.lightGray);
      // X axis time length 60 seconds domain axis
dynamically...
      ValueAxis axis = plot.getDomainAxis();
      axis.setAutoRange(true);
      axis.setFixedAutoRange(150000.0); // 60 seconds
      // Y axis
      ValueAxis range = plot.getRangeAxis();
      range.setRange(0, 1200);
      return result1;
   }
   // StrangeIP Graph
   private JFreeChart createChart2(final TimeSeriesCollection
dataset) {
      final JFreeChart result2 = ChartFactory.
```

```
createTimeSeriesChart(
                    "Dangerous IP", "Time", "Count", dataset,
true, true, false);
      result2.getLegend().setPosition(RectangleEdge.RIGHT);
      final XYPlot plot = result2.getXYPlot();
      plot.setBackgroundPaint(Color.white);
      plot.setDomainGridlinePaint(Color.lightGray);
      plot.setRangeGridlinePaint(Color.lightGray);
      // X axis time length 60 seconds domain axis
dynamically...
      ValueAxis axis = plot.getDomainAxis();
      axis.setAutoRange(true);
      axis.setFixedAutoRange(150000.0); // 60 seconds
      // Y axis
      ValueAxis range = plot.getRangeAxis();
      range.setRange(0, 400);
      return result2;
   }

   // Graph Thread
   class SetPacketNumber extends Thread {
      @Override
      public void run() {
            // TODO Auto-generated method stub
            int Snd = 0, Recv = 0, SnR = 0, Total = 0;
            int ipT = 0, tcpT = 0, udpT = 0, icmpT = 0,
synAckT = 0;
            while (true) {
                  try {
                        Thread.sleep(1000);
                  } catch (InterruptedException e) {
                        // TODO Auto-generated catch block
                        e.printStackTrace();
                  }
                  final Millisecond now = new Millisecond();
                  series1.add(now, Flooding.sendC);
                  series2.add(now, Flooding.recvC);
                  tcp.add(now, Flooding.synC);
                  udp.add(now, Flooding.udpC);
                  icmp.add(now, Flooding.icmpC);

                  synAck.add(now, Flooding.synAckC);
                  stip.add(now, Flooding.strangeC);
                  Snd = (int) (F.sendC);
                  Recv = (int) (F.recvC);
                  SnR = Snd + Recv;
                  Total += SnR;
                  tcpT = (int) (F.synC);
```

```
                    udpT = (int) (F.udpC);
                    icmpT = (int) (F.icmpC);
                    synAckT = (int) (F.synAckC);
                    ipT = (int) (F.strangeC);
                    String strSnd = Integer.toString(Snd);
                    String strRecv = Integer.toString(Recv);
                    String strSnR = Integer.toString(SnR);
                    String strTotal = Integer.toString(Total);
                    String stip = Integer.toString(ipT);
                    String tcp = Integer.toString(tcpT);
                    String udp = Integer.toString(udpT);
                    String synAck = Integer.toString(synAckT);
                    String icmp = Integer.toString(icmpT);
                    infoSend.setText(strSnd + "  /sec");
                    infoRecv.setText(strRecv + "  /sec");
                    infoSnR.setText(strSnR + "  /sec");
                    infoTotal.setText(strTotal+ " (packets)");
                    info_tcp.setText(tcp + "  /sec");
                    info_udp.setText(udp + "  /sec");
                    info_icmp.setText(icmp + "  /sec");
                    info_synAck.setText(synAck + "  /sec");
                    info_tf.setText(stip + "  /sec");
                    F.reset();
                }
            }
        }
        void printInformationFrame(Panel packetInformation) {
            packetInformation.setLayout(new BorderLayout());
            Panel p = new Panel();
            p.setLayout(new BorderLayout());
            packetInformation.add("North", new JLabel(
                    "<html><Font color = #F0381E> Traffic Board
</Font></html>",
                    JLabel.CENTER));
            packetInformation.add("Center", p);
            // Sending/Receiving Pack Quantity Panel.
            p.add("North", tPane5);
            Panel pTraffic = new Panel();
            pTraffic.setPreferredSize(new Dimension(205, 150));
            tPane5.addTab("Traffic", new ImageIcon("./Image/ddos.
jpg"), pTraffic);
            Panel mainTrafPl = new Panel();
            mainTrafPl.setLayout(new GridLayout(4, 2));
            pTraffic.add("Center", mainTrafPl);
            String[] labelTitle = new String[4];
            labelTitle[0] = "Send";
            labelTitle[1] = "Receive";
            labelTitle[2] = "Send & Receive";
```

```
        labelTitle[3] = "Total";
        mainTrafPl.add(new Label(labelTitle[0], Label.RIGHT));
        mainTrafPl.add(infoSend);
        mainTrafPl.add(new Label(labelTitle[1], Label.RIGHT));
        mainTrafPl.add(infoRecv);
        mainTrafPl.add(new Label(labelTitle[2], Label.RIGHT));
        mainTrafPl.add(infoSnR);
        mainTrafPl.add(new Label(labelTitle[3], Label.RIGHT));
        mainTrafPl.add(infoTotal);
        // DDoS Pack Quantity Panel.
        p.add("Center", tPane2);
        Panel pDos = new Panel();
        pDos.setPreferredSize(new Dimension(205, 170));
        tPane2.addTab("DoS", new ImageIcon("./Image/ddos.jpg"),
pDos);
        Panel mainDosPl = new Panel();
        mainDosPl.setLayout(new GridLayout(4, 2));
        pDos.add("Center", mainDosPl);
        String[] labelTitle2 = new String[4];
        labelTitle2[0] = "TCP flooding";
        labelTitle2[1] = "UDP flooding";
        labelTitle2[2] = "ICMP flooding";
        labelTitle2[3] = "Syn & Ack";
        mainDosPl.add(new Label(labelTitle2[0], Label.RIGHT));
        mainDosPl.add(info_tcp);
        mainDosPl.add(new Label(labelTitle2[1], Label.RIGHT));
        mainDosPl.add(info_udp);
        mainDosPl.add(new Label(labelTitle2[2], Label.RIGHT));
        mainDosPl.add(info_icmp);
        mainDosPl.add(new Label(labelTitle2[3], Label.RIGHT));
        mainDosPl.add(info_synAck);
        // Harmful Pack Quantity Panel.
        p.add("South", tPane4);
        Panel pSignature = new Panel();
        pSignature.setPreferredSize(new Dimension(205, 130));
        tPane4.addTab("Dangerous Packets", new ImageIcon("./
Image/caution.jpg"),
                    pSignature);
        Panel mainSignatureP = new Panel();
        mainSignatureP.setLayout(new GridLayout(2, 2));
        mainSignatureP.setPreferredSize(new Dimension(205, 50));
        pSignature.add(mainSignatureP);
        String[] labelTitle3 = new String[2];
        labelTitle3[0] = "China/HongKong";
        labelTitle3[1] = " ";// "HongKong";
        mainSignatureP.add(new Label(labelTitle3[0], Label.
RIGHT));
        mainSignatureP.add(info_tf);
```

```
        mainSignatureP.add(new Label(labelTitle3[1], Label.
RIGHT));

        // Packet information table

    }
}
```

Appendix 6

HOMESERVER

```
HomeServer.java
import dao.DBManager;
import dao.NameDAO;
import gnu.io.CommPortIdentifier;
import gnu.io.SerialPort;
import gnu.io.SerialPortEvent;
import gnu.io.SerialPortEventListener;
import service.NameManageService;
import service.UsageManageService;
import thread.TcpIpServer;
import thread.TestThread;
import java.io.InputStream;
import java.io.OutputStream;
import java.util.Enumeration;
public class HomeServer implements SerialPortEventListener {
    SerialPort serialPort;
    public static boolean on;
    /**
     * You should change to the serial port associated with
your Arduino.
     */
    private static final String PORT_NAMES[] = {
            "COM6", // Windows
    };
    /**
     * Input stream with a buffer to read data from the port.
     */
    private InputStream input;
    public OutputStream getOutput() {
        return output;
    }
    /**
     * Output stream for transferring data to the arduino via
```

```
port.
     */
    private OutputStream output;
    /**
     * Approximate time (2 seconds) to wait for the port to
open.
     */
    private static final int TIME_OUT = 2000;
    /**
     * The default baud rate for the port matches the speed of
the Serial. begin in Arduino.
     */
    private static final int DATA_RATE = 9600;
    public void initialize() {
        CommPortIdentifier portId = null;
        Enumeration portEnum = CommPortIdentifier.
getPortIdentifiers();
        // Find the identifier of the serial port that is
connected to your computer.
        while (portEnum.hasMoreElements()) {
            CommPortIdentifier currPortId =
(CommPortIdentifier) portEnum.nextElement();
            for (String portName : PORT_NAMES) {
                if (currPortId.getName().equals(portName)) {
                    portId = currPortId;
                    break;
                }
            }
        }
        // Terminate if the identifier is not found.
        if (portId == null) {
            System.out.println("Could not find COM port.");
            return;
        }
        try {
            // Serial port open, use class name as port
identification name for application.
            serialPort = (SerialPort) portId.open(this.
getClass().getName(),
                    TIME_OUT);
            // Set the port's parameters such as speed.
            serialPort.setSerialPortParams(DATA_RATE,
                    SerialPort.DATABITS_8,
                    SerialPort.STOPBITS_1,
                    SerialPort.PARITY_NONE);
            // Open stream for reading and writing via port.
            input = serialPort.getInputStream();
            output = serialPort.getOutputStream();
            // Register a listener that receives data from
```

```
Arduino.
            serialPort.addEventListener(this);
            serialPort.notifyOnDataAvailable(true);
        } catch (Exception e) {
            System.err.println(e.toString());
        }
    }
    /**
     * This method must be called when you stop using the port.
     * <p>
     * It prevents port locking on platforms such as Linux.
     */
    public synchronized void close() {
        if (serialPort != null) {
            serialPort.removeEventListener();
            serialPort.close();
        }
    }
    /**
     * Processes events for serial communication and reads and
outputs data.
     */
    public synchronized void serialEvent(SerialPortEvent
oEvent) {
        if (oEvent.getEventType() == SerialPortEvent.DATA_
AVAILABLE) {
            try {
                int available = input.available();
                byte chunk[] = new byte[available];
                input.read(chunk, 0, available);
                //output.write(123);
                System.out.print(new String(chunk));
                //SubStringService.getInstance().addUsage(new
String(chunk)); // 사용량 전송
            } catch (Exception e) {
                System.err.println(e.toString());
            }
        }
        // Ignore other event types. If you need events, add
them.
        // See SerialPortEvent source for other events
    }
    public static void main(String[] args) throws Exception {
        HomeServer main = new HomeServer();
        main.initialize();
        System.out.println("============= Connect DB
===============");
        if (DBManager.getInstance().getConnection() == null) //
DB connection
```

```
        {
            System.out.println("DB connection is not working
properly.");
            return;
        }
        System.out.println("==================== Connect
Server ======================");
        TcpIpServer server = new TcpIpServer(main.getOutput(),
1);
        server.start();
        System.out.println("==================== Name
management DB connection ========================");
        NameManageService.getInstance().init();
        NameDAO.load();
        System.out.println("========= Usage Management
===========");
        UsageManageService.getInstance().init();
        System.out.println("Started");
        TestThread tt = new TestThread();
        tt.start();
    }
}
///////////////////////////////////////////////////////////////
////////////////////////////////////
dao
///////////////////////////////////////////////////////////////
////////////////////////////////////
DBManager
///////////////////////////////////////////////////////////////
////////////////////////////////////
DBManager.java
package dao;
import properties.DBConfig;
import java.sql.*;
public class DBManager {
    private Connection con;
    private Statement stmt;
    private DBManager() {
    }
    private static class SingleHolder {
        public static DBManager single = new DBManager();
    }
    public static DBManager getInstance() {
        return SingleHolder.single;
    }
    public void close() {
        try {
            con.close();
        } catch (SQLException e) {
```

```
            e.printStackTrace();
        }
    }
    public Connection getConnection() {
        con = null;
        try {
            Class.forName("com.mysql.jdbc.Driver");
            con = DriverManager.getConnection("jdbc:mysql://
localhost/home_server",
                    DBConfig.id, DBConfig.db_pwd);
//            java.sql.Statement st = null;
//            ResultSet rs = null;
//            st = con.createStatement();
//            rs = st.executeQuery("SHOW DATABASES");
//
//            if (st.execute("SHOW DATABASES")) {
//                rs = st.getResultSet();
//            }
//
//            while (rs.next()) {
//                String str = rs.getNString(1);
//                System.out.println(str);
//            }
        } catch (SQLException sqex) {
            System.out.println("SQLException: " + sqex.
getMessage());
            System.out.println("SQLState: " + sqex.
getSQLState());
        } catch (Exception e) {
            return null;
        } finally {
            return con;
        }
    }
}
////////////////////////////////////////////////////////////////
/////////////////////////////////////
ElectricUsageDAO
////////////////////////////////////////////////////////////////
/////////////////////////////////////
ElectricUsageDAO.java
package dao;
import java.sql.Connection;
import java.sql.PreparedStatement;
import java.sql.ResultSet;
import java.sql.Timestamp;
/**
 * Created by Administrator on 2015-09-08.
 */
```

```
public class ElectricUsageDAO {
    private final static String INSERT = "INSERT INTO electric_
usage" +
            "(objId, useTime, name, e_usage) " +
            "VALUES (?, ?, ?, ?)";
    public static boolean insertUsage(int index, Timestamp
time, String name, double usage) {
        Connection con = null;
        try {
            con = DBManager.getInstance().getConnection();
            PreparedStatement stmt = con.
prepareStatement(INSERT);
            java.util.Date today = new java.util.Date();
            stmt.setInt(1, index);
            stmt.setTimestamp(2, time);
            stmt.setString(3, name);
            stmt.setDouble(4, usage);
            stmt.addBatch();
            stmt.executeBatch();
            stmt.close();
        } catch (Exception e) {
            e.printStackTrace();
            return false;
        } finally {
            DBManager.getInstance().close();
        }
        return true;
    }
    public static boolean isIndexUsed(int index) {
        Connection con = null;
        boolean used = true;
        try {
            con = DBManager.getInstance().getConnection();
            PreparedStatement s = con.prepareStatement("SELECT
count(index) as cnt FROM electric_usage WHERE ? = index");
            s.setInt(1, index);
            ResultSet rs = s.executeQuery();
            rs.next();
            used = rs.getInt("cnt") > 0;
        } catch (Exception e) {
            used = true;
        } finally {
            DBManager.getInstance().close();
        }
        return used;
    }
}
////////////////////////////////////////////////////////////////
///////////////////////////////////////
```

```
NameDAO
//////////////////////////////////////////////////////////////////
/////////////////////////////////////
NameDAO.java
package dao;
import service.NameManageService;
import java.sql.Connection;
import java.sql.PreparedStatement;
import java.sql.ResultSet;
import java.sql.Timestamp;
/**
 * Created by Administrator on 2015-09-08.
 */
public class NameDAO {
    private final static String INSERT = "INSERT INTO electric_
usage (index, name) VALUES (?, ?)";
    public static final String LOAD_QUERY = "SELECT * FROM
`name_table` WHERE `index`>0";
    public static void load() {
        Connection con = null;
        try {
            con = DBManager.getInstance().getConnection();
            PreparedStatement stmt = con.prepareStatement(LOAD_
QUERY);
            ResultSet rset = stmt.executeQuery();
            while (rset.next()) {
                int objId = rset.getInt("index");
                String name = rset.getString("name");
                System.out.println("[LOAD NameDAO] : \"" +
objId + "\" : " + name);
                NameManageService.getInstance().addName(objId,
name);
            }
        } catch (Exception e)
        {
            e.printStackTrace();
        } finally
        {
            DBManager.getInstance().close();
        }
    }
}
//////////////////////////////////////////////////////////////////
/////////////////////////////////////
model
//////////////////////////////////////////////////////////////////
/////////////////////////////////////
TimeWatt
//////////////////////////////////////////////////////////////////
```

```
////////////////////////////////////////
TimeWatt.java
package model;
/**
 * Created by Administrator on 2015-09-07.
 */
public class TimeWatt {
    private long time;
    private double watt;
    public TimeWatt(long time,double watt) {
        this.time = time;
        this.watt = watt;
    }
    public void setTime(long time) {
        this.time = time;
    }
    public void setWatt(double watt) {
        this.watt = watt;
    }
    public long getTime() {
        return time;
    }
    public double getWatt() {
        return watt;
    }
}
////////////////////////////////////////////////////////////////////
////////////////////////////////////////
UsageList
////////////////////////////////////////////////////////////////////
////////////////////////////////////////
UsageList.java
package model;
import java.util.ArrayList;
/**
 * Created by Administrator on 2015-09-10.
 */
public class UsageList {
    private static ArrayList<Integer> indexList;
    static {
        indexList = new ArrayList<Integer>();
    }
    public static boolean isInList(int index)
    {
        if(indexList.contains(index))
            return true;
        return false;
    }
    public static void addList(int index){
```

```
        indexList.add(index);
    }
}
////////////////////////////////////////////////////////////
//////////////////////////////////////
properties
////////////////////////////////////////////////////////////
//////////////////////////////////////
DBConfig
////////////////////////////////////////////////////////////
//////////////////////////////////////
DBConfig.java
package properties;
/**
 * Created by Administrator on 2015-09-08.
 */
public class DBConfig {
    public static final String id = "root";
    public static final String db_pwd = "apmsetup";
}
////////////////////////////////////////////////////////////
//////////////////////////////////////
DefaultConfig ///////////////////////////////////////////////
////////////////////////////////////////////////////////
DefaultConfig.java
package properties;
/**
 * Created by Administrator on 2015-09-07.
 */
public class DefaultConfig {
    // Time unit to receive input. (sec)
    public static int PERIOD_TIME = 5;
    public static int VOLTAGE = 220;
}
////////////////////////////////////////////////////////////
//////////////////////////////////////
service
////////////////////////////////////////////////////////////
//////////////////////////////////////
NameManageService
////////////////////////////////////////////////////////////
//////////////////////////////////////
NameManageService.java
package service;
import model.TimeWatt;
import thread.CalculateThread;
import java.util.ArrayList;
import java.util.Date;
import java.util.List;
```

```
import java.util.concurrent.ConcurrentHashMap;
/**
 * Created by Administrator on 2015-09-07.
 */
public class NameManageService {
    private ConcurrentHashMap<Integer, String> nameMap;
    @SuppressWarnings("synthetic-access")
    private static class SingletonHolder {
        protected static final NameManageService instance = new
NameManageService();
    }
    public static final NameManageService getInstance() {
        return SingletonHolder.instance;
    }
    public void init() {
        this.nameMap = new ConcurrentHashMap<Integer,
String>();
    }
    public void addName(int index, String name) {
        nameMap.put(index, name);
    }
    public String getName(int index) {
        System.out.println("TEST" + nameMap.size());
        if (!nameMap.containsKey(index))
            return "Name is blank.";
        return nameMap.get(index);
    }
}
////////////////////////////////////////////////////////////////
////////////////////////////////////////
SubStringService
////////////////////////////////////////////////////////////////
////////////////////////////////////////
SubStringService.java
package service;
import java.util.concurrent.ConcurrentHashMap;
/**
 * Created by Administrator on 2015-09-07.
 */
public class SubStringService {
    //TODO Separate incoming string management
    private StringBuilder sb;
//    private ConcurrentHashMap<String, StringBuilder> sbMap;
    @SuppressWarnings("synthetic-access")
    private static class SingletonHolder {
        protected static final SubStringService instance = new
SubStringService();
    }
    public static final SubStringService getInstance() {
```

```
        return SingletonHolder.instance;
    }
    public void init() {
        sb = new StringBuilder();
    }
    public void putString(String str) {
        sb.append(str);
        String tmp;
        if (str.indexOf("\\n") > 0) {
            String[] stringArray = new String(str).
split("\\n");
            sb.append(stringArray[0]);
            addUsage(sb.toString()); // Add usage
            sb = new StringBuilder();
            sb.append(stringArray[1]);
        }
    }
    public void addUsage(String str) {
        String[] stringArray = new String(str).split(":");
        UsageManageService.getInstance().addUsage(Integer.
parseInt(stringArray[0]), WattCalculateService.
calculate(Double.parseDouble(stringArray[1])));
    }
}
//////////////////////////////////////////////////////////////
////////////////////////////////////
UsageManageService
//////////////////////////////////////////////////////////////
////////////////////////////////////
UsageManageService.java
package service;
import model.TimeWatt;
import thread.CalculateThread;
import java.util.ArrayList;
import java.util.Date;
import java.util.List;
import java.util.concurrent.ConcurrentHashMap;
/**
 * Created by Administrator on 2015-09-07.
 */
public class UsageManageService {
    public static boolean on;
    public ConcurrentHashMap<Integer, List<TimeWatt>>
getUsageSec() {
        return usageSec;
    }
    public ConcurrentHashMap<Integer, List<TimeWatt>>
getUsageMin() {
        return usageMin;
```

```
    }
    private ConcurrentHashMap<Integer, List<TimeWatt>>
usageSec;
    private ConcurrentHashMap<Integer, List<TimeWatt>>
usageMin;
    @SuppressWarnings("synthetic-access")
    private static class SingletonHolder {
        protected static final UsageManageService instance =
new UsageManageService();
    }
    public static final UsageManageService getInstance() {
        return SingletonHolder.instance;
    }
    public void init() {
        this.usageSec = new ConcurrentHashMap<Integer,
List<TimeWatt>>();
        this.usageMin = new ConcurrentHashMap<Integer,
List<TimeWatt>>();
        calculateThread();
    }
    /**
     * @param index
     * @param watt      */
    public void addUsage(int index, double watt) {
        if (!usageSec.containsKey(index)) {
            usageSec.put(index, new ArrayList<TimeWatt>());
        }
        usageSec.get(index).add(new TimeWatt(System.
currentTimeMillis(), watt));
        Date d = new Date();
        System.out.println(d.toString() + " [
UsageManageService ] " + index + " : " + watt + "W 소모중");
    }
    public void calculateThread() {
        CalculateThread ct = new CalculateThread();
        ct.start();
    }
    public double get5MinWatt(int index) {
        double resultWatt = 0;
        for (TimeWatt watt : usageSec.get(index)) {
            resultWatt += watt.getWatt() * 5 / 3600;
        }
        usageSec.clear();
        return resultWatt;
    }
}
///////////////////////////////////////////////////////////////
////////////////////////////////////
WattCalculateService
```

```
////////////////////////////////////////////////////////////////
////////////////////////////////////
WattCalculateService.java
package service;
import properties.DefaultConfig;
/**
 * Created by Administrator
 */
public class WattCalculateService {
    /**
     * @param amp
     * @return
     */
    public static double calculate(double amp) {
        return amp * DefaultConfig.VOLTAGE;
    }
}
////////////////////////////////////////////////////////////////
////////////////////////////////////
thread
////////////////////////////////////////////////////////////////
////////////////////////////////////
CalculateThread
////////////////////////////////////////////////////////////////
////////////////////////////////////
CalculateThread.java
package thread;
import dao.ElectricUsageDAO;
import model.TimeWatt;
import model.UsageList;
import service.NameManageService;
import service.UsageManageService;
import java.sql.Timestamp;
import java.util.ArrayList;
import java.util.Date;
/**
 * Created by Administrator
 */
public class CalculateThread extends Thread {
    public void run() {
        while (true) {
            System.out.println("3-minute cycle thread execution
for power usage aggregation ");
            try {
                sleep(10 * 60 * 1000);
            } catch (InterruptedException e) {
                e.printStackTrace();
            }
            java.util.Date date = new java.util.Date();
```

```
            Timestamp time = new Timestamp(date.getTime());
            for (int index : UsageManageService.getInstance().
getUsageSec().keySet()) {
                if (!UsageManageService.getInstance().
getUsageMin().containsKey(index))
                    UsageManageService.getInstance().
getUsageMin().put(index, new ArrayList<TimeWatt>());
                double wh = UsageManageService.getInstance().
get5MinWatt(index);
                System.out.println("Three-minute aggregation
index : " + index + " / " + wh);
                UsageManageService.getInstance().getUsageMin().
get(index).add(new TimeWatt(System.currentTimeMillis(), wh));
                System.out.println(new Date().toString() + "
Usage : " + wh);
                ElectricUsageDAO.insertUsage(index, time,
NameManageService.getInstance().getName(index), wh);
            }
        }
    }
}
////////////////////////////////////////////////////////////////
//////////////////////////////////////////
TcpIpServer
////////////////////////////////////////////////////////////////
//////////////////////////////////////////
TcpIpServer.java
package thread;
import service.UsageManageService;
import java.io.DataInputStream;
import java.io.IOException;
import java.io.InputStream;
import java.io.OutputStream;
import java.net.ServerSocket;
import java.net.Socket;
import java.text.SimpleDateFormat;
import java.util.Date;
public class TcpIpServer implements Runnable {
    ServerSocket serverSocket;
    Thread[] threadArr;
    OutputStream output;
    public static void main(String[] args) {
    }
    public TcpIpServer(OutputStream output, int num) {
        try {

            serverSocket = new ServerSocket(7777);
            System.out.println(getTime() + " Server is
ready.");
```

```
            this.output = output;
            threadArr = new Thread[num];
        } catch (IOException e) {
            e.printStackTrace();
        }
    }
    public void start() {
        for (int i = 0; i < threadArr.length; i++) {
            threadArr[i] = new Thread(this);
            threadArr[i].start();
        }
    }
    @Override
    public void run() {
        while (true) {
            try {

                Socket socket = serverSocket.accept();
//                System.out.println(getTime() + " " + socket.
getInetAddress()
                InputStream in = socket.getInputStream();
                DataInputStream dis = new DataInputStream(in);
                String value;
                String result = "";
                while(dis.available() > 0 && (value = dis.
readLine()) != null) {
                    result += value;
                    System.out.println("[" + getTime() + "] " +
value);
                }
                if(result.indexOf("ON") > 0) {
                    UsageManageService.on = true;
                    System.out.println("TURN ON : 179(LED
stand)");
                    output.write(179);
                }
                else if(result.indexOf("OFF") > 0) {
                    UsageManageService.on = false;
                    System.out.println("TURN OFF : 179(LED
stand)");
                    output.write(120);
                }
                dis.close();
            } catch (Exception e) {
                e.printStackTrace();
            }
        }
    }
    static String getTime() {
```

```
        SimpleDateFormat f = new
SimpleDateFormat("[hh:mm:ss]");
        return "[HOME TCP SERVER] " + f.format(new Date());
    }
}
//////////////////////////////////////////////////////////////
////////////////////////////////////
TestThread
//////////////////////////////////////////////////////////////
////////////////////////////////////
TestThread.java
package thread;
import service.SubStringService;
import service.UsageManageService;
import java.util.Random;
/**
 * Created by Administrator on 2015-09-07.
 */
public class TestThread extends Thread {
    public TestThread() {
    }
    public void run() {
        int index = 0;
        while (true) {
            Random r = new Random();
            double randomValue = 0.10 + (0.09 - 0.10) * r.
nextDouble();
            if (!UsageManageService.on)
                randomValue = 0.0001 + (0.00001 - 0.0001) * r.
nextDouble();
            index++;
            SubStringService.getInstance().addUsage("192:" +
randomValue);
            try {
                sleep(5000);
            } catch (InterruptedException e) {
                e.printStackTrace();
            }
        }
    }
}
```

Appendix 7

RSSI (JAVA ANDROID AND PYTHON)

```
BleDevice
////////////////////////////////////////////////////////////////
/////////////////////////////////////
BleDevice.java
package so.stdc.ble;
import android.util.Log;
import java.util.Calendar;
import java.util.Date;
import java.util.List;
import java.util.Vector;
public class BleDevice
{
    private Date lastSeenTime = Calendar.getInstance().
getTime();
    private String name = "Unkown";
    private String address;
    private Vector<RssiSample> effectiveRssiSamples = new
Vector<RssiSample>();
    private double baseTXdBm = -58d;
    private double lossExponent = 2d;
    private double watchPosition = 0;
    private double watchSensitive = 2;
    private boolean watching = false;
    private Vector<RssiSample> allRssiSamples = new
Vector<RssiSample>();
    private double deltaThreshold = 7.5d;
    private long effectiveSampleTime = 2000;
    private boolean useTrendTrace = true;
    private double deltaRatio = 0d;
    public BleDevice(String address, String name, double rssi)
    {
        this.name = name;
        this.address = address;
```

```
        addRssi(rssi);
        checkValid();
    }
    public BleDevice(String address)
    {
        this.address = address;
        checkValid();
    }
    public BleDevice(String address, String name, RssiSample
rssiSample)
    {
        this.name = name;
        this.address = address;
        addRssi(rssiSample);
        checkValid();
    }
    public double getWatchPosition()
    {
        return watchPosition;
    }
    public void setWatchPosition(double watchPosition)
    {
        this.watchPosition = watchPosition;
    }
    public double getDeltaRatio()
    {
        return deltaRatio;
    }
    public void setDeltaRatio(double deltaRatio)
    {
        this.deltaRatio = deltaRatio;
    }
    public double getDeltaThreshold()
    {
        return deltaThreshold;
    }
    public long getEffectiveSampleTime()
    {
        return effectiveSampleTime;
    }
    public boolean isUseTrendTrace()
    {
        return useTrendTrace;
    }
    public void setDeltaThreshold(double deltaThreshold)
    {
        this.deltaThreshold = deltaThreshold;
    }
    public void setEffectiveSampleTime(long
```

```
effectiveSampleTime)
    {
        this.effectiveSampleTime = effectiveSampleTime;
    }
    public void setUseTrendTrace(boolean useTrendTrace)
    {
        this.useTrendTrace = useTrendTrace;
    }
    private boolean checkValid()
    {
        if (address == null)
        {
            throw new IllegalStateException("BleDevice.address
must be not null");
        }
        return true;
    }
    public double getWatchSensitive()
    {
        return watchSensitive;
    }
    public void setWatchSensitive(double watchSensitive)
    {
        this.watchSensitive = watchSensitive;
    }
    public boolean isWatching()
    {
        return watching;
    }
    public void setWatching(boolean watching)
    {
        this.watchPosition = getEstimatedDistance();
        this.watching = watching;
    }
    public void setName(String name)
    {
        this.name = name;
    }
    public String getAddress()
    {
        return address;
    }
    public String getName()
    {
        return name;
    }
    public double getBaseTXdBm()
    {
        return baseTXdBm;
```

```
    }
    public void setBaseTXdBm(double baseTXdBm)
    {
        this.baseTXdBm = baseTXdBm;
    }
    public double getLossExponent()
    {
        return lossExponent;
    }
    public void setLossExponent(double lossExponent)
    {
        this.lossExponent = lossExponent;
    }
    public Date getLastSeenTime()
    {
        return lastSeenTime;
    }
    public Vector<RssiSample> getAllRssiSamples()
    {
        return allRssiSamples;
    }
    public double getEstimatedDistance()
    {
        double rssi = getRssi();
        return Math.pow(10d, (baseTXdBm - rssi) / (10 *
lossExponent));
    }
    private synchronized void addRssi(RssiSample rssiSample)
    {
        allRssiSamples.add(rssiSample);
        if (allRssiSamples.size() > 100)
        {
            allRssiSamples.remove(0);
        }
        List<RssiSample> rssiSamples = this.allRssiSamples;
        if (rssiSamples.size() > 5)
        {
            rssiSamples = this.allRssiSamples.subList(this.
allRssiSamples.size() - 6, this.allRssiSamples.size() - 1);
        }
        if (rssiSamples.size() > 1)
        {
            double delta = getRssi() - rssiSamples.
get(rssiSamples.size() - 1).getRssi();
            double averageDelta = 0d;
            for (int index = 0; index < rssiSamples.size() - 2;
index++)
            {
                averageDelta += (rssiSamples.get(index).
```

```
getRssi() - rssiSamples.get(index + 1).getRssi());
            }
            averageDelta = averageDelta / (rssiSamples.size()
- 1);
            if (averageDelta == 0.0d) {
                if (delta == 0.0d) {
                    deltaRatio = 0.0d;
                } else
                {
                    deltaRatio = deltaThreshold *
deltaThreshold;
                }
            } else {
                deltaRatio = delta / averageDelta;
            }
//            Log.d("delta", "average delta: " + averageDelta);
//            Log.d("delta", "delta: " + delta);
//            Log.d("delta", "delta ratio: " + deltaRatio);
            if (0 <= deltaRatio && deltaRatio < deltaThreshold)
            {
                double avg = 0;
                int start = effectiveRssiSamples.size() - 5;
                start = (start < 0)? 0:start;
                int count=0;
                for (int index = start; index <
effectiveRssiSamples.size(); index++)
                {
                    avg += effectiveRssiSamples.
elementAt(index).getRssi();
                    count++;
                }
                avg = (avg + rssiSample.getRssi()) / (count+1);
                rssiSample.setRssi(avg);
                effectiveRssiSamples.add(rssiSample);
                if (effectiveRssiSamples.size() > 20)
                {
                    effectiveRssiSamples.remove(0);
                }
//                Log.d("delta", "effective: " + avg);
//                Log.d("delta", "effective SIZE: " +
effectiveRssiSamples.size());
            }
        }
        lastSeenTime = rssiSample.getSampledDate();
    }
    public void addRssi(double rssi)
    {
        RssiSample rssiSample = new RssiSample(rssi, Calendar.
getInstance().getTime());
```

```
        addRssi(rssiSample);
    }
    public synchronized void addRssi(BleDevice rssi)
    {
        addRssi(new RssiSample(rssi.getRssi(), Calendar.
getInstance().getTime()));
//          Log.d("RSSI", "add sample: " + rssi.getRssi());
//          Log.d("RSSI", "sample size: " + allRssiSamples.
size());
//          Log.d("RSSI", "rssi: " + getRssi());
    }
    public synchronized double getRssi()
    {
        long sampleTime = Calendar.getInstance().
getTimeInMillis() - effectiveSampleTime;
        if (useTrendTrace)
        {
            if (effectiveRssiSamples.size() <= 0)
            {
                if (allRssiSamples.size() <= 0)
                {
                    return 0d;
                } else
                {
                    double sum = 0;
                    int count = 0;
                    for (RssiSample rssi : allRssiSamples)
                    {
                        if (rssi.getSampledDate().getTime() >
sampleTime)
                        {
                            sum += rssi.getRssi();
                            count++;
                        }
                    }
                    return sum / count;
                }
            }
            double sum = 0;
            int count = 0;
            for (RssiSample rssi : effectiveRssiSamples)
            {
                if (rssi.getSampledDate().getTime() >
sampleTime)
                {
                    sum += rssi.getRssi();
                    count++;
                }
            }
```

```
            if (count == 0)
            {
                for (RssiSample rssi : effectiveRssiSamples)
                {
                    sum += rssi.getRssi();
                }
                count = effectiveRssiSamples.size();
            }
            return sum / count;
        }
        else
        {
            double sum = 0;
            int count = 0;
            for (RssiSample rssi : allRssiSamples)
            {
                if (rssi.getSampledDate().getTime() >
sampleTime)
                {
                    sum += rssi.getRssi();
                    count++;
                }
            }
            return sum / count;
        }
    }
    @Override
    public boolean equals(Object object)
    {
        if (object==null) return false;
        if (this.getClass() != object.getClass()) return false;
        return getAddress().equals(((BleDevice) object).
getAddress());
    }
    @Override
    public String toString()
    {
        return address;
    }
}
////////////////////////////////////////////////////////////////
//////////////////////////////////////
BleDeviceList
////////////////////////////////////////////////////////////////
//////////////////////////////////////
BleDeviceList.java
package so.stdc.ble;
import java.util.*;
public class BleDeviceList  implements List<BleDevice>
```

```
{
    public static BleDeviceList instance;
    private long lastSeenCheckInterval = 2000;
    private List<BleDevice> devices = new Vector<BleDevice>();
    private BleDeviceList()
    {
        new Timer().schedule(new LastSeenCheckTask(),
lastSeenCheckInterval);
    }
    public static BleDeviceList getInstance()
    {
        if (instance == null)
        {
            instance = new BleDeviceList();
        }
        return instance;
    }
    @Override
    public void add(int location, BleDevice device)
    {
        if (devices.contains(device))
        {
            devices.get(devices.indexOf(device)).
addRssi(device);
        }
        else
        {
            devices.add(location, device);
        }
    }
    @Override
    public boolean add(BleDevice device)
    {
        if (devices.contains(device))
        {
            devices.get(devices.indexOf(device)).
addRssi(device);
            return true;
        }
        else
        {
            return devices.add(device);
        }
    }
    @Override
    public boolean addAll(int location, Collection<? extends
BleDevice> devices)
    {
        List<BleDevice> tempDevices = new Vector<BleDevice>(0);
```

```
        for (BleDevice device: devices)
        {
            if (this.devices.contains(device))
            {
                this.devices.get(this.devices.indexOf(device)).
addRssi(device);
            }
            else
            {
                tempDevices.add(device);
            }
        }
        return tempDevices.addAll(location, tempDevices);
    }
    @Override
    public boolean addAll(Collection<? extends BleDevice>
devices)
    {
        return addAll(devices.size(), devices);
    }
    @Override
    public void clear()
    {
        devices.clear();
    }
    @Override
    public boolean contains(Object device)
    {
        return devices.contains(device);
    }
    @Override
    public boolean containsAll(Collection<?> devices)
    {
        return this.devices.containsAll(devices);
    }
    @Override
    public boolean equals(Object object)
    {
        return devices.equals(object);
    }
    @Override
    public BleDevice get(int location)
    {
        return devices.get(location);
    }
    @Override
    public int hashCode()
    {
        return devices.hashCode();
```

```
    }
    @Override
    public int indexOf(Object object)
    {
        return devices.indexOf(object);
    }
    @Override
    public boolean isEmpty()
    {
        return devices.isEmpty();
    }
    @Override
    public Iterator<BleDevice> iterator()
    {
        return devices.iterator();
    }
    @Override
    public int lastIndexOf(Object object)
    {
        return devices.lastIndexOf(object);
    }
    @Override
    public ListIterator<BleDevice> listIterator()
    {
        return devices.listIterator();
    }
    @Override
    public ListIterator<BleDevice> listIterator(int location)
    {
        return devices.listIterator(location);
    }
    @Override
    public BleDevice remove(int location)
    {
        return devices.remove(location);
    }
    @Override
    public boolean remove(Object object)
    {
        return devices.remove(object);
    }
    @Override
    public boolean removeAll(Collection<?> collection)
    {
        return devices.removeAll(collection);
    }
    @Override
    public boolean retainAll(Collection<?> collection)
    {
```

```
        return devices.retainAll(collection);
    }
    @Override
    public BleDevice set(int location, BleDevice device)
    {
        if (devices.contains(device))
        {
            BleDevice existDevice = devices.get(devices.
indexOf(device));
            existDevice.addRssi(device);
            return existDevice;
        }
        else
        {
            return devices.set(location, device);
        }
    }
    @Override
    public int size()
    {
        return devices.size();
    }
    @Override
    public List<BleDevice> subList(int start, int end)
    {
        BleDeviceList bleDeviceList = new BleDeviceList();
        bleDeviceList.addAll(devices.subList(start, end));
        return bleDeviceList;
    }
    @Override
    public synchronized Object[] toArray()
    {
        Object[] array = new Object[devices.size()];
        for(int index = 0; index < devices.size(); index++)
        {
            array[index] = devices.get(index);
        }
        return array;
    }
    @Override
    public synchronized  <T> T[] toArray(T[] array)
    {
        T[] tempArray;
        if (array.length >= devices.size())
        {
            tempArray = array;
        }
        else
        {
```

```
            tempArray = (T[])new Object[devices.size()];
        }
        for(int index = 0; index < devices.size(); index++)
        {
            tempArray[index] = (T)devices.get(index);
        }
        return tempArray;
    }
    public long getLastSeenCheckInterval()
    {
        return lastSeenCheckInterval;
    }
    public void setLastSeenCheckInterval(long
lastSeenCheckInterval)
    {
        if (this.lastSeenCheckInterval <= 0 &&
lastSeenCheckInterval > 0)
        {
            new Timer().schedule(new LastSeenCheckTask(),
lastSeenCheckInterval);
        }
        this.lastSeenCheckInterval = lastSeenCheckInterval;
    }
    public void destroy()
    {
        lastSeenCheckInterval = 0;
    }
    private class LastSeenCheckTask extends TimerTask
    {
        @Override
        public synchronized void run()
        {
            while(true)
            {
                List<BleDevice> removeList = new
Vector<BleDevice>();
                for (BleDevice device : devices)
                {
                    if (Calendar.getInstance().
getTimeInMillis() - device.getLastSeenTime().getTime() >
lastSeenCheckInterval)
                    {
                        removeList.add(device);
                    }
                }
                devices.removeAll(removeList);
                if (lastSeenCheckInterval > 0)
                {
                    //new Timer().schedule(new
```

```
LastSeenCheckTask(), lastSeenCheckInterval);
                    try
                    {
                        Thread.sleep(lastSeenCheckInterval);
                    } catch (InterruptedException e)
                    {
                        e.printStackTrace();
                    }
                }
                else
                {
                    break;
                }
            }
        }
    }
}
//////////////////////////////////////////////////////////////////
/////////////////////////////////////
RssiSample
//////////////////////////////////////////////////////////////////
/////////////////////////////////////
RssiSample.java
package so.stdc.ble;
import java.util.Date;
public class RssiSample
{
    private Date sampledDate;
    private double rssi;
    public RssiSample(double rssi, Date sampledDate)
    {
        this.rssi = rssi;
        this.sampledDate = sampledDate;
    }
    public Date getSampledDate()
    {
        return sampledDate;
    }
    public void setSampledDate(Date sampledDate)
    {
        this.sampledDate = sampledDate;
    }
    public double getRssi()
    {
        return rssi;
    }
    public void setRssi(double rssi)
    {
        this.rssi = rssi;
```

```
    }
}
////////////////////////////////////////////////////////////////
////////////////////////////////////////////////
////////////////////////////////////////////////////////////////
////////////////////////////////////////
C and C++ PLC
////////////////////////////////////////////////////////////////
////////////////////////////////////////
//MainServer//
////////////////////////////////////////////////////////////////
////////////////////////////////////////
#define DEBUGLEVEL 1
#define CAN_RESEND 0
#include <iostream>
#include <fstream>
#include <cstdlib>
#include <time.h>
#include "lib/Manager/DeviceManager.h"
#include "lib/Manager/WebPageManager.h"
#include "lib/CountingServer.h"
using namespace std;
int main(int argc, char* argv[])
{
   int ThisMonthWatt = 0;
   int PrevThisMonthWatt = 0;
   int LastMonthWatt = 0;
   devicemanager DM;
   webpagemanager WM;
   countingServer CS;
   pthread_t ServerThread;
   mail RecvMail;
   time_t timer;
   struct tm *t;
   time(&timer);
   t = localtime(&timer);
   cout << "2014 Capstone Smart Home Project" << endl;
   cout << "Team Pikachu - Main Server" << endl;
   cout << "Author : Jun Ho Huh" << endl;
   cout << "UpTime : " << t->tm_year + 1900 << "-" << t->tm_mon
+1 << "-" << t->tm_mday << "  " << t->tm_hour << ":" << t->tm_
min << ":" << t->tm_sec << endl;
   if(DM.Initialize() && CS.Initialize("192.168.0.7", 11550) ==
0)
   {
      pthread_create(&ServerThread, NULL, &CS.Start, NULL);
      pthread_detach(ServerThread);
      while(1)
      {
```

```
            DM.InBox.PushMail(1000, 2005, "ReqWatt");
            sleep(1);
            RecvMail = DM.OutBox.PopMail(1000);
            if(RecvMail.To == 1000 && RecvMail.From == 2005)
// received from WattMeter
            {
                  ThisMonthWatt = atoi(RecvMail.Data.c_str());
                  if(PrevThisMonthWatt != ThisMonthWatt)
                  {
                        WM.Write(ThisMonthWatt,
LastMonthWatt);
                  }
                  PrevThisMonthWatt = ThisMonthWatt;
            }
            sleep(1);
            if(CS.HomePeople == 0) // home is empty
            {
                  // turn off all devices
                  DM.InBox.PushMail(1000, 3000, "Off");
                  //CS.HomePeople = 1;
            }
            else
            {
                  //turn on all deivces
                  DM.InBox.PushMail(1000, 3000, "On");
                  //CS.HomePeople = 0;
            }
            sleep(1);
      }
   }
}
////////////////////////////////////////////////////////////////
/////////////////////////////////////
BBang2
////////////////////////////////////////////////////////////////
/////////////////////////////////////
#ifndef BBang2_socket
   #define BBang2_scoket
   #include <arpa/inet.h>
   struct socket_t
   {
      int FD;
      struct sockaddr_in Addr;
      unsigned int AddrSize;
      int SerialNum;
      socket_t()
      {
            FD = 0;
            AddrSize = sizeof(Addr);
```

```
            SerialNum = 1000; // initial serialnumber is
server number.
      }
   };
#endif
/////////////////////////////////////////////////////////////
////////////////////////////////////
BBang2_string
/////////////////////////////////////////////////////////////
////////////////////////////////////
#ifndef BBang2_string
   #define BBang2_string
   #include <string>
   using namespace std;
   string toString(int number)
   {
      string number_string = "";
      char ones_char;
      int ones = 0;
      while(true){
            ones = number % 10;
            switch(ones){
                  case 0: ones_char = '0'; break;
                  case 1: ones_char = '1'; break;
                  case 2: ones_char = '2'; break;
                  case 3: ones_char = '3'; break;
                  case 4: ones_char = '4'; break;
                  case 5: ones_char = '5'; break;
                  case 6: ones_char = '6'; break;
                  case 7: ones_char = '7'; break;
                  case 8: ones_char = '8'; break;
                  case 9: ones_char = '9'; break;
                  //default : ErrorHandling("Trouble
converting number to string.");
            }
            number -= ones;
            number_string = ones_char + number_string;
            if(number == 0){
                  break;
            }
            number = number/10;
      }
      return number_string;
   }
   string* StringSplit(string strOrigin, string strTok)
   {
      unsigned int cutAt;
      int index = 0;
      string* strResult = new string[10];
```

```
        while ((cutAt = strOrigin.find_first_of(strTok)) !=
strOrigin.npos)
        {
                if (cutAt > 0)
                {
                        strResult[index++] = strOrigin.substr(0,
cutAt);
                }
                strOrigin = strOrigin.substr(cutAt+1);
        }
        if(strOrigin.length() > 0)
        {
                strResult[index++] = strOrigin.substr(0, cutAt);
        }
        return strResult;
    }
#endif
////////////////////////////////////////////////////////////////
/////////////////////////////////////
BBang2_uart
////////////////////////////////////////////////////////////////
/////////////////////////////////////
//-- to communication using rs232
#ifndef BUFF_SIZE
   #define BUFF_SIZE 128
#endif
#ifndef BAUDRATE
   #define BAUDRATE 9600
#endif
#ifndef NON_BLOCK
   #define NON_BLOCK 0
#endif
#ifndef BLOCK
   #define BLOCK 1
#endif
#ifndef BBang2_uart
   #define BBang2_uart
#include <iostream>
using namespace std;
   #include <sys/types.h>
   #include <termios.h>
   #include <fcntl.h>
   #include <unistd.h>
   #include <stdio.h>
   #include <string.h>
   class uart
   {
      private :
```

```
            const char* tty;
            int FD;
            struct termios IO;
            char Buffer[BUFF_SIZE];;
      public :
            bool Initialize(const char* tty, int BlockMode);
            int ReadLine(char* Buffer);
            int WriteLine(const char* Msg);
};
bool::uart::Initialize(const char* tty, int BlockMode)
{
      this->tty = tty;
      if(BlockMode == BLOCK)
            FD = open(tty, O_RDWR | O_NOCTTY);
      else if(BlockMode == NON_BLOCK)
            FD = open(tty, O_RDWR | O_NOCTTY | O_NONBLOCK);
      else
            return false;
      if(FD == -1)
      {
            perror("UART :: ");
            return false;
      }
      switch(BAUDRATE) //baudRate
      {
      case 9600 :
            IO.c_cflag = B9600;
            break;
      case 19200 :
            IO.c_cflag = B19200;
            break;
      case 38400 :
            IO.c_cflag = B19200;
            break;
      }
      IO.c_cflag |= CS8;
      IO.c_cflag |= CLOCAL;
      IO.c_cflag |= CREAD;
      IO.c_iflag = 0;
      IO.c_oflag = 0;
      IO.c_lflag = 0;
      IO.c_cc[VTIME] = 0;
      IO.c_cc[VMIN] = 1;
      tcflush(FD, TCIFLUSH);
      tcsetattr(FD, TCSANOW, &IO);
      return true;
}

int uart::ReadLine(char* Buffer)
```

```
    {
        int line = 0;
        char Input;
        memset(Buffer, 0, BUFF_SIZE);
        read(FD, &Input, sizeof(Input));
        while(Input != '+')
        {
                if(line > 128)
                {
                        memset(Buffer, 0, BUFF_SIZE);
                        return 0;
                }
                Buffer[line] = Input;
                line++;
                read(FD, &Input, sizeof(Input));
        }
        return line;
    }
    int uart::WriteLine(const char* Msg)
    {
        for(unsigned int i =0; i < strlen(Msg); i++)
        {
                write(FD, Msg+i, 1);
        }
        return strlen(Msg+1);
    }
#endif
////////////////////////////////////////////////////////////////
///////////////////////////////////////
Manager
////////////////////////////////////////////////////////////////
///////////////////////////////////////
DeviceManager
////////////////////////////////////////////////////////////////
///////////////////////////////////////
//------------------------------------//
//-------- Device Manager --------//
//------------------------------------//
//-- this connect and control devices via uart
//-- use between server and devices protocol (BSDP)
//-- From(int) To(int) User(int) Command(string) +(end
charater)
//-- each devices have uniq serial number, so DM can
distinguish between one device from another
#ifndef BUFF_SIZE
    #define BUFF_SIZE 128
#endif
#ifndef CAN_RESEND
    #define CAN_RESEND 1
```

```
#endif
#ifndef DeviceManager
   #define DeviceManager
   #include "../BBang2/BBang2_uart.h"
   #include "../BBang2/BBang2_string.h"
   #include "MailBox.h"
   #include "LogManager.h"
   #include <stdlib.h>
   #include <iostream>
   #include <string>
   #include <list>
   using namespace std;
   //----------------------------//
   //----- Device Class -----//
   //---------------------------//
   class device
   {
   private :
      string Name;
      int SerialNum;
      list<string> FuncList; // list of usable function.
   public :
      device()
      {
            SerialNum = 0;
      }
      device(int SerialNum, string Name)
      {
            this->SerialNum = SerialNum;
            this->Name = Name;
      }
      string getName(){ return Name; }
      int getSerialNum(){ return SerialNum; }
      bool isUsableFunc(string Function)
      {
            for(list<string>::iterator Pos = FuncList.begin();
Pos != FuncList.end(); Pos++)
            {
                  if(Pos->compare(Function) ==0) // Find !!
                        return true;
            }
            return false; // not found T_T
      }
   };
   //-------------------------------//
   //-- Device Manager Class --//
   //-------------------------------//
   class devicemanager
   {
```

```
    private :
        pthread_t RecvThread;
        pthread_t SendThread;
        list<device> DeviceList;
        bool SearchDevice(int SerialNum);
        bool ConnectDevice(int SerialNum);
        static void* RecvCommand(void* null);
        static void* SendCommand(void* null);
    public :
        static uart Uart;
        static logmanager Log;
        static mailbox InBox;
        static mailbox OutBox;
        static bool Busy; // is enable send command ?

        static int MailCount;
        static int PrevMailCount;
        bool Initialize();
    };
    bool devicemanager::Initialize()
    {
        if(!Uart.Initialize("/dev/ttyUSB0", BLOCK))
        {
                Log.Write("DM::Initialize() => Failed", DEBUG_
LEVEL::ERROR);
                return false;
        }
        pthread_create(&RecvThread, NULL,
&devicemanager::RecvCommand, NULL);
        pthread_detach(RecvThread);
        pthread_create(&SendThread, NULL,
&devicemanager::SendCommand, NULL);
        pthread_detach(SendThread);
        Log.Write("DM::Initialize()", DEBUG_LEVEL::DEBUG);
        return true;
    }
    bool devicemanager::ConnectDevice(int SerialNum)
    {
        return true;
    }
    bool devicemanager::SearchDevice(int SerialNum)
    {
        list<device>::iterator Pos;
        for(Pos = DeviceList.begin(); Pos != DeviceList.end();
Pos++)
        {
                if(Pos->getSerialNum() == SerialNum)
                {
                        Log.Write("DM::SearchDevice("+toString(Seria
```

```
lNum)+") => True", DEBUG_LEVEL::DEBUG);
                  return true;
            }
      }
      Log.Write("DM::SearchDevice("+toString(SerialNum)+") =>
Failed", DEBUG_LEVEL::DEBUG);
      return false;
   }
   void* devicemanager::SendCommand(void* null)
   {
      char Buffer[BUFF_SIZE];
      mail newMail;
      mail prevMail;
      int BusyCounter;
      while(1)
      {
            //-- chaek mail box.
            if(!InBox.isEmpty())
            {
                  if(!Busy)
                  {
                        newMail = InBox.PopMail();
                        if(MailCount != PrevMailCount)
                  {
                              prevMail.ResendCount++;
                              if((prevMail.ResendCount <= 2)
&& (CAN_RESEND == 1))
                              {
                                    Log.Write("Resend mail
- From:"+toString(prevMail.From)+" To:"+toString(prevMail.To)+"
Data:"+prevMail.Data, DEBUG_LEVEL::WARN);
                                    InBox.PushMail(prevMail);
                              }
                              else
                              {
                                    Log.Write("Trash mail
- From:"+toString(prevMail.From)+" To:"+toString(prevMail.To)+"
Data:"+prevMail.Data, DEBUG_LEVEL::WARN);
                              }
                              PrevMailCount = MailCount;
                        }
                        if(newMail.To != 100 && newMail.To !=
2000 && newMail.To != 3000)
{
                              MailCount++;
                              prevMail = newMail;
                        }
                        Log.Write("Send Mail
From:"+toString(newMail.From)+" To:"+toString(newMail.To)+"
```

```
Data:"+newMail.Data, DEBUG_LEVEL::INFO);
                                //-- Send Mail
                                sprintf(Buffer, "1000%d %d %s +",
newMail.To, newMail.From, newMail.Data.c_str());
                                Uart.WriteLine("1000 1111 1111 begin
+");
                                Uart.WriteLine(Buffer);
                        }
                        else
                        {
                                BusyCounter++;
                                if(BusyCounter > 10)
                                {
                                        Busy = false;
                                        BusyCounter = 0;
                                        Log.Write("plc line is blocked
by network error.", DEBUG_LEVEL::WARN);
                                }
                        }
                }
        }
        pthread_exit(NULL);
    }
    void* devicemanager::RecvCommand(void* null)
    {
        char Buffer[BUFF_SIZE];
        string Msg;
        string* tmpStr;
        while(1)
        {
                if(Uart.ReadLine(Buffer) > 0)
                {
                        Msg = Buffer;
                        tmpStr = StringSplit(Msg, " "); // [From]
[To] [User] [Command]
                        if(tmpStr[1].compare("1111") == 0) // this
Msg is broadcasted message
                        {
                                if(tmpStr[3].compare("begin") ==0)
                                        Busy = true;
                                else if(tmpStr[3].compare("end") ==
0)
                                        Busy = false;
                        }
                        else if(tmpStr[1].compare("1000") == 0) //
this Msg is delivered from somewhere to server
                        {

                                mail newMail(atoi(tmpStr[0].c_str()),
```

```
atoi(tmpStr[2].c_str()), tmpStr[3]);
                         OutBox.PushMail(newMail);
                         MailCount--;
                         Log.Write("Receive Mail
From:"+toString(newMail.From)+" To:"+toString(newMail.To)+"
Data:"+newMail.Data, DEBUG_LEVEL::INFO);
                    }
               }
          }
          pthread_exit(NULL);
     }
     //-- DM's static member initialization.
     uart devicemanager::Uart;
     logmanager devicemanager::Log;
     bool devicemanager::Busy;
     int devicemanager::MailCount = 0;
     int devicemanager::PrevMailCount = 0;
     mailbox devicemanager::InBox;
     mailbox devicemanager::OutBox;
#endif
////////////////////////////////////////////////////////////////
/////////////////////////////////////
LogManager
////////////////////////////////////////////////////////////////
/////////////////////////////////////
//--------------------//
//-- Log Manager --//
//--------------------//
#ifndef DEBUGLEVEL
     #define DEBUGLEVEL 1
#endif
#ifndef LogManager
     #define LogManager
     #include <fstream>
     #include <string>
     #include <time.h>
     #include "../BBang2/BBang2_string.h"
     using namespace std;
     struct DEBUG_LEVEL
     {
          static int ERROR;
          static int WARN;
          static int DEBUG;
          static int INFO;
     };
     class logmanager
     {
     private :
          fstream Log;
```

```
        time_t timer;
    public :
        void Write(string str, int level);
    };
    void logmanager::Write(string str, int level)
    {
        if(level >= DEBUGLEVEL)
        {
                time(&timer);
                struct tm *t = localtime(&timer);
                string FileName = "data/" + toString(t->tm_year +
1900) + "-" + toString(t->tm_mon) + "-" + toString(t->tm_mday);
                string LogStr;
                do{ Log.open(FileName.c_str(), ios::out |
ios::app); }while( Log <= 0 );
                switch(level)
                {
                case 0:
                        LogStr = "INFO(" + toString(t->tm_hour) +
":" + toString(t->tm_min) + ":" + toString(t->tm_sec) + ")
" + str;
                        break;
                case 1:
                        LogStr = "DEBUG(" + toString(t->tm_hour) +
":" + toString(t->tm_min) + ":" + toString(t->tm_sec) + ")
" + str;
                        break;
                case 2:
                        LogStr = "WARN(" + toString(t->tm_hour) +
":" + toString(t->tm_min) + ":" + toString(t->tm_sec) + ")
" + str;
                        break;
                case 3:
                        LogStr = "ERROR(" + toString(t->tm_hour) +
":" + toString(t->tm_min) + ":" + toString(t->tm_sec) + ")
" + str;
                        break;
                }
                Log << LogStr << endl;
                cout << LogStr << endl;
                Log.close();
        }
    }
    int DEBUG_LEVEL::ERROR = 3;
    int DEBUG_LEVEL::WARN = 2;
    int DEBUG_LEVEL::DEBUG = 1;
    int DEBUG_LEVEL::INFO = 0;
#endif
////////////////////////////////////////////////////////////////
```

```
//////////////////////////////////////
MailBox
//////////////////////////////////////////////////////////////////////////
//////////////////////////////////////
//-----------------//
//-- Mail Box --//
//-----------------//
//for the server and DM communication
#ifndef Mail
   #define Mail
   #include <list>
   #include <string>
   #include <pthread.h>
   using namespace std;
   struct mail
   {
      mail()
      {
            this->From = 0;
            this->To = 0;
            this->ResendCount = 0;
      }
      mail(int From, int To, string Data)
      {
            this->From = From;
            this->To = To;
            this->Data = Data;
            this->ResendCount = 0;
      }
      int From;
      int To;
      string Data;
      int ResendCount;
   };
   class mailbox
   {
   private :
      list<mail> MailBox;
      pthread_mutex_t var = PTHREAD_MUTEX_INITIALIZER;
      list<mail>::iterator SearchMail(int SerialNum);
   public :
      int Size();
      bool isEmpty();
      void PushMail(mail newMail);
      void PushMail(int From, int To, string Data);
      mail PopMail(int SerialNum);
      mail PopMail();
   };
   int mailbox::Size()
```

```
{
    return MailBox.size();
}
bool mailbox::isEmpty()
{
    return MailBox.empty();
}
void mailbox::PushMail(mail newMail)
{
    pthread_mutex_lock(&var);
    MailBox.push_front(newMail);
    pthread_mutex_unlock(&var);
}
void mailbox::PushMail(int From, int To, string Data)
{
    mail newMail(From, To, Data);
    pthread_mutex_lock(&var);
    MailBox.push_front(newMail);
    pthread_mutex_unlock(&var);
}
mail mailbox::PopMail()
{
    pthread_mutex_lock(&var);
    mail rtnMail = (mail)MailBox.back();
    MailBox.pop_back();
    pthread_mutex_unlock(&var);
    return rtnMail;
}
mail mailbox::PopMail(int SerialNum)
{
    pthread_mutex_lock(&var);
    mail rtnMail;
    list<mail>::iterator Pos = SearchMail(SerialNum);
    if(Pos != MailBox.end())
    {
          rtnMail = *Pos;
          MailBox.erase(Pos);
    }
    pthread_mutex_unlock(&var);
    return rtnMail;
}
list<mail>::iterator mailbox::SearchMail(int SerialNum)
{
    list<mail>::iterator Pos;
    for(Pos = MailBox.begin(); Pos != MailBox.end(); Pos++)
    {
          if(Pos->To == SerialNum)
                return Pos;
    }
```

```
      return (Pos = MailBox.end());
   }
#endif
////////////////////////////////////////////////////////////////
/////////////////////////////////////
WebPageManager
////////////////////////////////////////////////////////////////
/////////////////////////////////////
//--------------------//
//-- PageManager --//
//--------------------//
#ifndef PageManager
   #define PageManager
   #include <fstream>
   #include <string>
   #include <time.h>
   #include "../BBang2/BBang2_string.h"
   #include "LogManager.h"
   using namespace std;
   class webpagemanager
   {
   private :
      fstream WebPage;
      time_t timer;
      logmanager Log;
   public :
      void Read();
      void Write(int ThisMonthWatt, int LastMonthWatt);
   };
   void webpagemanager::Read()
   {
      char Buffer[100];
      WebPage.open("index.html", ios::in);
      while(!WebPage.eof())
      {
            WebPage.getline(Buffer, 100);
            cout << Buffer << endl;
      }
      Log.Write("WP::Read()", DEBUG_LEVEL::DEBUG);
      WebPage.close();
   }
   void webpagemanager::Write(int ThisMonthWatt, int
LastMonthWatt)
   {
      WebPage.open("index.html", ios::out);
      WebPage << "<html>" << endl;
      WebPage << "<head>" << endl;
      WebPage << "<title>Pikachu:: Power Usage </title>" <<
endl;
```

```
        WebPage << "</head>" << endl;
        WebPage << "<script>" << endl;
        WebPage << "var ThisMonthWatt = " << ThisMonthWatt <<
";" << endl;
        WebPage << "var LastMonthWatt = " << LastMonthWatt <<
";" << endl;
        WebPage << "function Display()" << endl;
        WebPage << "{" << endl;
        WebPage << "document.write('This
Month:'+ThisMonthWatt+'kWh</br>');" << endl;
        WebPage << "document.write('Last
Month:'+LastMonthWatt+'kWh</br>');" << endl;
        WebPage << "}" << endl;
        WebPage << "</script>" << endl;
        WebPage << "<body onLoad = Display();>" << endl;
        WebPage << "</body>" << endl;
        WebPage << "</html>" << endl;
        Log.Write("WP::("+toString(ThisMonthWatt)+ ", " +
toString(LastMonthWatt) + ")", DEBUG_LEVEL::DEBUG);
        WebPage.close();
    }
#endif
/////////////////////////////////////////////////////////////////
//////////////////////////////////////
Server
/////////////////////////////////////////////////////////////////
//////////////////////////////////////
#ifndef BUFF_SIZE
   #define BUFF_SIZE 128
#endif
#ifndef LISTEN_SIZE
   #define LISTEN_SIZE 20
#endif
#ifndef Server
   #define Server
   #include "../BBang2/BBang2_string.h"
   #include "../Device/DeviceManager.h"
   #include <iostream>
   #include <string.h>
   #include <pthread.h>
   #include <string>
   #include <list>
   #include <unistd.h>
   #include <sys/types.h>
   #include <arpa/inet.h>
   #include <sys/socket.h>
   using namespace std;
   struct socket_t
   {
```

```
        int FD;
        struct sockaddr_in Addr;
        unsigned int AddrSize;
        int SerialNum;
        socket_t()
        {
               FD = 0;
               AddrSize = sizeof(Addr);
               SerialNum = 1000; // initial serialnumber is
server number.
        }
    };
    class server
    {
    private :
        socket_t ServerSocket;
        devicemanager DM;
        pthread_t UserThread;
        static void* User(void* Socket); // process user's
demmand
        void SendMsg(socket_t Target, string Msg);
        string SendMailAndReceive(int From, int To, string Msg);
    public :
        int Initialize(string ServerIP, int Port); // Server and
DM init.
        int Start(); // Server receive start;
    };
    int server::Initialize(string ServerIP, int Port)
    {
        memset(&ServerSocket.Addr, 0, ServerSocket.AddrSize);
        ServerSocket.Addr.sin_family = AF_INET;
        ServerSocket.Addr.sin_port = htons(Port);
        ServerSocket.Addr.sin_addr.s_addr = inet_
addr(ServerIP.c_str());
        DM.Initialize();
        ServerSocket.FD = socket(PF_INET, SOCK_STREAM, 0);
        if(ServerSocket.FD == -1)
               return -1;
        if(bind(ServerSocket.FD, (struct sockaddr*)&ServerSocket.
Addr, ServerSocket.AddrSize) == -1)
               return -2;
        if(listen(ServerSocket.FD, LISTEN_SIZE) == -1)
               return -3;
        return 0;
    }
    int server::Start()
    {
        socket_t ClientSocket;
        int ClientCount = 0;
```

```
        while(1)
        {
                ClientSocket.FD = accept(ServerSocket.FD, (struct
sockaddr*)&ClientSocket.Addr, &ClientSocket.AddrSize);
                if(ClientSocket.FD == -1)
                        cout << "Server :: Socket Err" << endl;
                else // new client appear Du-Dung.
                {
                        if(ClientCount > LISTEN_SIZE)
                                ClientCount = 0;
                        ClientSocket.SerialNum = 4000 + ClientCount;
// giving out serial number to client
                        ClientCount++;
                        pthread_create(&UserThread, NULL,
&server::User, (socket_t*)&ClientSocket);
                        pthread_detach(UserThread);
                        cout << "Server :: new Client connected =>
IP : " << inet_ntoa(ClientSocket.Addr.sin_addr) << "Serial : "
<< ClientSocket.SerialNum << endl;
                }
        }
        return 0;
    }
    void* server::User(void* Socket)
    {
        socket_t UserSocket = Socket;
        int line = 0;
        char InputBuffer[BUFF_SIZE];
        string* Buffer;
        while(UserSocket.FD > 0) // while socket alive.
        {
                line = read(UserSocket.FD, InputBuffer, BUFF_
SIZE);
                //Buffer = InputBuffer;
                Buffer = StringSplit(InputBuffer, " "); // [To
- device serial num] [Command - string]
                if(Buffer[1].compare("Init") == 0) // Connect new
device
                {
                        if(DM.SearchDevice(atoi(Buffer[0].c_str())))
// already exist
                                SendMsg(UserSocket, "AOK");
                        else
                        {
                                SendMsg(UserSocket,
SendMailAndReceive(UserSocket.SerialNum, atoi(Buffer[0].c_
str()), Buffer[1]));
                        }
                }
```

```
        }
        pthread_exit(NULL);
    }
    void server::SendMsg(socket_t Target, string Msg)
    {
    }
    string server::SendMailAndReceive(int From, int To, string
Msg)
    {
        mail ReceivedMail;
        DM.InBox.PushMail(From, To, Msg); // Send Command to
device
        while(ReceivedMail.From != From) // wait until mail
receive.
        {
                ReceivedMail = DM.OutBox.PopMail(From);
        }
        //SendMsg(UserSocket, ReceivedMail.Data); //Send data
received from device to client
        return ReceivedMail.Data;
    }
#endif
////////////////////////////////////////////////////////////////
//////////////////////////////////////
CountingServer
////////////////////////////////////////////////////////////////
//////////////////////////////////////
#ifndef BUFF_SIZE
    #define BUFF_SIZE 128
#endif
#ifndef LISTEN_SIZE
    #define LISTEN_SIZE 20
#endif
#ifndef Server
    #define Server
    #include "BBang2/BBang2_string.h"
    #include "BBang2/BBang2_socket.h"
    #include "Manager/LogManager.h"
    #include <iostream>
    #include <string.h>
    #include <pthread.h>
    #include <string>
    #include <list>
    #include <unistd.h>
    #include <sys/types.h>
    #include <arpa/inet.h>
    #include <sys/socket.h>
    using namespace std;
    class countingServer
```

```
    {
    private :
        static void* User(void* Socket); // process user's
demmand
        void SendMsg(socket_t Target, string Msg);
        string SendMailAndReceive(int From, int To, string Msg);
    public :
        static logmanager Log;
        static socket_t ServerSocket;
        static pthread_t UserThread;
        static int HomePeople;
        int Initialize(string ServerIP, int Port); // Server and
DM init.
        static void* Start(void* null); // Server receive start;
    };
    int countingServer::Initialize(string ServerIP, int Port)
    {
        memset(&ServerSocket.Addr, 0, ServerSocket.AddrSize);
        ServerSocket.Addr.sin_family = AF_INET;
        ServerSocket.Addr.sin_port = htons(Port);
        ServerSocket.Addr.sin_addr.s_addr = inet_
addr(ServerIP.c_str());
        ServerSocket.FD = socket(PF_INET, SOCK_STREAM, 0);
        if(ServerSocket.FD == -1)
        {
                Log.Write("Server::socket open failed", DEBUG_
LEVEL::ERROR);
                return -1;
        }
        if(bind(ServerSocket.FD, (struct sockaddr*)&ServerSocket.
Addr, ServerSocket.AddrSize) == -1)
        {
                Log.Write("Server::bind failed", DEBUG_
LEVEL::ERROR);
                return -2;
        }
        if(listen(ServerSocket.FD, LISTEN_SIZE) == -1)
        {
                Log.Write("Server::listen failed", DEBUG_
LEVEL::ERROR);
                return -3;
        }
        Log.Write("Server::Initialize()", DEBUG_LEVEL::DEBUG);
        return 0;
    }
    void* countingServer::Start(void* null)
    {
        socket_t ClientSocket;
        int ClientCount = 0;
```

```
      while(1)
      {
            ClientSocket.FD = accept(ServerSocket.FD, (struct
sockaddr*)&ClientSocket.Addr, &ClientSocket.AddrSize);
            if(ClientSocket.FD == -1)
                  Log.Write("Server::accept failed. IP = " +
string(inet_ntoa(ClientSocket.Addr.sin_addr)), DEBUG_
LEVEL::WARN);
            else // new client appear Du-Dung.
            {
                  if(ClientCount > LISTEN_SIZE)
                        ClientCount = 0;
                  ClientSocket.SerialNum = 4001 + ClientCount;
// giving out serial number to client
                  ClientCount++;
                  pthread_create(&UserThread, NULL,
&countingServer::User, (socket_t*)&ClientSocket);
                  pthread_detach(UserThread);
                  Log.Write("Server:" + string(inet_
ntoa(ClientSocket.Addr.sin_addr))+": Connected", DEBUG_
LEVEL::DEBUG);
            }
      }
      pthread_exit(NULL);
   }
   void* countingServer::User(void* Socket)
   {
      socket_t UserSocket = *(socket_t*)Socket;
      int line = 0;
      char InputBuffer[BUFF_SIZE];
      string* Buffer;
      while(UserSocket.FD > 0)
      {
            line = read(UserSocket.FD, InputBuffer, BUFF_
SIZE);
            if(line > 0)
            {
                  Buffer = StringSplit(InputBuffer, " "); //
[To - Device SerialNumber] [Command]
                  if(Buffer[0].compare("1000") == 0) // to
Server
                  {
                        if(Buffer[1].compare("In") == 0) //
someone coming back to home
                        {
                              HomePeople++;
                              Log.
Write("Server::"+string(inet_ntoa(UserSocket.Addr.sin_addr))+":
HomePeople++", DEBUG_LEVEL::DEBUG);
```

```
                              }
                              else if(Buffer[1].compare("Out") ==
0) // someone leave from home
                              {
                                     HomePeople--;
                                     if(HomePeople < 0)
                                            HomePeople = 0;
                                     Log.
Write("Server::"+string(inet_ntoa(UserSocket.Addr.sin_addr))+":
HomePeople--", DEBUG_LEVEL::DEBUG);
                              }
                              else if(Buffer[1].compare("Bye") ==
0) // end connection
                              {
                                     close(UserSocket.FD);
                                     Log.
Write("Server::"+string(inet_ntoa(UserSocket.Addr.sin_addr))+":
Disconnected", DEBUG_LEVEL::DEBUG);
                              }
                       }
                       else // to devices
                       {
                       }
                }
         }
         pthread_exit(NULL);
   }
   logmanager countingServer::Log;
   socket_t countingServer::ServerSocket;
   pthread_t countingServer::UserThread;
   int countingServer::HomePeople = 0;
#endif
/////////////////////////////////////////////////////////////////
/////////////////////////////////////
Server
/////////////////////////////////////////////////////////////////
/////////////////////////////////////
device
/////////////////////////////////////////////////////////////////
/////////////////////////////////////
device_manager
/////////////////////////////////////////////////////////////////
/////////////////////////////////////
#include "device_manager.h"
void deviceManager::Initialize()
{
   Uart.Initialize("/dev/ttyUSB0"); //you can change this tty,
if you use another one or have some problem.
}
```

```
bool deviceManager::ConnectDevice(int SerialNum)
{
   char Buffer[BUFF_SIZE];
   char* DeviceSerialNum;
   char* DeviceName;
   if(deviceManager::Send(SerialNum, "Init") > 0)
   {
      if(deviceManager::Recv(Buffer) > 0)
      {
            DeviceSerialNum = strtok(Buffer, " "); // tokenize
device serial number
            strtok(NULL, " "); // next token is server devcie
serial number (10000)
            DeviceName = strtok(NULL, " "); // tokenize device
name;
            device newDevice(DeviceName,
atoi(DeviceSerialNum));
            DeviceList.push_front(newDevice);
            cout << "newDevice Connected. => " << DeviceName
<< endl;
            return true;
      }
   }
   return false;
}
int deviceManager::Send(int SerialNum, const char* Command)
{
   char Buffer[BUFF_SIZE];
   sprintf(Buffer, "10000%d %s +", SerialNum, Command);
   return Uart.WriteLine(Buffer);
}
//-- receive message from somewhere.
//-- this function wait 1s for receive something.
//-- if not received anything, it return 0.
//-- if uart run blocking mode, it works just receive function
but it makes endless loop.
int deviceManager::Recv(char* Buffer)
{
   int line = 0;
   while(line <= 0)
      line = Uart.ReadLine(Buffer);
   return line;
}
char* deviceManager::SendCommand(int SerialNum, const char*
Command)
{
   char Buffer[BUFF_SIZE];
   if(deviceManager::Send(SerialNum, Command) > 0)
   {
```

```
        if(deviceManager::Recv(Buffer) > 0)
        {
              return Buffer;
        }
    }
    return NULL;
}
bool deviceManager::SerchDevice(int SerialNum)
{
    for(list<device>::iterator Pos = DeviceList.begin(); Pos !=
DeviceList.end(); Pos++)
    {
        if(Pos->getSerialNum() == SerialNum) { return true; }
    }
    return false;
}
////////////////////////////////////////////////////////////////
////////////////////////////////////////
device_manager
////////////////////////////////////////////////////////////////
////////////////////////////////////////
//-- it controls devices via rs232.
#ifndef DEVICEMANAGER
#define DEVICEMANAGER
#include "../type.h"
#include "uart.h"
#include <iostream>
#include <stdio.h>
#include <stdlib.h>
#include <string.h>
#include <list>
using namespace std;
class deviceManager
{
private :
    uart Uart;
    list<device> DeviceList;
    int Send(int SerialNum, const char* Command); // send
command to device.
    int Recv(char* Buffer); // receive message from device.
thread.
    void SendBegin();
    void SendEnd();
public :
    void Initialize(); // baudrate = 9600 (fixed)
    bool ConnectDevice(int SerialNum); // new device connect.
    char* SendCommand(int SerialNum, const char* Command);
    bool SerchDevice(int SerialNum);
};
```

```
#endif
//////////////////////////////////////////////////////////////
///////////////////////////////////////
uart
//////////////////////////////////////////////////////////////
///////////////////////////////////////
#include "uart.h"
void uart::Initialize(const char* TTY)
{
   tty = TTY;
   //FD = open(tty, O_RDWR | O_NOCTTY | O_NONBLOCK);
   FD = open(tty, O_RDWR | O_NOCTTY);
   if(FD == -1)
   {
      perror("Uart Error");
   }
   memset(&IO, 0, sizeof(IO));
   memset(Buffer, 0, 1024);
   IO.c_cflag = B9600; //baudRate
   IO.c_cflag |= CS8; // Data bit is 8bit
   IO.c_cflag |= CLOCAL;
   IO.c_cflag |= CREAD;
   IO.c_iflag = 0;
   IO.c_oflag = 0;
   IO.c_lflag = 0;
   IO.c_cc[VTIME] = 0;
   IO.c_cc[VMIN] = 1;
   tcflush(FD, TCIFLUSH);
   tcsetattr(FD, TCSANOW, &IO);
}
int uart::WriteLine(char* Msg)
{
   for(unsigned int i = 0; i < strlen(Msg); i++)
   {
      write(FD, Msg + i, 1);
   }
   return strlen(Msg) + 1;
}
int uart::ReadLine(char* Buffer)
{
   int line = 0;
   int count = 0;
   char Input;
   memset(Buffer, 0, BUFF_SIZE);
   read(FD, &Input, sizeof(Input));
   while(Input != '+')
   {
      if(count > 128)
            return 0;
```

```
        Buffer[line] = Input;
        line++;
        count++;
        read(FD, &Input, sizeof(Input));
    }
    return line;
}
////////////////////////////////////////////////////////////////
/////////////////////////////////////
uart
////////////////////////////////////////////////////////////////
/////////////////////////////////////
//-- between server and device protocol (SADP)
//-- ==> from to msg
//-- ex) 00000(Server) 00001(some avr device) req(request)
//
//-- initial uart baudrate is 9600;
//-- PLC Module (NC-PLCD) baudrate is 9600. (fixed)
//
//-- all of device have uniq serial number. it makes classify
each devices.
//-- device name is not uniq. so, user can change the devcie
name.
//-- ex) LightDevice => Room1_Light
//-- server serial number is 00000 (fixed)
//-- wattmeter is 00001 (fixed)
#ifndef BUFF_SIZE
   #define BUFF_SIZE 128
#endif
#ifndef NAME_SIZE
   #define NAME_SIZE 128
#endif
#ifndef UART
#define UART
#include <stdio.h>
#include <stdlib.h>
#include <string.h>
#include <unistd.h>
#include <sys/types.h>
#include <termios.h>
#include <fcntl.h>
class uart
{
private :
   const char* tty;
   int FD;
   struct termios IO;
   char Buffer[BUFF_SIZE];
public :
```

```
   void Initialize(const char* TTY);
   int ReadLine(char* Buffer);
   int WriteLine(char* Msg);
};
class device
{
private :
   char Name[NAME_SIZE]; // Name size = 255
   int SerialNum; // for BSDP
public :
   device(char* Name, int SerialNum)
   {
            strcpy(this->Name, Name);
            this->SerialNum = SerialNum;
   }
   char* getName() { return Name; }
   char* setName(char* newName) { strcpy(Name, newName); return
Name; }
   int getSerialNum() { return SerialNum; }
   //int setSerialNum(int newSerialNum) { SerialNum =
newSerialNum; return SerialNum; }
};
#endif
////////////////////////////////////////////////////////////////
/////////////////////////////////////
server
////////////////////////////////////////////////////////////////
/////////////////////////////////////
server 1
////////////////////////////////////////////////////////////////
/////////////////////////////////////
#include "server.h"
//-- server initialize process is composed 2step.
//-- 1st step, socket open, binding, define listening size and
general settings.
//-- 2nd step, accept new client and compare which is user or
devices.
void server::Initialize(const char* ServerIP, int Port)
{
   memset(&ServerSocket.Addr, 0, ServerSocket.AddrSize);
   ServerSocket.Addr.sin_family = AF_INET; // kind of network
service == ethernet
   ServerSocket.Addr.sin_port = htons(Port); // port number
   ServerSocket.Addr.sin_addr.s_addr = inet_addr(ServerIP); //
server ip
   DM.Initialize(); // device initialize
   if(server::Init_Step1())
      server::Init_Step2();
}
```

```
bool server::Init_Step1()
{
   ServerSocket.FD = socket(PF_INET, SOCK_STREAM, 0); // socket
open
   if(ServerSocket.FD == -1)
   {
      perror("Initialize");
      return false;
   }
   if(bind(ServerSocket.FD, (struct sockaddr*)&ServerSocket.
Addr, ServerSocket.AddrSize) == -1)
   {
      perror("Server Error");
      return false;
   }
   if(listen(ServerSocket.FD, LISTEN_SIZE) == -1)
   {
      perror("Server Error");
      return false;
   }
   return true;
}
bool server::Init_Step2() // after step1, accept new client.
{
   socket_t ClientSocket;
   while(1)
   {
      ClientSocket.FD = accept(ServerSocket.FD, (struct
sockaddr*)&ClientSocket.Addr, &ClientSocket.AddrSize);
      if(ClientSocket.FD == -1)
      {
            perror("connection error");
      }
      else // new client appear Du dung !!
      {
            int line = 0; // length of input buffer
            memset(Buffer, 0, BUFF_SIZE);
            cout << "--new connection-- " << inet_
ntoa(ClientSocket.Addr.sin_addr) << endl;
            line = read(ClientSocket.FD, Buffer, BUFF_SIZE);
            if(line > 0) // now playing - nickelcome :
something in ur buffer
            {
                  //--------------------------//
                  //-- connect new user --//
                  //------------------------//
                  if(strcmp(Buffer, "*User*") == 0)
                  {
                        memset(Buffer, 0, BUFF_SIZE);
```

```
                        pthread_create(&RecvThread, NULL,
&server::User, (socket_t*)&ClientSocket);
                        pthread_detach(RecvThread);
                        server::SendMsg(ClientSocket, "AOK");
                        cout << inet_ntoa(ClientSocket.Addr.
sin_addr) << ": User connected" << endl;
                  }
                  //---------------------------------//
                  //-- connect invalid socket --//
                  //---------------------------------//
                  else
                  {
                        cout << inet_ntoa(ClientSocket.Addr.
sin_addr) << ": invalid connection: " << Buffer << endl;
                        server::Close(ClientSocket);
                  }
            }
      }
   }
   return true;
}
//-- process user request
//-- such as requset watt data, turn on the light and so on.
void* server::User(void* UserSocket)
{
   socket_t userSocket = *(socket_t*)UserSocket;
   int line = 0;
   char Input[BUFF_SIZE];
   char TmpBuffer[BUFF_SIZE];
   char* SerialNum;
   char* Command;
   while(userSocket.FD > 0)
   {
      line = read(userSocket.FD, Input, BUFF_SIZE);
      if(line > 0)
      {
            //-- all of command form is "serialnuber command"
            SerialNum = strtok(Input, " "); // Tokenize
SerialNumber
            Command = strtok(NULL, " ");
            if(DM.SerchDevice(atoi(SerialNum))) // is device
exist ?
            {
                  if(strcmp(Command, "ReqWatt") == 0) //
request amount of uesd watt
                  {
                        memset(TmpBuffer, 0, BUFF_SIZE);
                        strcpy(TmpBuffer,DM.
SendCommand(atoi(SerialNum), "ReqWatt")); // WattMeter serial
```

```
number must 00001
                        if(TmpBuffer != NULL)
                        {
                                cout << inet_ntoa(userSocket.
Addr.sin_addr) << ": ReqWatt"  << endl;
                                server::SendMsg(userSocket,
TmpBuffer);
                        }
                        else
                        {
                                cout << inet_ntoa(userSocket.
Addr.sin_addr) << ": ReqWatt Failed"  << endl;
                                server::SendMsg(userSocket,
"Failed");
                        }
                }
                else if(strcmp(Command, "On") ==0 ) // turn
on device
                {
                        memset(TmpBuffer, 0, BUFF_SIZE);
                        strcpy(TmpBuffer,
DM.SendCommand(atoi(SerialNum), "On"));
                        if(strcmp(TmpBuffer, "AOK") == 0)
                        {
                                cout << inet_ntoa(userSocket.
Addr.sin_addr) << ": On => " << SerialNum << endl;
                                server::SendMsg(userSocket,
"AOK");
                        }
                        else
                        {
                                cout << inet_ntoa(userSocket.
Addr.sin_addr) << ": On Failed => " << SerialNum << endl;
                                server::SendMsg(userSocket,
"Failed");
                        }
                }
                else if(strcmp(Command, "Off") == 0)
                {
                        memset(TmpBuffer, 0, BUFF_SIZE);
                        cout << atoi(SerialNum) << endl;
                        strcpy(TmpBuffer,
DM.SendCommand(atoi(SerialNum), "Off"));
                        if(strcmp(TmpBuffer, "AOK") == 0)
                        {
                                cout << inet_ntoa(userSocket.
Addr.sin_addr) << ": Off => " << SerialNum << endl;
                                server::SendMsg(userSocket,
"AOK");
```

```
                        }
                        else
                        {
                               cout << inet_ntoa(userSocket.
Addr.sin_addr) << ": Off Failed => " << SerialNum << endl;
                               server::SendMsg(userSocket,
"Failed");
                        }
                  }
            }
            else if(strcmp(Command, "Add") == 0) // add new
device
            {
                  if(DM.ConnectDevice(atoi(SerialNum)))
                  {
                        cout << inet_ntoa(userSocket.Addr.
sin_addr) << ": Add device => " << SerialNum << endl;
                        server::SendMsg(userSocket, "AOK");
                  }
                  else
                  {
                        cout << inet_ntoa(userSocket.Addr.
sin_addr) << ": Add device failed => " << SerialNum << endl;
                        server::SendMsg(userSocket,
"Failed");
                  }
            }
            else // nai. not exist.
            {
                  server::SendMsg(userSocket, "ERR");
            }
      }
      memset(Input, 0, BUFF_SIZE);
      memset(TmpBuffer, 0, BUFF_SIZE);
   }
   pthread_exit(NULL);
}
int server::SendMsg(socket_t Target, const char* Msg)
{
   int line;
   line = write(Target.FD, Msg, strlen(Msg));
   if(line < 0) // send failed
   {
      line = write(Target.FD, Msg, strlen(Msg)); // retry
      if(line < 0)
      {
            cout << inet_ntoa(Target.Addr.sin_addr) << ": msg
send failed." << endl;
            server::Close(Target);
```

```
            return -1;
        }
    }
    return line;
}
int server::Close(socket_t Target)
{
    cout << inet_ntoa(Target.Addr.sin_addr) << ": disconnected."
<< endl;
    return close(Target.FD); // not find. => target is not
device, just user.
}
//////////////////////////////////////////////////////////////
////////////////////////////////////
server 2
//////////////////////////////////////////////////////////////
////////////////////////////////////
#ifndef SERVER
#define SERVER
#include "../type.h"
#include "../device/device_manager.h"
#include <iostream>
#include <stdio.h>
#include <stdlib.h>
#include <string.h>
#include <unistd.h>
#include <sys/types.h>
#include <arpa/inet.h>
#include <sys/socket.h>
#include <pthread.h>
#include <list>
using namespace std;
struct socket_t
{
    int FD;
    struct sockaddr_in Addr;
    unsigned int AddrSize;
    socket_t()
    {
        FD = 0;
        AddrSize = sizeof(Addr);
    }
};
class server
{
private :
    socket_t ServerSocket;
    pthread_t MainThread; // accept thread
    pthread_t RecvThread; // user recv msg thread
```

```
   char Buffer[BUFF_SIZE];
   bool Init_Step1(); // Socket Open >> bind >> listen
   bool Init_Step2(); // accept >> recvThread
   static int SendMsg(socket_t Target, const char* Msg);
   static int Close(socket_t Target); // disconnect tartget
socket
   static void* WattMeter(void* DeviceSocket); // amount of
used watt
   static void* GeneralDevice(void* DeviceSocket); // turn on,
turn off, amount of used watt
   static void* User(void* UserSocket);
public :
   static deviceManager DM;
   void Initialize(const char* ServerIP, int Port);
};
#endif
///////////////////////////////////////////////////////////////
////////////////////////////////////
HomeAutomation_AVR
///////////////////////////////////////////////////////////////
/////////////////////////////////////
General_Device
///////////////////////////////////////////////////////////////
/////////////////////////////////////
//--- PLC Between Server and Device Protocol (BSDP) ---//
//-- "From To Command +"
//-- ex) "10000 20000 ReqWatt +"
//
//-- From => int
//-- To => int
//-- Command => char*
//-- + => end of command
//
//--- PLC Addr Map ---//
//-- Server => 10000
//-- All Device (include Server) => 11111
//-- this is kind of broad casting.
//-- measure devices => 2xxxx
//-- general devices => 3xxxx
#define BUFF_SIZE 128
#define SERIALNUM 3005
#define F_CPU 16000000UL
#define BAUDRATE 9600
#include <avr/io.h>
#include <avr/interrupt.h>
#include <util/delay.h>
#include <stdio.h>
#include <stdlib.h>
#include <string.h>
```

```
#include "lib/atmega128_PLC_Network.hpp"
#include "lib/atmega128_uart.hpp"
//-- Device Constant --//
char SerialNum[BUFF_SIZE];
char Name[BUFF_SIZE];
//-- Buffer Variable --//
char Buffer[BUFF_SIZE];
int BuffIndex = 0;
//-- UART --//
UART Uart;
int main(void)
{
   //-- Device initialize --//
   DDRA = 0xff;
   DDRG = 0xff;
   plc.Initialize(1);
   //Uart.initialize(0);
   itoa(SERIALNUM, SerialNum, 10); // serial number, int to
char
   strcpy(Name, "WattMeter"); // device name

   //-- Device Work !!! --//
   //-- read watt date from WattMeter, and store.
   //-- if data is bigger than int type range, send data to
server
    while(1)
    {
        PORTG = 0x02;
        _delay_ms(300);
        PORTG = 0x00;
        _delay_ms(300);
    }
}
void ProcessCommand(char* Buffer)
{
   int From;
   int To;
   int User;
   char* Command;

   From = atoi(strtok(Buffer, " "));
   To = atoi(strtok(NULL, " "));
   User = atoi(strtok(NULL, " "));
   Command = strtok(NULL, " ");

   if(To == SERIALNUM || To == 3000 || To == 11111)
   {
      if(strcmp(Command, "begin") == 0)
      {
```

```
            plc.setBusy(true);
      }
      else if(strcmp(Command, "end") == 0)
      {
            plc.setBusy(false);
      }
      else if(strcmp(Command, "Init") == 0)
      {
            plc.setBusy(false);
            plc.SendMsg(From, User, Name);
      }
      //-- Here, this device's command
      else if(strcmp(Command, "On") == 0)
      {
            plc.setBusy(false);
            PORTA = 0xff;
            if(To == SERIALNUM)
                  plc.SendMsg(From, User, "AOK");
      }
      else if(strcmp(Command, "Off") == 0)
      {
            plc.setBusy(false);
            PORTA = 0x00;
            if(To == SERIALNUM)
                  plc.SendMsg(From, User, "AOK");
      }
   }
}
////////////////////////////////////////////////////////////
//////////////////////////////////////
atmega128_PLC_Network
////////////////////////////////////////////////////////////
//////////////////////////////////////
//-- Atmega128 PLC Network
#ifndef BUFF_SIZE
   #define BUFF_SIZE 128
#endif
#ifndef F_CPU
   #define F_CPU 16000000UL
#endif
#ifndef SERIALNUM
   #define SERIALNUM 2000
#endif
#ifndef BAUDRATE
   #define BAUDRATE 9600 // NC_PLCD baudrate is 9600 (fixed)
#endif
#ifndef atmega128_PLC_Network
   #define atmega128_PLC_Network
   #include <util/delay.h>
```

```
#include "atmega128_uart.hpp"
class PLC
{
private:
    UART Uart;
    int UartNumber;
    bool Busy; // plc line busy (true = can't use)

    void WriteLine(int TargetDevice,int UserNum, const char*
Command)
    {
          char Buffer[BUFF_SIZE];
          sprintf(Buffer, "%d %d %d %s +", SERIALNUM,
TargetDevice, UserNum, Command);
          while(Busy); // if line is busy, wait.
          Uart.send(this->UartNumber, Buffer);
          _delay_ms(1);
    }

public:
    PLC()
    {
          UartNumber = 0;
          Busy = false;
    }

    //bool isBusy(){return Busy;}
    bool setBusy(bool Busy){this->Busy = Busy; return this-
>Busy;}

    void Initialize(int UartNumber)
    {
          this->UartNumber = UartNumber;
          Uart.initialize(UartNumber);
    }

    void SendMsg(int TargetDevice, int UserNum, const char*
Msg)
    {
          this->WriteLine(1111, 1111, "begin");
          this->WriteLine(TargetDevice, UserNum, Msg);
          this->WriteLine(1111, 1111, "end");
    }
};

//-- Global Functions and Variables --//
//-- to interaction and work, plc use ISR.

//-------------------------------//
```

```
   //-- MUST DEFINE IN MAIN !!!!!! --//
   //--------------------------------//
   //-- compare command list and input command
   //-- if input command is valid command, process the command
   //-- Between Server and Device Protocol
   //-- From(int) To(int) Command(char) +
   //-- end of command => '+'
   void ProcessCommand(char* Buffer); // <======= RIGHT HERE !

   char ISR_Buffer[BUFF_SIZE];
   int BufferIndex; // must initialize in main.
   PLC plc;

   ISR(USART1_RX_vect)
   {
      PORTG = 0x01; // input led
      if(BufferIndex > BUFF_SIZE) // Buffer overflow
      {
            memset(ISR_Buffer, 0, BUFF_SIZE);
            BufferIndex = 0;
      }
      ISR_Buffer[BufferIndex] = UDR1;
      BufferIndex++;
      if(ISR_Buffer[BufferIndex -1] == '+') // if end of
command
      {
            ProcessCommand(ISR_Buffer);
            memset(ISR_Buffer, 0, BUFF_SIZE);
            BufferIndex = 0;
      }
      PORTG = 0x00;
   }

#endif
//////////////////////////////////////////////////////////////
////////////////////////////////////
atmega128_uart
//////////////////////////////////////////////////////////////
////////////////////////////////////
//-- Atmega128 UART
//-- v1.2
#ifndef atmega128_uart
   #define atmega128_uart
   #ifndef F_CPU
      #define F_CPU 16000000UL
   #endif
   #ifndef BAUDRATE
      #define BAUDRATE 9600 // If you do not specify a
Overlate default is 9600
```

```
#endif
#define MYUBRR (((F_CPU / (BAUDRATE * 16UL))) -1)
#include <avr/io.h>
#include <avr/interrupt.h>

class UART
{
public:
    void initialize(int UartNumber)
    {
            cli();
            if(UartNumber == 0)
            {
                    UBRR0H = (unsigned char)(MYUBRR>>8);
                    UBRR0L = (unsigned char)(MYUBRR);
                    UCSR0B = (1<<RXEN0)|(1<<TXEN0)|(1<<RXCIE0);

            }
            else if(UartNumber == 1)
            {
                    UBRR1H = (unsigned char)(MYUBRR>>8);
                    UBRR1L = (unsigned char)(MYUBRR);
                    UCSR1B = (1<<RXEN1)|(1<<TXEN1)|(1<<RXCIE1);
            }
            sei();
    }

    int send(int UartNumber, const char* Msg)
    {
            if(UartNumber == 0)
                    return uart0_send(Msg);
            else
                    return uart1_send(Msg);
    }

    int uart0_send(const char* Msg)
    {
            int i = 0;
            for(i = 0; Msg[i] != 0x00; i++)
            {
                    while(!(UCSR0A & (1<<UDRE0)));
                    UDR0 = Msg[i];
            }

            return i;
    }

    int uart1_send(const char* Msg)
    {
```

```
            int i = 0;
            for(i = 0; Msg[i] != 0x00; i++)
            {
                  //while(!(UCSR1A & (1<<UDRE1)));
                  while(!(UCSR1A & 0x20));
                  UDR1 = Msg[i];
            }
            return i;
      }
   };

#endif
//////////////////////////////////////////////////////////////
////////////////////////////////////
WattMater_Counter
//////////////////////////////////////////////////////////////
////////////////////////////////////
//-- WattMeter Device Addon.
//-- receive pulse from watt meter and count pulse.
#define F_CPU 8000000UL
#define BAUDRATE 9600
#define MYUBRR (((F_CPU / (BAUDRATE * 16UL))) -1)
#include <avr/io.h>
#include <util/delay.h>
int prev;
int main(void)
{
   DDRA = 0x00; // Pulse input.
   DDRB = 0xff;
   PORTA = 0xff;
   //-- init Uart => output
   UBRRH = (unsigned char)(MYUBRR>>8);
   UBRRL = (unsigned char)(MYUBRR);
   UCSRB = (1<<RXEN)|(1<<TXEN)|(1<<RXCIE);
    while(1)
    {
        if(PINA == 0xFE) // Incoming pulses
      {
            if(prev == 0) // Chattering prevention !
            {
                  prev = 1;
                  UDR = '1';
            }
      }
      else if(PINA == 0xFF)
            prev = 0;
    }
}
//////////////////////////////////////////////////////////////
```

```
/////////////////////////////////////
WattMeter_Device
////////////////////////////////////////////////////////////////
/////////////////////////////////////
//--- PLC Between Server and Device Protocol (BSDP) ---//
//-- "From To Command +"
//-- ex) "10000 20000 ReqWatt +"
//
//-- From => int
//-- To => int
//-- Command => char*
//-- + => end of command
//
//--- PLC Addr Map ---//
//-- Server => 10000
//-- All Device (include Server) => 11111
//-- this is kind of broad casting.
//-- measure devices => 2xxxx
//-- general devices => 3xxxx
#define BUFF_SIZE 128
#define SERIALNUM 2005
#define F_CPU 16000000UL
#define BAUDRATE 9600
#include <avr/io.h>
#include <avr/interrupt.h>
#include <util/delay.h>
#include <stdio.h>
#include <stdlib.h>
#include <string.h>
#include "lib/atmega128_PLC_Network.hpp"
#include "lib/atmega128_uart.hpp"
//-- Device Constant --//
char SerialNum[BUFF_SIZE];
char Name[BUFF_SIZE];
//-- Buffer Variable --//
char Buffer[BUFF_SIZE];
int BuffIndex = 0;
//-- Used amount of watt data --//
char WattBuffer[BUFF_SIZE];
int CurrentWatt = 0;
//-- UART --//
UART Uart;
int main(void)
{
   //-- Device initialize --//
   DDRG = 0xff;
   plc.Initialize(1);
   Uart.initialize(0);
   itoa(SERIALNUM, SerialNum, 10); // serial number, int to
```

```
char
   strcpy(Name, "WattMeter"); // device name

   //-- Device Work !!! --//
   //-- read watt date from WattMeter, and store.
   //-- if data is bigger than int type range, send data to
server
    while(1)
    {
      PORTG = 0x02;
      _delay_ms(75);
      PORTG = 0x00;
      _delay_ms(300);
    }
}
ISR(USART0_RX_vect) // Incoming pulses => watt ++
{
   if(UDR0 == '1')
   {
      CurrentWatt++;
      itoa(CurrentWatt, WattBuffer, 10);
   }
}
void ProcessCommand(char* Buffer)
{
   int From;
   int To;
   int User;
   char* Command;

   From = atoi(strtok(Buffer, " "));
   To = atoi(strtok(NULL, " "));
   User = atoi(strtok(NULL, " "));
   Command = strtok(NULL, " ");

   if(To == SERIALNUM || To == 11111)
   {
      if(strcmp(Command, "begin") == 0)
      {
            plc.setBusy(true);
      }
      else if(strcmp(Command, "end") == 0)
      {
            plc.setBusy(false);
      }
      else if(strcmp(Command, "Init") == 0)
      {
            plc.setBusy(false);
            plc.SendMsg(From, User, Name);
```

```
        }
        //-- Here, this device's command
        else if(strcmp(Command, "ReqWatt") == 0)
        {
              plc.setBusy(false);
              plc.SendMsg(From, User, WattBuffer);
        }
    }
}
//////////////////////////////////////////////////////////////////
//////////////////////////////////////
atmega128_PLC_Network
//////////////////////////////////////////////////////////////////
//////////////////////////////////////
//-- Atmega128 PLC Network
//-- Protocol defined for communications between a plurality of
PLC modules
#ifndef BUFF_SIZE
   #define BUFF_SIZE 128
#endif
#ifndef F_CPU
   #define F_CPU 16000000UL
#endif
#ifndef SERIALNUM
   #define SERIALNUM 2000
#endif
#ifndef BAUDRATE
   #define BAUDRATE 9600 // NC_PLCD baudrate is 9600 (fixed)
#endif
#ifndef atmega128_PLC_Network
   #define atmega128_PLC_Network
   #include <util/delay.h>
   #include "atmega128_uart.hpp"
   class PLC
   {
   private:
      UART Uart;
      int UartNumber;
      bool Busy; // plc line busy (true = can't use)

      void WriteLine(int TargetDevice,int UserNum, const char*
Command)
      {
            char Buffer[BUFF_SIZE];
            sprintf(Buffer, "%d %d %d %s +", SERIALNUM,
TargetDevice, UserNum, Command);
            while(Busy); // if line is busy, wait.
            Uart.send(this->UartNumber, Buffer);
            _delay_ms(1);
```

```
        }

    public:
        PLC()
        {
                UartNumber = 0;
                Busy = false;
        }

        //bool isBusy(){return Busy;}
        bool setBusy(bool Busy){this->Busy = Busy; return this-
>Busy;}

        void Initialize(int UartNumber)
        {
                this->UartNumber = UartNumber;
                Uart.initialize(UartNumber);
        }

        void SendMsg(int TargetDevice, int UserNum, const char*
Msg)
        {
                this->WriteLine(1111, 1111, "begin");
                this->WriteLine(TargetDevice, UserNum, Msg);
                this->WriteLine(1111, 1111, "end");
        }
    };

    //-- Global Functions and Variables --//
    //-- to interaction and work, plc use ISR.

    //--------------------------------//
    //-- MUST DEFINE IN MAIN !!!!!! --//
    //--------------------------------//
    //-- compare command list and input command
    //-- if input command is valid command, process the command
    //-- Between Server and Device Protocol
    //-- From(int) To(int) Command(char) +
    //-- end of command => '+'
    void ProcessCommand(char* Buffer); // <======= RIGHT HERE !

    char ISR_Buffer[BUFF_SIZE];
    int BufferIndex; // must initialize in main.
    PLC plc;

    ISR(USART1_RX_vect)
    {
        PORTG = 0x01; // input led
        if(BufferIndex > BUFF_SIZE) // Buffer overflow
```

```
        {
                memset(ISR_Buffer, 0, BUFF_SIZE);
                BufferIndex = 0;
        }
        ISR_Buffer[BufferIndex] = UDR1;
        BufferIndex++;
        if(ISR_Buffer[BufferIndex -1] == '+') // if end of
command
        {
                ProcessCommand(ISR_Buffer);
                memset(ISR_Buffer, 0, BUFF_SIZE);
                BufferIndex = 0;
        }
        PORTG = 0x00;
    }

#endif
/////////////////////////////////////////////////////////////////
/////////////////////////////////////
atmega128_uart
/////////////////////////////////////////////////////////////////
/////////////////////////////////////
//-- Atmega128 UART
//-- v1.2
#ifndef atmega128_uart
    #define atmega128_uart
    #ifndef F_CPU
        #define F_CPU 16000000UL
    #endif
    #ifndef BAUDRATE
        #define BAUDRATE 9600 // If you do not specify a
Overlate default is 9600
    #endif
    #define MYUBRR (((F_CPU / (BAUDRATE * 16UL))) -1)
    #include <avr/io.h>
    #include <avr/interrupt.h>

    class UART
    {
    public:
        void initialize(int UartNumber)
        {
                cli();
                if(UartNumber == 0)
                {
                        UBRR0H = (unsigned char)(MYUBRR>>8);
                        UBRR0L = (unsigned char)(MYUBRR);
                        UCSR0B = (1<<RXEN0)|(1<<TXEN0)|(1<<RXCIE0);
```

```
            }
            else if(UartNumber == 1)
            {
                  UBRR1H = (unsigned char)(MYUBRR>>8);
                  UBRR1L = (unsigned char)(MYUBRR);
                  UCSR1B = (1<<RXEN1)|(1<<TXEN1)|(1<<RXCIE1);
            }
            sei();
      }

      int send(int UartNumber, const char* Msg)
      {
            if(UartNumber == 0)
                  return uart0_send(Msg);
            else
                  return uart1_send(Msg);
      }

      int uart0_send(const char* Msg)
      {
            int i = 0;
            for(i = 0; Msg[i] != 0x00; i++)
            {
                  while(!(UCSR0A & (1<<UDRE0)));
                  UDR0 = Msg[i];
            }
            return i;
      }

      int uart1_send(const char* Msg)
      {
            int i = 0;
            for(i = 0; Msg[i] != 0x00; i++)
            {
                  //while(!(UCSR1A & (1<<UDRE1)));
                  while(!(UCSR1A & 0x20));
                  UDR1 = Msg[i];
            }
            return i;
      }
   };
#endif
////////////////////////////////////////////////////////////////
//////////////////////////////////////////
atmega128_uart
////////////////////////////////////////////////////////////////
//////////////////////////////////////////
//-- Atmega128 UART
//-- v1.1
```

```
#ifndef atmega128_uart
   #define atmega128_uart
   #ifndef F_CPU
      #define F_CPU 16000000UL
   #endif
   #ifndef BAUDRATE
      #define BAUDRATE 9600
   #endif
   #define MYUBRR (((F_CPU / (BAUDRATE * 16UL))) -1)
   #include <avr/io.h>
   #include <avr/interrupt.h>

   class UART
   {
   public:
      bool initialize(int UartNumber)
      {
            cli();
            UBRR1H = (unsigned char)(MYUBRR>>8);
            UBRR1L = (unsigned char)(MYUBRR);
            UCSR1B = (1<<RXEN1)|(1<<TXEN1)|(1<<RXCIE1);
            sei();

            return true;
      }

      int send(int UartNumber, const char* Msg)
      {
            if(UartNumber == 0)
                  return uart0_send(Msg);
            else
                  return uart1_send(Msg);
      }

      int uart0_send(const char* Msg)
      {
            int i = 0;
            for(i = 0; Msg[i] != 0x00; i++)
            {
                  while(!(UCSR0A & (1<<UDRE0)));
                  UDR0 = Msg[i];
            }

            return i;
      }

      int uart1_send(const char* Msg)
      {
            int i = 0;
```

```
            for(i = 0; Msg[i] != 0x00; i++)
            {
                  while(!(UCSR1A & (1<<UDRE1)));
                  UDR1 = Msg[i];
            }

            return i;
      }
   };

#endif
```

About the Author

Jun-Ho Huh is an Assistant Professor in the Department of Software, Catholic University of Pusan, South Korea. He is Associate Editor at Human-Centric Computing and Information Sciences (HCIS), Springer, Berlin Heidelberg with FTRA (SCOPUS/ESCI indexed), Associate Editor at Journal of Information Processing Systems (JIPS), Korea Information Processing Society (SCOPUS/ESCI indexed), and Submission Editor at Journal of Multimedia Information System, Korea Multimedia Society. He finished the Cooperative Marine Science and Engineering Program, Texas A&M University at Galveston, United States of America in Aug. 2006. He received B.S. in Science Degree from Department of Major of Applied Marine Sciences (Currently Faculty of Marine Biomedical Sciences), B.S. in Engineering Degree (Double Major) from Department of Major of Computer Engineering from Jeju National University at Ara, Jeju, Republic of Korea in Aug. 2007. M.A. in Education Degree from Department of Major of Computer Science Education, Graduate School of Education, Pukyoug National University at Daeyeon, Busan, Republic of Korea in Aug. 2012. He received the Ph.D. in Engineering Degree from Department of Major of Computer Engineering, Graduate School, Pukyoug National University at Daeyeon, Busan, Republic of Korea in Feb. 2016.He received the Best Paper Award from Korea Multimedia Society Eight times (Nov. 2014, May. 2015, Nov. 2015, May, 2016. Oct, 2016, May, 2017: Triple). and Undergraduate Student Paper Bronze Medal (Corresponding Author) Awarded, Korea Information Processing Society (Apr. 28, 2017), and he received Best Paper Award The 10th 2016 International Interdisciplinary Workshop Series from HSST (Aug. 2016). Also, he received Best Paper Award The 16th International Conference on Control, Automation and Systems (Oct. 2016), ICROS with IEEE Xplore. Currently he is Associate Editor at Human-centric Computing and Information Sciences (HCIS), Springer Berlin Heidelberg (SCOPUS/ESCI indexed). Also, he is Associate Editor at Journal of Information Processing

Systems (JIPS), Korea Information Processing Society (SCOPUS/ESCI indexed) and Submission Editor at Journal of Multimedia Information System, Korea Multimedia Society. He is Senior Research Engineer of SUNCOM Co., Republic of Korea (Aug. 2015- Jun. 2016) and Clinical Assistant Professor of E-Green Remote Continuing Education, Seoul, Republic of Korea (Mar. 2016- Feb. 2017). He was Research Professor of Dankook University at Jukjeon, Yongin, Republic of Korea (July. 2016- Sep. 2016). Currently he is Assistant Professor of Department of Software, Catholic University of Pusan, Republic of Korea (Dec. 2016-). His research interests are Green IT, Smart Grid, Network Security, IoT, Curriculum of Computer.

Index

www.ingramcontent.com/pod-product-compliance
Lightning Source LLC
LaVergne TN
LVHW080345011026
849161LV00009B/2192